Shelf-Life Evaluation of Foods

Second Edition

Edited by

C.M.D. Man
Senior Lecturer
Division of Food Science & Biotechnology
School of Applied Science
South Bank University
London, England

and

A.A. Jones
Group Technical Manager
Rayner & Co. Ltd.
London, England

AN ASPEN PUBLICATION®

Aspen Publishers, Inc.
Gaithersburg, Maryland
2000

The authors have made every effort to ensure the accuracy of the information herein. However, appropriate information sources should be consulted, especially for new or unfamiliar procedures. It is the responsibility of every practitioner to evaluate the appropriateness of a particular opinion in the context of actual clinical situations and with due considerations to new developments. The author, editors, and the publisher cannot be held responsible for any typographical or other errors found in this book.

Library of Congress Cataloging-in-Publication Data

Shelf-life evaluation of foods/C.M.D. Man, Adrian Jones, [editors].—2nd ed.
p. cm.
ISBN 0-8342-1782-1
1. Food—Storage. 2. Food—Preservation. 3. Food—Shelf-life dating. I. Man, C.M.D.
II. Jones, Adrian (Adrian A.)
TP373.3.S485 2000
641.4—dc21
00-023883

Orders: (800) 638-8437
Customer Service: (800) 234-1660

Editorial Services: Nora Fitzpatrick
Library of Congress Catalog Card Number: 00-023883
ISBN: 0-8342-1782-1

Printed in the United States of America

1 2 3 4 5

Contents

Contributors

R.H. Dale
Formerly Eden Vale, Dale Farm Dairy Group, Northern Foods plc, Nr. Northwich, Cheshire, United Kingdom

P. Dalgaard
Ministry of Food, Agriculture and Fisheries, Danish Institute for Fisheries Research, Dept. of Seafood Research, Technical University of Denmark, Lyngby, Denmark

M.J. Ellis
Formerly H J Heinz Co Ltd., Kitt Green, Wigan, Lancashire, United Kingdom

M.R. Goddard
Unilever Research Colworth Laboratory, Colworth House, Sharnbrook, Bedford, United Kingdom

J.A.K. Howarth
Kellogg Company of Great Britain Ltd., Park Road, Stretford, Manchester, United Kingdom

F. Johansson
SIK, The Swedish Institute for Food Research, PO Box 5401, S-402 29, Göteborg, Sweden

A.A. Jones
Rayner & Co Ltd., Edmonton, London, England

H.P. Jones
Formerly Lyons Bakeries (UK) Ltd., Fish Dam Lane, Carlton, Barnsley, South Yorkshire, United Kingdom

L. Leistner
Formerly the Federal Centre for Meat Research, Kulmbach, Germany

A. Leufvén
SIK, The Swedish Institute for Food Research, Göteborg, Sweden

M. Lewis
Formerly Eden Vale, Dale Farm Dairy Group, Northern Foods plc, Nr. Northwich, Cheshire, United Kingdom

C.M.D. Man
Division of Food Science & Biotechnology, School of Applied Science, South Bank University, London, England

A.V. Martin
Formerly Yorkreco, Nestec (York) Ltd., York, United Kingdom

R.P. Singh
Department of Biological and Agricultural Engineering, University of California, Davis, California

P.J. Smedley
Rayner & Co Ltd., Edmonton, London, United Kingdom

U. Stöllman
SIK, The Swedish Institute for Food Research, Göteborg, Sweden

H. Symons
Formerly American Frozen Food Institute, McLean, Virginia

S.J. Walker
Campden and Chorleywood Food Research Association, Chipping Campden, Gloucestershire, United Kingdom

R.J. Winger
Institute of Food, Nutrition and Human Health, Massey University, Palmerston North, New Zealand

Preface to Second Edition

A gratifying level of demand meant that the first edition of this book had to be reprinted in 1996. And, it would appear, based on the various reviews that were brought to our attention, that the book has indeed fulfilled its objective in providing a convenient and useful reference to the shelf-life evaluation of foods. We were therefore rather pleased when Aspen Publishers, Inc., asked us to continue the work of putting together a second edition for this book, originally published by Blackie Academic and Professional, an imprint of Chapman and Hall, London.

Safe and wholesome food products of consistent quality have their origin in the proper and correct evaluation of their shelf lives during product development. The objective of this edition, therefore, is essentially the same as that of the previous edition. We were very fortunate to be able to consult nearly all the contributors to the first edition in the preparation of this new edition. Consequently, many original chapters have been retained and updated; a few were substantially revised and 4 completely new ones have been added, making a total of 16 chapters. In keeping with the aims of the previous edition, we have tried to include sufficient information, both in respect of the principles and the practice of shelf-life evaluation, for the experienced workers as well as the newcomers. As before, constructive comments from readers are welcome, whose interest in the subject and support of our endeavor are deeply appreciated.

Work such as this is the result of team efforts. We thank all the contributors, past and present, for helping to make this edition a reality. We dedicate this volume to them. We thank the publishers, too, for their enthusiasm and encouragement, and in particular, Christine Havens-Hafer, Development Editor at Aspen Publishers, Inc., whose professionalism, exceptional patience, wise counsel, and unfailing support have sustained us throughout what, at times, must seem like an everlasting project. Finally, we would like to thank our families once again for their love, care, and understanding while we were busy trying to extend the shelf life of this book.

C.M.D.M.
A.A.J.

PART I

The Principles

Chapter 1

Scientific Principles of Shelf-Life Evaluation

R.P. Singh

1.1 INTRODUCTION

Foods are perishable by nature. Numerous changes take place in foods during processing and storage. It is well known that conditions used to process and store foods may adversely influence the quality attributes in foods. Upon storage for a certain period of time, one or more quality attributes of a food may reach an undesirable state. At that instant, the food is considered unsuitable for consumption and it is said to have reached the end of its shelf life. In this chapter, certain major modes of food deterioration will be examined. These modes of deterioration will be expressed mathematically as rate equations. Reaction kinetics and various models that are used in practice to express quality changes in foods will be discussed. At the end of this chapter, a brief discussion of sensors that indicate time and temperature exposures of foods and their use in monitoring shelf life of foods will be presented.

1.2 MAJOR MODES OF FOOD DETERIORATION

During storage and distribution, foods are exposed to a wide range of environmental conditions. Environmental factors such as temperature, humidity, oxygen, and light can trigger several reaction mechanisms that may lead to food degradation. As a consequence of these mechanisms, foods may be altered to such an extent that they are either rejected by the consumer or they may become harmful to the person consuming them. It is therefore imperative that a good understanding of different reactions that cause food deterioration is gained prior to developing specific procedures for the evaluation of the shelf life of foods. Physical, chemical, and microbiological changes are the leading causes of food deterioration. A brief discussion of these changes is presented in this section. For more extended coverage of these topics, the reader is referred to books by Fennema (1996), Coultate (1996), Taub and Singh (1998), and Labuza (1982) as well as later chapters of this book.

1.2.1 Physical changes

Physical changes are caused by mishandling of foods during harvesting, processing, and distribution; these changes lead to a reduced shelf life of foods. Bruising of fruits and vegetables during harvesting and post-harvest handling leads to the development of rot. Crushing of dried snack foods during distribution seriously affects their consumer acceptability. Tuberous and leafy vegetables lose water when kept in atmospheres with low humidity and they subsequently wilt. Dried foods when kept in high humidity may pick up moisture and become soggy. In the case of frozen foods, fluctuating temperatures are often destructive to foods. For example, fluctuating temperatures cause recrystalization of ice cream leading to an undesirable sandy texture. Freezer burn is a major quality defect in frozen foods that is induced from the exposure of frozen foods to fluctuating temperatures. If the fluctuations are large and the food undergoes a phase change, then there are more prominent undesirable changes, for example, changes caused by thawing and refreezing of foods. Similarly, phase changes involving melting and solidifying of fats are detrimental to the quality of candies and other lipid-containing confectionery items.

1.2.2 Chemical changes

During the processing and storage of foods, several chemical changes occur that involve the internal food components and the external environmental factors. These changes may cause food deterioration and reduce the shelf life. The most important chemical changes are associated with enzymic action; oxidative reactions, particularly lipid oxidation that alters the flavor of many lipid-containing foods; and non-enzymic browning that causes changes in appearance.

At favorable temperatures, such as room temperature, many enzymic reactions proceed at rapid rates altering the quality attributes of foods. For example, fruits upon cutting tend to brown rapidly at room temperature due to the reaction of phenolase with the cell constituents that are released upon cutting of the tissue in the presence of oxygen. Enzymes such as lipoxygenase, if not denatured during the blanching process, can affect food quality even at subfreezing temperatures. In addition to temperature, other environmental factors such as oxygen, water, and pH induce deleterious changes in foods that are catalyzed by enzymes.

The presence of unsaturated fatty acids in foods is a prime reason for the development of rancidity during storage as long as oxygen is available. While development of off-flavors is markedly noticeable in rancid foods, the generation of free radicals during the autocatalytic process leads to several other undesirable reactions, for example, loss of vitamins, alteration of color, and degradation of proteins.

The rates of lipid oxidation are influenced by several factors. As indicated previously in this chapter, the environmental temperature is an important variable. The presence of oxygen in the immediate vicinity of the food leads to increased rates of oxidation. Similarly, water plays an important role; lipid oxidation occurs at high rates at very low water activities. In determining the shelf life of foods that contain lipids, the important reaction mechanisms and their rates of reactions must be known. In addition to lipid oxidation, there are other chemical reactions that are induced by light such as loss of vitamins and browning of meats. The influence of light on riboflavin loss in milk stored in high density polyethylene bottles is shown in Figure 1–1 (Singh et al., 1975).

Non-enzymic browning is a major cause of quality change and degradation of nutritional content in many foods. This type of browning reaction occurs due to the interaction between reducing sugars and amino acids. These reactions result in the loss of protein solubility, darkening of lightly colored dried products and the development of bitter flavors. Environmental factors such as temperature, water activity, and pH have an influence on non-enzymic browning.

1.2.3 Microbiological changes

In determining the influence of microorganisms on the shelf life of foods, the rates of microbial growth as a function of various environmental factors must be known. Microbes have the ability to multiply at high rates when favorable conditions are present. Prior to harvest, fruits and vegetables have generally good defense mechanisms against microbial attacks; however, after separation from the plant, they can easily succumb to microbial proliferation. Similarly, meat upon slaughter is susceptible to rapid microbial deterioration. Microbial proliferation in cottage cheese stored at three temperatures is shown in Figure 1–2 (Shellhammer and Singh, 1991). Microbial growth in foods results in food spoilage with the development of undesirable sensory characteristics and in certain cases the food may become unsafe for consumption.

The pathogenicity of certain microorganisms is a major safety concern in processing and handling of foods. Upon ingestion, microorganisms

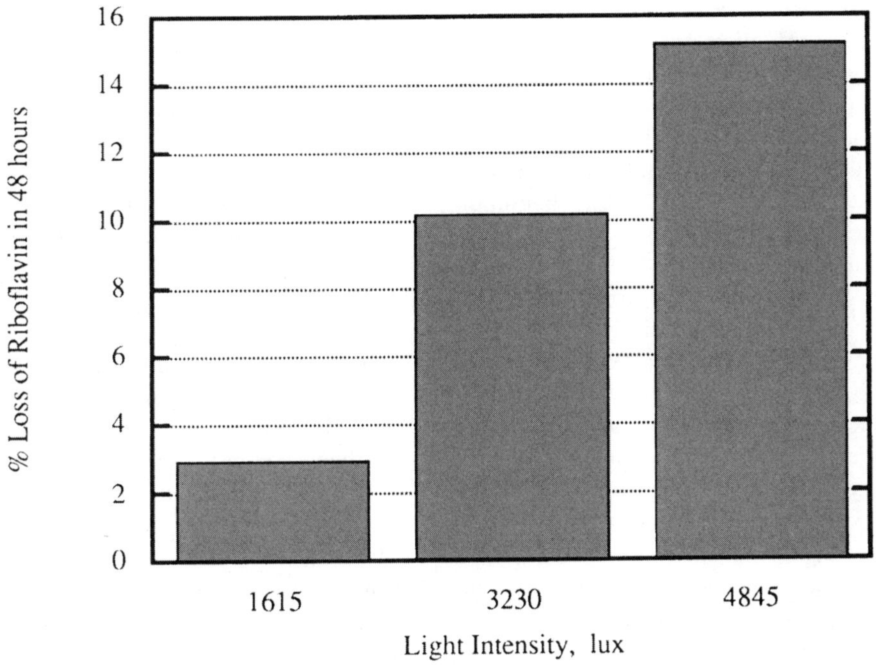

Figure 1–1 Loss of riboflavin in whole milk stored in blow molded high density polyethylene bottles under light at 4.4°C.

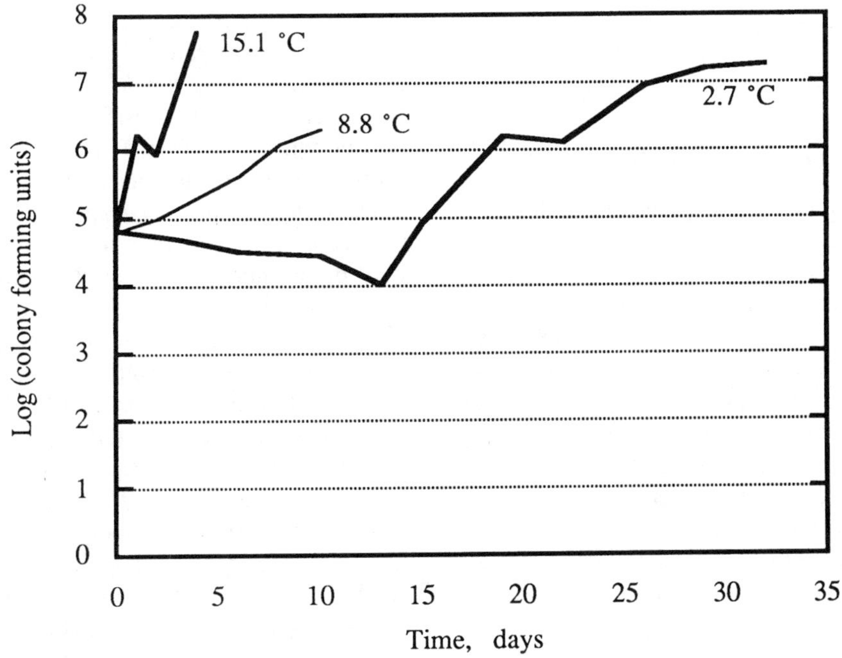

Figure 1–2 Standard plate count for cottage cheese stored at three constant temperatures.

such as *Salmonella* species and *Escherichia coli* strains cause infection while others such as *Clostridium botulinum* and *Staphylococcus aureus* produce chemicals in foods that are toxic to humans. Botulism is severe food poisoning because of the high potency of the toxin and the high fatality rates associated with it. Food poisoning due to *Staphylococcus* is more frequent although it is not as severe as botulism. The presence of molds and their growth on a food product may result in undesirable appearance and off-flavors. Under certain favorable conditions mold spores may also produce chemicals that are toxic to humans. During food processing and storage, special procedures are employed to minimize or prevent the occurrence of pathogenic and spoilage microorganisms.

Various measures commonly used to minimize or prevent microbial growth in foods include altering the environmental temperature, such as either lowering the temperature to slow the growth processes or raising the temperature to destroy the microbes; removing water such that it is unavailable to the growing microbes; controlling the environmental gases such as lowering oxygen or increasing carbon dioxide concentration; reducing or stopping the microbial growth by adding acid; and promoting fermentation, thus lowering the pH.

1.3 EVALUATION OF FOOD QUALITY

A common practice employed to evaluate the shelf life of a given food product is to determine changes in selected quality characteristics over a period of time. One may consider quality of a food as a gross measure of the food deterioration occurring in a food item. However, it should be recognized that the term "quality" is meant to encompass several quality attributes or characteristics. From a consumer's standpoint, the sensory expectations derived from the presence (or absence) of desirable (or undesirable) characteristics of a given food determine the quality of a product. Therefore, a food product noted for its high quality has more of the desirable characteristics.

Empirical or analytical techniques may be used to quantify the quality attributes of food. For example, enumeration of microbes and determination of chemical components of a product are analytical techniques, whereas, given the assurance of product safety, the use of human subjects to monitor changes in the magnitudes of quality characteristics constitutes an empirical technique. Considerable research has been conducted on quality evaluation of foods with an aim to determine shelf life (Jul, 1984; Van Arsdel et al., 1969; Singh et al., 1986; Labuza, 1982; Taub and Singh, 1998).

In the case of frozen foods, their shelf life has been described by the use of various terms. For example, practical storage life (PSL) is defined as "the period of frozen storage after freezing of an initially high quality product during which the organoleptic quality remains suitable for consumption or the process intended" (IIR, 1986). This definition requires the use of a number of sensory evaluation procedures and statistical analysis techniques. Most of these procedures rely on the use of sensory difference tests (discrimination methods), and the shelf-life failure is determined with strict statistical criteria. Shelf-life failure is often identified by "Just Noticeable Difference" (JND). JND is the "earliest time when a difference between the quality of test and control samples can be detected by trained sensory panels" (Van Arsdel et al., 1969). The use of this criterion is not restricted to one specific quality attribute or characteristic, rather it represents the overall quality of a product.

Using statistical analysis, probability of a difference between the test product and the control may be assigned at any predetermined confidence level. It may even be related to the noticeable difference between the test and the control samples that may be of commercial significance. The use of a difference test, such as JND, provides information on how long a product may be stored, for example, the PSL of a product. Readers are referred to Table 15–1 (Chapter 15) for PSL values of some common foods. While this is useful information on shelf life of different foods, it fails to provide a measure of the rate at which the overall quality is changing.

There is considerable evidence in the literature that temperature plays a major role in causing changes in food quality during storage. Higher storage temperatures generally lead to increased quality deterioration. In the past, there have been several attempts to use mathematical techniques and models to describe changes in food quality as influenced by storage temperature. Some of the earlier interest in this area was stimulated by observations that fluctuating storage temperatures caused more detrimental change in frozen foods than storage at constant temperature. Hicks (1944) modeled the changes in food quality by assuming an exponential relationship between the rate of quality change and temperature. He used an *effective mean temperature* to represent a diurnal temperature fluctuation. The effective mean temperature was a constant value that produced the same change in quality as a regular sinusoidal temperature change. Other temperature functions, such as square and sawtooth functions, were modeled by Schwimmer et al. (1955). Using mixed linear and exponential relationships between temperature and rate of quality change, Schubert (1977) also developed an effective mean temperature for the periodic temperature fluctuations. He used an empirical function containing three experimentally derived parameters to describe the effect of temperature on the rate of quality change. Employing chemical kinetic theory, Labuza (1979) used random variable temperature conditions to obtain effective temperature. These approaches have simplified the incorporation of fluctuating temperatures in calculating quality changes in foods.

Use of the chemical kinetic approach to model changes in food quality was suggested by Kwolek and Bookwalter (1971). The kinetic approach and the Arrhenius relationship that describes the influence of temperature on the reaction rate constants were also promoted by Saguy and Karel (1980). Lai and Heldman (1982) developed methods to determine the value of the activation energy of food quality losses from shelf-life data at known temperatures. A computer-aided method to simulate changes in food quality during storage of frozen foods was used by Singh (1976). Wells et al. (1987) discussed a graphical representation of the temperature history and food quality relationships. Ross (1998) provided a mathematical approach to model quality loss in foods during storage. In managing stocks of perishable inventory, the use of food quality attributes was elaborated by Wells and Singh (1997). Taoukis et al. (1997) provided examples of kinetic modeling of quality attributes for shelf-life predictions.

1.3.1 Reaction kinetics

Chemical kinetics involves the study of the rates and mechanisms by which a chemical species converts to another. The rate of a chemical reaction is determined by the mass of a product produced or the reactant consumed per unit time. In general, the rates of reactions may be determined by monitoring the concentration of either the reactants or the products of the reactions. The mechanism of a reaction, on the other hand, is more difficult to determine since it involves the sequence of steps that produce the overall result. Mechanisms of reactions are determined only in simple systems.

A general form of kinetic model may be obtained by considering the following chemical reaction:

$$aA + bB \underset{k_b}{\overset{k_f}{\leftrightarrow}} cC + dD \qquad (1.1)$$

where A and B are reactants; C and D are the products; *a*, *b*, *c*, and *d* are stoichiometric coefficients for the reactants and products; and k_f and k_b are forward and backward reaction rate constants. The rate at which a reactant, for example reactant A, changes is given by

$$-\frac{d[A]}{dt} = k_f[A]^{\alpha}[B]^{\beta} - k_b[C]^{\gamma}[D]^{\delta} \qquad (1.2)$$

where [A], [B], [C], and [D] are the concentrations of reactants (mass per unit volume); α, β, γ, and δ are the reaction orders with respect to each product or reactant; and *t* is time. Equations (1.1) and (1.2) are for a general case, since these equations are unsolvable because of too many unknowns; therefore, simplifying procedures are used. For example, the reaction conditions may be chosen such that either the forward or the backward reaction is predominant. If the concentration of reactant B in equation (1.1) is kept very high, then the change in concentration of B is negligible. As a consequence, the backward rate constant will be considerably smaller than the forward reaction rate constant. For this case, the rate of reaction can be represented by

$$-\frac{d[A]}{dt} = k'_f[A]^n \qquad (1.3)$$

where k'_f is the pseudo forward rate constant and n is the reaction order. The rate of reaction is described mainly by the concentration of a species. Due to the complex nature of foods, it is difficult to determine the actual mechanisms of the intermediate reactions that lead to a particular change in quality. The following approach is commonly used to analyze general quality changes in foods.

A general rate expression may be written as follows for a quality attribute Q:

$$\pm\frac{d[Q]}{dt} = k[Q]^n \qquad (1.4)$$

where ± refers to either decreasing or increasing value of the attribute Q, k is the pseudo forward rate constant, and n is the observed order of reaction. It is assumed that the environmental factors such as temperature, humidity and light, and concentrations of other components are kept constant.

Equation (1.4) may be rewritten more specifically for a quality attribute that is decreasing with time and following an nth order of reaction,

$$-\frac{d[Q]}{dt} = k[Q]^n \qquad (1.5)$$

1.3.1.1 Zero order reaction

Consider a quality attribute Q, which decreases during the storage period as shown in Figure 1–3. A closer examination of the linear plot reveals that the rate of loss of a quality attribute is constant throughout the storage period and it does not depend on the concentration of Q.

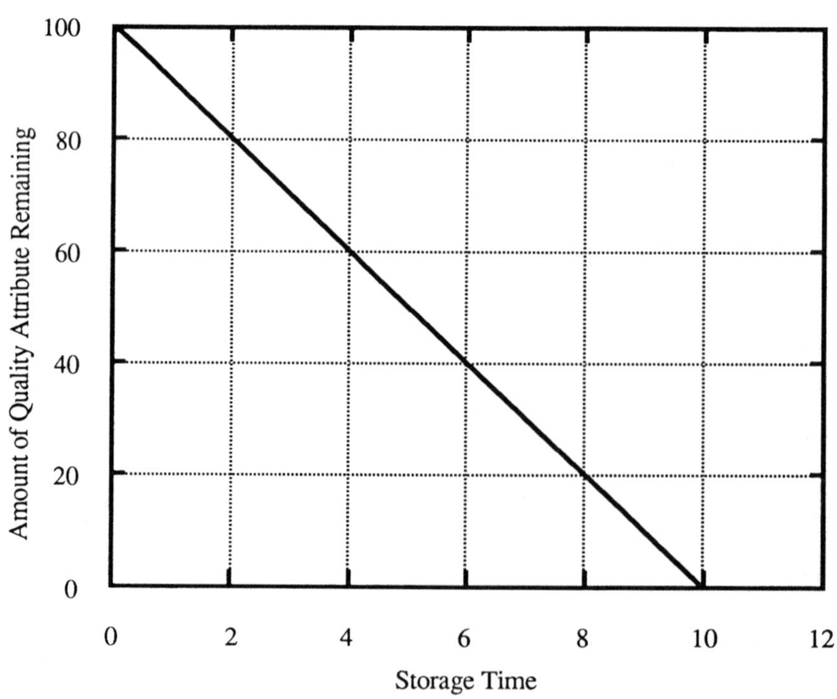

Figure 1–3 Decrease in a quality attribute of a food during storage (zero order reaction).

This assumption has been widely used in the food science literature. This linear plot represents a zero order reaction therefore, substituting $n = 0$ in equation (1.5) gives

$$-\frac{d[Q]}{dt} = k \tag{1.6}$$

Assuming that $[Q]_0$ represents some initial value of a quality attribute and $[Q]$ is the amount of that attribute left after time t, then separating the variables in equation (1.6) and setting up the limits of integration,

$$-\int_{[Q]_0}^{[Q]} d[Q] = k\int_0^t dt \tag{1.7}$$

Integrating

$$-\,|\,[Q]|_{[Q]_0}^{[Q]} = kt \tag{1.8}$$

or

$$[Q]_0 - [Q] = kt \tag{1.9}$$

The left-hand side of equation (1.9) denotes the extent of reaction when the reaction order, n, is equal to 0. Using symbol ξ for the extent of reaction, the equation becomes

$$\xi = kt \tag{1.10}$$

If the end of shelf life, t_s, is noted by the quality attribute reaching a certain level, say $[Q]_e$, then

$$[Q]_e = [Q]_0 - kt_s \tag{1.11}$$

Therefore, the shelf life, t_s, may be calculated as

$$t_s = \frac{[Q]_0 - [Q]_e}{k} \tag{1.12}$$

The use of zero order rate equation (1.6) is useful in describing such reactions as enzymic degradation, non-enzymic browning and lipid oxidation that lead to development of rancid flavors.

1.3.1.2 First order reaction

Consider Figure 1–4 where a quality attribute, Q, decreases in an exponential manner with storage time. The rate of loss of a quality attribute is dependent on the amount of quality attribute remaining; this implies that as time proceeds and the quality attribute decreases so does the rate of reaction. This exponential plot between quality attribute and time represents a first order reaction, $n = 1$, and equation (1.5) is modified as follows:

$$-\frac{d[Q]}{dt} = k[Q] \tag{1.13}$$

To integrate the preceding equation, the variables are separated,

$$-\frac{d[Q]}{[Q]} = k dt \tag{1.14}$$

Assuming that $[Q]_0$ represents some initial value of a quality attribute and $[Q]$ is the amount of that attribute left after time t, and assigning the limits of integration in equation (1.14), the following is obtained:

$$-\int_{[Q]_0}^{[Q]} \frac{d[Q]}{[Q]} = k\int_0^t dt \tag{1.15}$$

Integrating

$$-\,|\ln[Q]|_{[Q]_0}^{[Q]} = k\,|\,t\,|_0^t \tag{1.16}$$

or

$$\ln[Q]_0 - \ln[Q] = kt \tag{1.17}$$

Therefore, for a first order reaction, the extent of reaction is

$$\xi = \ln[Q]_0 - \ln[Q] \tag{1.18}$$

Rewriting equation (1.17),

$$\ln\frac{[Q]}{[Q]_0} = -kt \tag{1.19}$$

where $[Q]$ is the amount of quality attribute left at time t.

At the end of shelf life, t_s, for a certain final level of quality attribute, Q_e, equation (1.19) can also be written as

$$\ln\frac{Q_e}{Q_0} = -kt_s \tag{1.20}$$

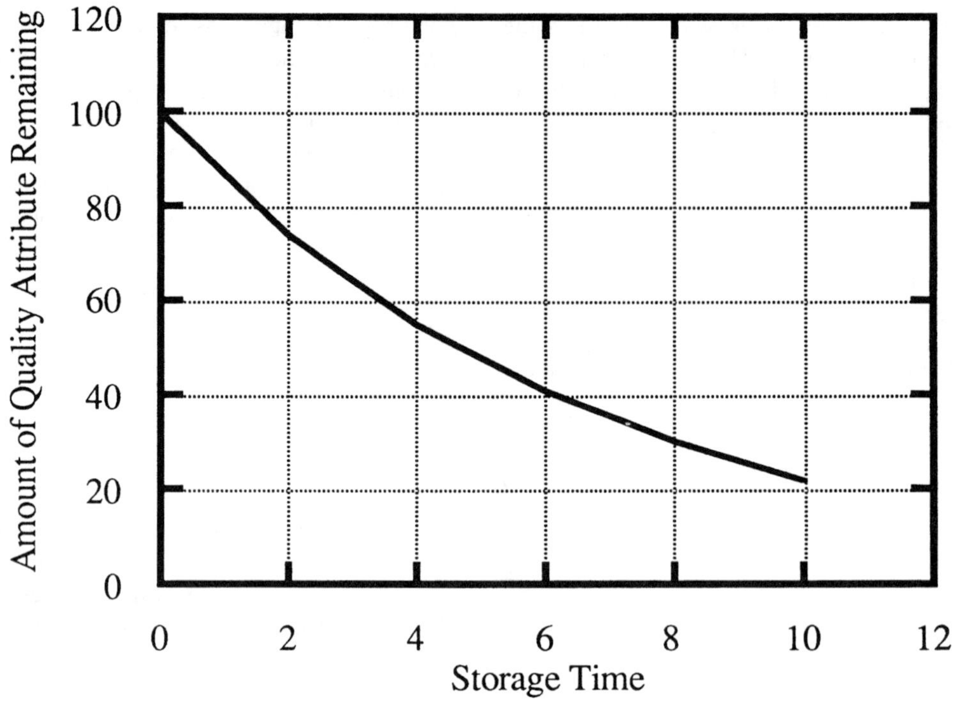

Figure 1–4 Decrease in a quality attribute of a food during storage (first order reaction).

or

$$t_s = \frac{\ln \frac{Q_0}{Q_e}}{k} \quad (1.21)$$

Sometimes, it is desirable to know the time for half life of a reaction. To obtain the half-life time, equation (1.21) may be modified by substituting $Q_e = 0.5\ Q_0$,

$$t_{\frac{1}{2}} = \frac{0.693}{k} \quad (1.22)$$

The types of food deterioration reactions that show first order losses include vitamin and protein losses and microbial growth.

In analyzing experimental data on the rate of change of a given quality attribute, the following two issues are important. First, the analytical precision of measuring attribute level affects the value obtained for rate constant. In Table 1–1 an estimate of percentage error in reaction rate constant, k, is given for different levels of analytical precision (Benson, 1960). As seen in this table, as analytical precision decreases, and when the change in reactant species monitored is small, there is a greater percentage error in the calculated value of the reaction rate constant.

The second important issue in kinetic analysis is the change in reactant species monitored. In Figure 1–5, two plots are shown, one for zero order reaction and the other for a first order reaction. In this case, the overlap between zero order and first order plots, for up to 55% reduction of the quality attribute, indicates that either zero or first order model can be used to describe the change in the attribute. However, beyond 55% reduction, there is a considerable difference in the level of quality attribute predicted by the zero order and first order models. Therefore, if the experimental data on the rate of change of a

Table 1–1 An estimate of percentage error in reaction rate constant, *k*, as a function of analytical precision and change in reactant species monitored

Analytical precision (%)	*Change in reactant species monitored*						
	1%	*5%*	*10%*	*20%*	*30%*	*40%*	*50%*
±0.1	14	2.8	1.4	0.7	0.5	0.4	0.3
±0.5	70	14	7	3.5	2.5	2	1.5
±1.0	>100	28	14	7	5	4	3
±2.0	>100	56	28	14	10	8	6
±5.0	>100	>100	70	35	25	20	15
±10.0	>100	>100	>100	70	50	40	30

Source: Benson, 1960.

quality attribute are available only for less than 55% reduction in the quality of an attribute, then a simple zero order reaction model would be sufficient. However, use of that model to estimate changes greater than 55% reduction of the quality attribute can lead to serious errors if the correct reaction order was in fact first order. Thus, caution must be exercised in determining the reaction order. Labuza (1984) has illustrated some of the common pitfalls that should be avoided in using statistical evaluation of rate of loss of quality.

Most reactions showing losses in food quality may be described by zero or first order; there are

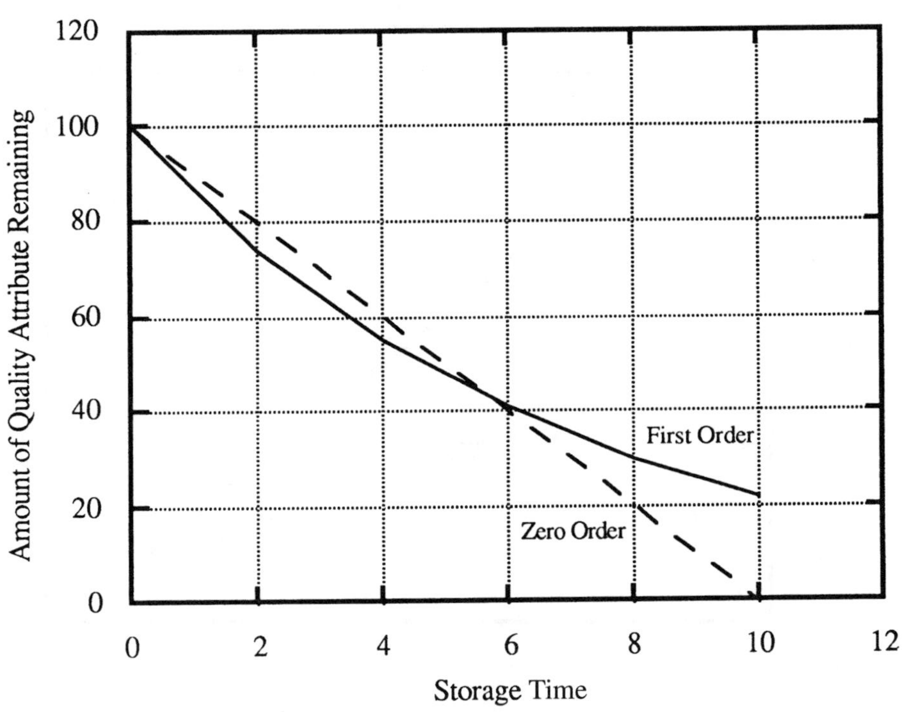

Figure 1–5 Decrease in a quality attribute of a food following two different orders of reactions.

some studies in the literature that indicate use of other orders. A second order reaction was used by Singh and Heldman (1976) to describe vitamin C degradation in canned infant foods. In studying oxygen uptake in lipid oxidation kinetics, Labuza (1971) found a half order reaction with respect to oxygen for relatively pure lipids. The order of reaction increased to first order when antioxidants were added.

1.3.1.3 Temperature effects

The influence of temperature on the reaction rate may be described by using the Arrhenius relationship, as follows:

$$k = k_0 e^{\left[-\frac{E_A}{RT}\right]} \quad (1.23)$$

where k_0 is the pre-exponential factor, E_A is the activation energy, R is the ideal gas constant, and T is the temperature (absolute scale).

The temperature dependence of the reaction rate can be shown graphically as in Figure 1–6. On a semilogarithmic axis, a linear plot is obtained between rate constant and the inverse of absolute temperature. The slope of the straight line gives the activation energy. A steeper slope implies that the reaction rate constant is influenced to a larger extent by a change in temperature.

Another parameter that is often used in the literature to describe the relationship between temperature and reaction rate constant is the Q_{10} value. Q_{10} is defined as follows:

$$Q_{10} = \frac{\text{reaction rate at temperature } (T + 10)^\circ\text{C}}{\text{reaction rate at temperature } T^\circ\text{C}} \quad (1.24)$$

It can be shown that Q_{10} and activation energy E_A are related by the following expression (Singh and Heldman, 1993):

$$\log Q_{10} = \frac{E_A}{2.303R}\left[\frac{10}{T_A(T_A + 10)}\right] \quad (1.25)$$

1.3.2 Determination of kinetic parameters

In order to describe the influence of temperature on the reaction rates, such as by using the Arrhenius expression, it is necessary to know the

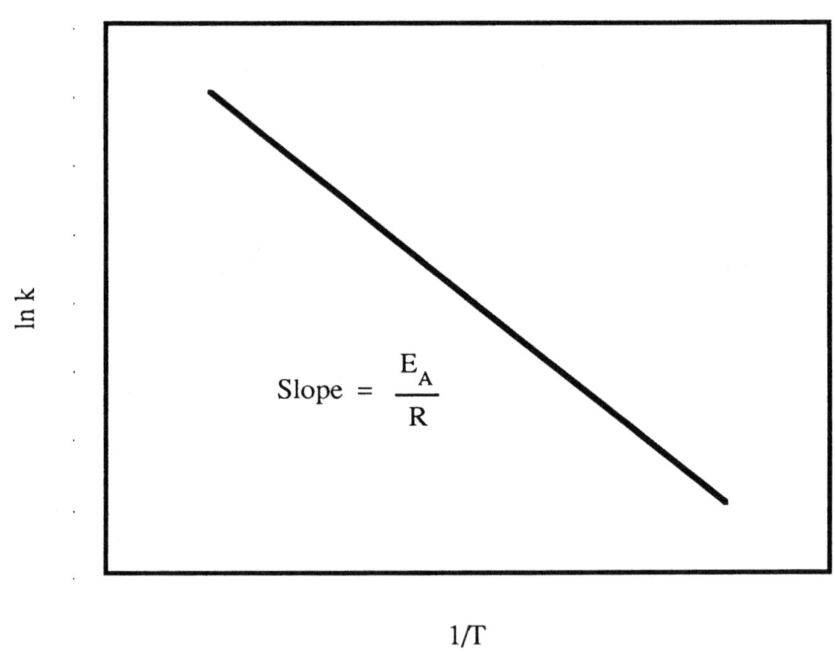

Figure 1–6 Influence of temperature on the reaction rate constant. Slope = E_A/R.

values of kinetic parameters such as rate constant and activation energy. There are various methods that have been used for determining the kinetic parameters. The most common methods are presented as follows.

(a) Differential method

Equation (1.5),

$$-\frac{d[Q]}{dt} = k[Q]^n$$

may be linearized as

$$\ln\left(-\frac{d[Q]}{dt}\right) = n\ln[Q] + \ln k \quad (1.26)$$

In this method, first, the $[Q]$ vs time data are plotted. Then the values of the slope at different $[Q]$ are obtained. The slope values are $-d[Q]/dt$. Next, the slope values are converted as $\ln(-d[Q]/dt)$ and they are plotted against $\ln[Q]$. Using linear regression, the best linear fit is obtained. The slope of the linear plot gives the order of the reaction, and the intercept gives $\ln k$, yielding the value of k.

(b) Fit of experimental data to a linear equation

In this procedure, the concentration of a quality attribute $[Q]$ is measured at different times of storage. The $[Q]$ vs time data are then plotted either on a simple graph paper or an attempt is made to fit the data to a linear equation. If the $[Q]$ vs time data show a good fit to a linear equation, then the data indicate a zero order reaction, as shown by a linear relationship of equation (1.9). On the other hand, if the $\ln[Q]$ vs time fits a linear equation, then the reaction is considered to be of a first order as shown by the relationship of equation (1.17).

(c) Time to complete half life of a reaction

From equation (1.9), for 50% extent of the reaction, when $n = 0$,

$$0.5[Q]_0 = kt_{1/2} \quad (1.27)$$

or

$$\frac{t_{1/2}}{[Q]_0} = \text{constant} \quad (1.28)$$

when $n = 1$, for 50% extent of the reaction, according to equation (1.22)

$$t_{1/2} = \text{constant} \quad (1.29)$$

In this procedure, a storage study is carried out by increasing the initial concentration of a reactant $[Q]$ to different levels, and the half life for each value of initial concentration, $[Q]_0$, is determined. The data obtained are then fit either into equation (1.28) or (1.29). Depending on the best fit, the order of the reaction is determined.

When kinetic data are collected at different temperatures, there are two additional procedures that are most commonly employed to determine the kinetic parameters. The linear regression method involves plotting the log of rate constant against the reciprocal of (absolute) temperature. It is necessary to obtain rate constants for at least three different temperatures. Since this plot should be a straight line, linear regression procedures are used to determine the slope and intercept, thus obtaining values for the pre-exponential factor and the activation energy in equation (1.23).

A non-linear regression method suggested by Arabshahi and Lund (1985), Cohen and Saguy (1985) and Haralampu et al. (1985) is used to determine the activation energy directly from the concentration or level of a quality attribute. This method precludes the determination of rate constants. The advantage of this method lies in the use of original data points related to the variation of quality attribute in calculating the desired kinetic parameters. Standard computer routines such as SAS NONLIN (SAS Institute Inc, 1982) and BMDP 3R (University of California, 1983) are available for non-linear regression.

The expression describing change in quality attribute in a non-linear form for a zero order reaction is as follows:

$$Q_{ij} = Q_0 - k_0 t_{ij} \exp\left(-\frac{E_A}{RT}\right) \quad (1.30)$$

and for a first order reaction, the expression is

$$Q_{ij} = Q_0 \exp\left[-k_0 t_{ij} \exp\left(-\frac{E_A}{RT_j}\right)\right] \quad (1.31)$$

where the subscripts i and j indicate the time and temperature of the corresponding quality attribute measurement.

1.3.3 Shelf-life plots

Labuza (1982) presented a simple approach to determine the effect of temperature on food quality. His method is particularly suited for situations when the available data are meager, for example in cases when the experimental data are available only as time needed to reach a certain value of a quality attribute. This is true for many published studies on shelf life in the literature. A brief introduction to developing shelf-life plots is as follows.

For a given reaction order the following can be written,

$$k = \frac{\text{amount lost at time } t_s}{t_s} \qquad (1.32)$$

where the numerator, the amount lost at time t_s, for a first order reaction = ln Q_0/Q_e, and for a zero order reaction = $Q_0 - Q_e$.

Taking logarithm of both sides of equation (1.32),

$$\ln k = \ln (\text{amount lost at time } t_s) - \ln t_s \qquad (1.33)$$

The Arrhenius expression, equation (1.23), may be rewritten as

$$\ln k = \ln k_0 - \frac{E_A}{RT} \qquad (1.34)$$

Equating right-hand sides of equations (1.33) and (1.34), and rearranging to give

$$-\ln t_s = -\ln (\text{amount lost at time } t_s) + \ln k_0 - E_A/RT \qquad (1.35)$$

According to the preceding equation, a plot of t_s with respect to $1/T$ on a semilog paper should give a straight line (Figure 1–7); the activation energy can then be obtained from the slope of the line, $+E_A/2.303R$. For a small temperature range, less than ± 20°C, t_s may be plotted directly against T without significant error (Labuza, 1982) as shown in Figure 1–8. Then

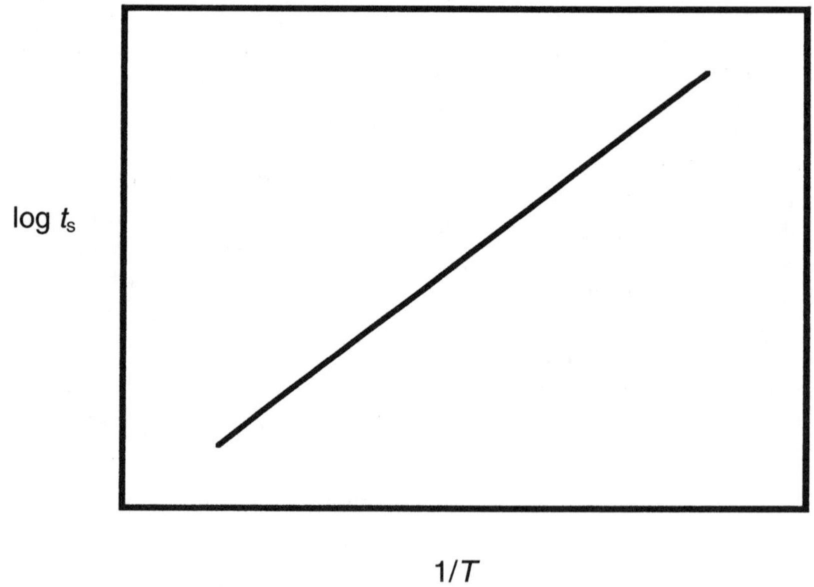

Figure 1–7 A shelf-life plot using inverse of absolute temperature for x-axis.

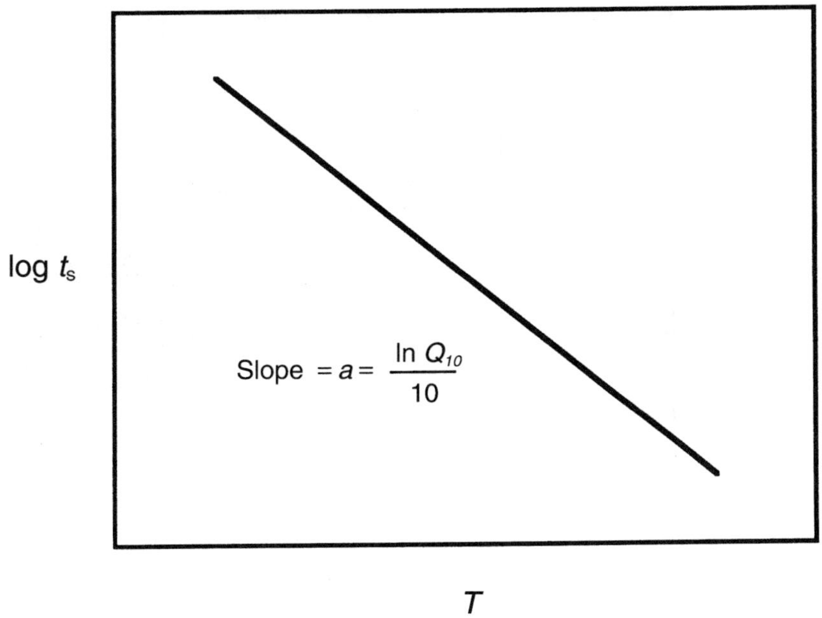

Figure 1–8 A shelf-life plot. Slope = $a = \ln Q_{10}/10$.

the equation for the straight line shown in Figure 1–8 may be written as

$$t_s = t_0 e^{-aT} \quad (1.36)$$

where t_0 is the shelf life at a reference temperature, a is the slope of the line, and T is the temperature difference between the temperature at which the shelf life, t_s, is desired and the reference temperature.

For a zero order reaction, shelf life and rate constant are inversely proportional. Equation (1.24) may be rewritten as

$$Q_{10} = \frac{\text{Shelf life at } T°C}{\text{Shelf life at } (T+10)°C} \quad (1.37)$$

Then from equations (1.36) and (1.37), the following is obtained:

$$Q_{10} = e^{10a} \quad (1.38)$$

or

$$a = \frac{\ln Q_{10}}{10} \quad (1.39)$$

The preceding equation is useful in designing accelerated shelf-life tests. An example is shown in Figure 1–9. If the shelf life at 25°C is 20 months, a series of lines may be drawn through the point representing 25°C and 20 months for different Q_{10} values as shown in Figure 1–9. If the accelerated tests are to be conducted at 40°C, then the accelerated shelf-life times for different Q_{10} values may be obtained from this plot. Thus, if $Q_{10} = 4$, the shelf life at 40°C is 2.5 months, if $Q_{10} = 8$, the shelf life at 40°C is 0.88 months. The same procedure can be easily implemented on a computer with a spreadsheet program. In Figure 1–10, a spreadsheet program and results for the preceding example are shown.

1.4 USE OF SENSORS TO MONITOR SHELF LIFE OF FOODS

In recent years there has been an increased interest in sensors that can be used to monitor quality of stored foods. Many developmental efforts have focused on the time–temperature indicators. As the name suggests, these sensors provide information related to the temperature

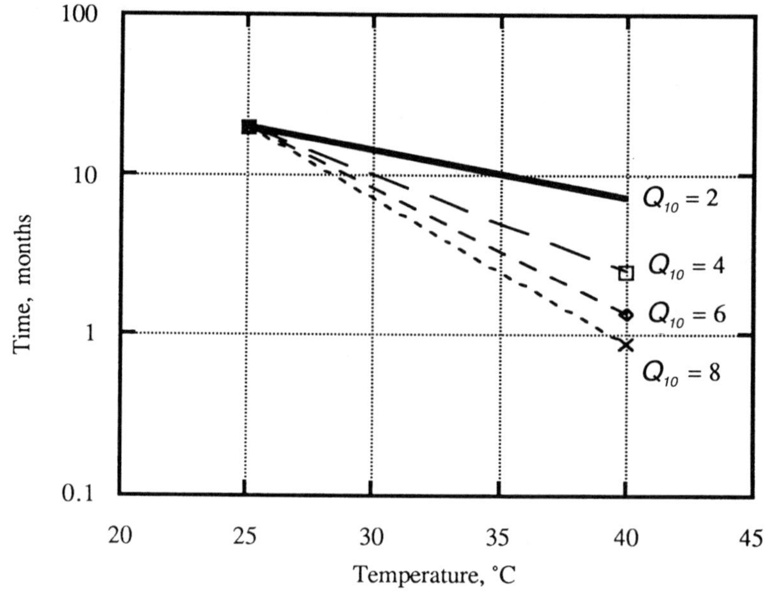

Figure 1–9 Accelerated shelf-life plots for different Q_{10} values.

history of the products to which the sensors are attached. Since temperature is one of the most important environmental factors that influences a number of quality attributes in foods, there is a recognized merit in knowing the temperature exposure of a food consignment during storage and distribution. For the indicators to be most useful for this purpose, they must be irreversible. Preferably, they should provide an integrated response of both time and temperature. During the last few decades, a number of patents have been issued related to devices that can indicate the integrated effect of time and temperature (Schoen and Byrne, 1972; Taoukis et al., 1991b; Kramer

	A	B	C	D	E	F
1	*Ti*	25	C			
2	*T*	40	C			
3	*Q*10	2				
4	*a*	.069	=LN(B3)/10			
5	shelf life at *Ti*	20				
6	shelf life at *T*	7.071	=B5*EXP(-B4*(B2-B1))			
7						
8						
9						
10						
11						
12						
13						

Enter reference temperature *Ti* in cell B1, temperature *T* at which shelf life is desired in cell B2, *Q*10 in cell B3, and shelf life at temperature *Ti* in cell B5. Enter formulas as shown in cells B4 and B6. The results will be shown in cell B6.

Figure 1–10 A spreadsheet to calculate accelerated shelf life at a higher temperature.

and Farquhar, 1976; Farquhar, 1977; Byrne, 1976; Olley, 1976; Wells and Singh, 1992). These devices provide an indication of temperature history using a variety of different principles, such as mechanical, enzymic, or chemical change; development of color; or movement of color along a scale.

Time–temperature indicators may be classified as either critical temperature indicators, partial history indicators, or full history indicators (Taoukis et al., 1991a; Singh et al., 1984).

1.4.1 Critical temperature indicators

Critical temperature indicators give a response only when the temperature either goes above or falls below a certain pre-selected temperature. These indicators do not integrate time with temperature; therefore, they fail to provide the temperature history effect. These types of indicators are useful to indicate if certain temperature was reached during the exposure period, e.g., whether a freeze-sensitive product was exposed to freezing temperature. This type of information is useful if quality attributes of the product are sensitive to some critical temperatures. The MonitorMark™ Thaw Indicator and MonitorMark™ Freeze Indicator (manufactured by Packaging Division/3M™, St Paul, Minnesota) are examples of critical temperature indicators.

1.4.2 Partial history indicators

The partial history indicators provide a response only when the temperature exceeds (or is below) a certain predetermined value. An example is the MonitorMark™ Extended Response Indicator (manufactured by Packaging Division/3M™, St Paul, Minnesota). This type of indicator, described in detail by Manske (1976, 1983), contains a pad saturated with a chemical mixture. The chemical diffuses along a porous wick after the pad is brought into contact with the wick and only when the temperature exceeds the melting point of the chemical. Since the diffusion process is temperature dependent, then for temperatures greater than the melting point of the selected chemical, one obtains a time–temperature integrated response from this type of indicator. Different chemicals with their corresponding melting points can be used for different applications. The performance of these indicators and their applications have been reported by Taoukis and Labuza (1989a, 1989b), Taoukis et al. (1990), and Fu et al. (1991b). These types of indicators have been used in the shipment of pharmaceutical products to tropical regions by the World Health Organization.

A more recent 3M™ sensor is marketed as MonitorMark™ Threshold Indicator. If the temperature exposure exceeds a critical threshold temperature, there is the appearance of blue color in the indicator's circular window. An activation strip is used to prevent indicator response during shipment and storage of the indicator prior to its application to a food package. Upon installation, the activation strip is removed and the indicator begins to respond to the surrounding temperature. Indicators for five threshold temperatures are available, namely, –15, –5, 1, 17, and 26°C. The indicator is 31.8 mm × 19.1 mm × 1.6 mm (approximate size measurements), and it contains a porous wick layer over a reservoir pad containing a blue-dyed chemical that has a desired melt point. The film activation strip separates the wick from the reservoir. The wick is covered with a paperboard layer containing a circular viewing window and a clear film protective over-layer. On the underside of the sensor, there is a layer of pressure-sensitive adhesive for attaching it to a package surface. Prior to installation, the indicator must be preconditioned by placing it in a conditioning temperature chamber that is typically 5–10°C below the indicator response temperature.

1.4.3 Full history indicators

Full history indicators provide a continuous response based on changes in the temperature; the response of these indicators is at a rapid rate at higher temperatures and at a slow rate at lower temperatures. Examples of full history time–temperature indicators include LifeLines Freshness Monitor (manufactured by LifeLines Technologies, Morristown, New Jersey), and

IPOINT Time/Temperature Monitor (manufactured by I-POINT Biotechnologies A.B., Malmo, Sweden, currently discontinued). Improved versions of I-POINT have now become available with new names VITSAB® TTI (Cox Recorders, Belmont, North Carolina) and MonitorMark™ Smart Label (3M™, St Paul, Minnesota). The principle of IPOINT Time/Temperature Monitor is described by Blixt and Tiru (1977), Blixt (1983) and Blixt et al. (1977). This indicator involves two pouches, one containing a lipolytic enzyme and the other containing a lipid substrate. The indicator is activated by breaking the seal between the two pouches and mixing their contents. An enzymic hydrolysis of the lipid substrate results in a decrease of pH with a consequent change in color. Different indicator response times can be obtained by changing the concentration of the enzyme-substrate. The activation energies of these indicators can be made to vary from 15 to 35 kcal/mole, which covers a broad range suitable for many quality degradation reactions during storage and distribution. This indicator has been extensively studied by Taoukis and Labuza (1989a) and Wells and Singh (1988c).

The Lifelines™ Fresh-Scan indicator (manufactured by LifeLines Technology Inc., Morristown, New Jersey) uses a temperature-dependent polymerization reaction (Patel and Yang, 1983; Patel et al., 1977; Patel and Yee, 1980; Fields and Prusik, 1986). In this type of indicator disubstituted diacetylene crystals are allowed to polymerize. As the polymer is formed, there is a distinct change in color. The indicator is in an active state upon manufacturing; therefore, it requires special care during shipment prior to its attachment to a food package. The range of activation energy is 20 to 24 kcal/mole, useful for many food storage applications. The application of this indicator has been explored for a number of foods (Wells and Singh, 1988b, 1988c; Taoukis, 1989; Taoukis and Labuza, 1989a).

The MonitorMark™ Smart Label is a temperature-sensitive sensor that changes color from off-white to black over time and darkens more rapidly at higher temperatures. The sensor is applied to food packages at the processor's location. It is meant to show consumers when packages have been exposed to higher than recommended temperatures for unduly long periods.

The usefulness of these sensors in monitoring quality changes depends on how well they can mimic changes in quality. Studies conducted to date show that these indicators are suitable to follow changes in selected quality attributes due to temperature. In a comprehensive study conducted with 12 different foods (Wells and Singh, 1988b), selected quality attributes were monitored both sensorily and analytically. It was found that a number of these changes were due to temperature exposure. This study also showed that the indicator response of several models could be used to monitor changes in the selected quality attributes. For example, in case of frozen strawberries, a decrease in firmness (an attribute measured using sensory panels) and a decrease in measured amount of ascorbic acid were well correlated with the response of IPOINT model 2340 indicator as shown in Figure 1–11. Similarly, the development of coconut flavor in UHT sterilized milk was strongly correlated with the response of LifeLines Freshness indicator as seen in Figure 1–12. Published studies on correlation of time–temperature indicator response with changes in quality attributes in various foods include hamburgers (Singh and Wells, 1985; Chen et al., 1989); frozen strawberries (Singh and Wells, 1987); tomatoes, lettuce, UHT milk, fruitcake (Wells and Singh, 1988b); pasteurized whole milk (Mistry and Kosikowski, 1983; Cherng and Zall, 1989; Grissius et al., 1987); ice cream (Dolan et al., 1985); refrigerated ready-to-eat salads (Campbell, 1986); refrigerated orange juice (Chen and Zall, 1987b); pasteurized cream (Chen and Zall, 1987a); and cottage cheese (Shellhammer and Singh, 1991; Chen and Zall, 1987a).

The mathematical basis of how the full history indicators mimic a given quality attribute is amply presented in the literature (Wells and Singh, 1992, 1988a). Kinetic models are useful to describe changes in food quality along with the response of time–temperature indicators (Wells and Singh, 1992; Fu et al., 1991a). Briefly, their

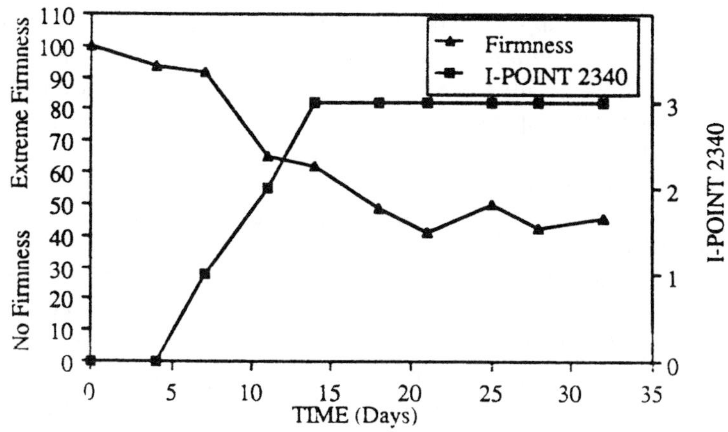

Figure 1–11 A comparison of mean response of I-POINT model 2340 (■) and mean sensory score of tomato firmness (▲) at 15°C (Wells and Singh, 1988b).

approach involved using indicator response at a constant reference temperature and the activation energy of the indicator to predict a constant temperature equivalent of indicator response change for any interval between successive indicator inspections. The constant temperature equivalent was then used to predict the amount of a food quality attribute expected during the same interval. The mathematical derivations that describe this approach for a first order reaction model were presented by Wells and Singh (1988a) and for a zero order reaction by Wells and Singh (1992). The equivalent temperature used by these authors is different from the *effective mean temperature* used by Hicks (1944) and Schwimmer et al. (1955). The equivalent temperature is calculated on the basis of time–temperature indicator and the elapsed time between indicator inspections. If the time intervals between the successive inspections are small or the temperature of storage is constant, then the equivalent temperature predicted by the time–

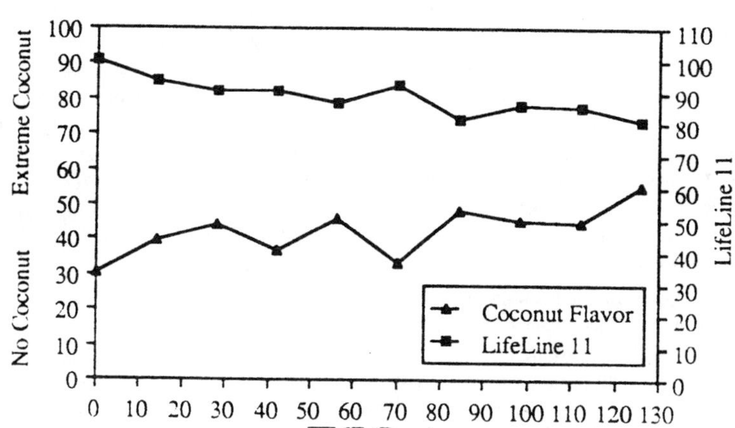

Figure 1–12 A comparison of mean response of LifeLine model 11 (■) and mean sensory score of coconut flavor (▲) in UHT sterilized milk stored at 5°C (Wells and Singh, 1988b).

temperature indicator will be approximately equal to the *effective mean temperature* calculated from the change in food quality. The advantage of using the equivalent temperature approach is that it does not require the food quality to be observed experimentally or predicted theoretically based on temperature history recorded by a data acquisition system.

The time–temperature indicators have proven to be useful aids in monitoring the quality of foods during storage and distribution. The information provided by the indicators can be implemented in computer-aided quality based inventory management systems (Singh and Wells, 1989; Singh, 1989; Wells and Singh, 1989, 1998; Wright and Taub, 1998). It is important to note that a time–temperature indicator provides a history of temperature changes in the immediate vicinity of the indicator. Mathematical models have been proposed to extend this local temperature history to that of the entire unit package used for a food with known thermal properties (Malcata, 1990); these models need to be validated for industrial conditions.

1.5 CONCLUSIONS

In studying shelf life of foods, it is important to measure the rate of change of a given quality attribute. As shown in this chapter, these rates can be modeled using simple zero order or first order kinetics. The shelf-life plots are useful in incorporating the effect of temperature on changes in food quality. The principles of shelf-life evaluation discussed in this chapter should assist a reader in the development of predictive capabilities. This quantitative treatment is necessary to improve and maintain the storage and distribution segment of the food chain.

REFERENCES

Arabshahi, A. and Lund, D.B. (1985) Considerations in calculating kinetic parameters from experimental data. *J. Food Process Eng.*, **7**, 239–251.

Benson, S.W. (1960) *Foundation of Chemical Kinetics*, McGraw-Hill, New York, NY.

Blixt, K.G. (1983) The I-point TTM—a versatile biochemical time-temperature integrator. IIR Commission C2 Preprints, *Proc. 16th Intl. Cong. Refrig.*, Intl. Inst. of Refrig., Paris. p. 629.

Blixt, K.G. and Tiru, M. (1977) An enzymatic time/temperature device for monitoring the handling of perishable commodities. *Dev. Biol. Std.*, **36**, 237.

Blixt, K.G. Tornmarck, S.I.A., Juhlin, R., Salenstedt, K.R. and Tiru, M. (1977) Enzymatic substrate composition absorbed on a carrier. U.S. Patent 4,043,871.

Byrne, C.H. (1976) Temperature indicators—the state of the art. *Food Technol.*, **30**(6), 66.

Campbell, L.A. (1986) Use of a time-temperature indicator in monitoring quality of refrigerated salads. *M.S. Thesis*. Michigan State University, East Lansing, MI.

Chen J.H. and Zall, R.R. (1987a) Packaged milk, cream and cottage cheese can be monitored for freshness using polymer indicator labels. *Dairy Food Sanit.*, **7**, 402.

Chen, J.H. and Zall, R.R. (1987b) Refrigerated orange juice can be monitored for freshness using a polymer indicator label. *Dairy Food Sanit.*, **7**, 280.

Chen, H.C., Singh, R.P. and Reid, D.R. (1989) Quality changes in hamburger meat during frozen storage. *Int. J. Refrig.* **12**, 88–94.

Cherng, Y.S. and Zall, R.R. (1989) Use of time temperature indicators to monitor fluid milk movement in commercial practice. *Dairy Food Environ. Sanit.*, **9**, 439.

Cohen E., and Saguy, I. (1985) Statistical evaluation of Arrhenius model and its applicability in prediction of food quality losses. *J. Food Proc. Preserv.*, **9**, 273–290.

Coultate, T.P. (1996) *Food, the Chemistry of Its Components*. Royal Society of Chemistry, London, England.

Dolan, K.D., Singh, R.P. and Wells, J.H. (1985) Evaluation of time-temperature related quality changes in ice cream during storage. *J. Food Proc. Preserv.*, **9**, 253.

Farquhar, J.W. (1977) Time-temperature indicators in monitoring the distribution of frozen foods. *J. Food Qual.*, **1**(2), 119.

Fennema, O.R. (1996) *Food Chemistry*, Marcel Dekker, New York, NY.

Fields, S.C. and Prusik, T. (1986) Shelf life estimation of beverage and food products using bar coded time-temperature indicator labels, in *The Shelf Life of Foods and Beverages* (ed G. Charalambous), Elsevier Science Publishers, Amsterdam, the Netherlands, 23.

Fu, B., Taoukis, P.S. and Labuza, T.P. (1991a) Predictive microbiology for monitoring spoilage of dairy products with time temperature indicators. *J. Food Sci.*, **56**(5), 1209–1215.

Fu, B., Taoukis, P.S. and Labuza, T.P. (1991b) Theoretical design of a variable activation energy time-temperature integrator for prediction of food or drug shelf life. *Drug. Dev. Ind. Pharm.*, **18**(8), 829–850.

Grissius, R., Wells, J.H., Barrett, E. and Singh, R.P. (1987) Correlation of time-temperature indicator response with microbial growth in pasteurized milk. *J. Food Proc. Preserv.*, **11**, 309–324.

Haralampu, S.G., Saguy, I. and Karel, M. (1985) Estimation of Arrhenius parameters using three least squares methods. *J. Food Process. Preserv.*, **9**, 129–143.

Hicks, E.W. (1944) Note on the estimation of the effect of diurnal temperature fluctuations on reaction rates in stored foodstuffs and other materials. *J. CSIRO,* Australia **17**, 111–114.

International Institute of Refrigeration (IIR) (1986*) Recommendations for the Processing and Handling of Frozen Foods*, 3rd ed, IIR, Paris, France.

Jul, M. (1984) *The Quality of Frozen Foods*. Academic Press, New York, NY.

Kramer, A. and Farquhar, J. (1976) Testing of time-temperature indicating and defrost devices. *Food Technol.*, **30**(2), 50.

Kwolek, W.F. and Bookwalter, G.N. (1971) Predicting storage stability from time-temperature data. *Food Technol.,* **25**(10), 1025,1026,1028,1029,1031,1037.

Labuza, T.P. (1971) Nutrient losses during drying and storage of dehydrated foods. *Crit. Rev. Food Technol.,* **3**, 355.

Labuza, T.P. (1979) A theoretical comparison of losses in food under fluctuating temperature sequences. *J. Food Sci.,* **44**(4), 1162–1168.

Labuza, T.P. (1982) *Shelf-Life Dating of Foods*, Food and Nutrition Press, Westport, CT.

Labuza, T.P. (1984) Application of chemical kinetics to deterioration of foods. *J. Chem. Educ.*, **61**, 348.

Lai, D. and Heldman, D.R. (1982) Analysis of kinetics of quality changes in frozen foods. *J. Food Proc. Eng.,* **6**, 179–200.

Malcata, F.X. (1990) The effect of internal thermal gradients on the reliability of surface mounted full-history time-temperature indicators. *J. Food Proc. Preserv.*, **14**, 481.

Manske, W.J. (1976) Selected time interval indicating device. U.S. Patent 3,954,011.

Manske, W.J. (1983) The application of controlled fluid migration to temperature limit and time temperature integrators. C2 Preprints, *16 Intl. Cong. Refrig.*, Intl. Inst. of Refrig., Paris, France. p. 797.

Mistry, V.V. and Kosikowski, F.F. (1983) Use of time-temperature indicators as quality control devices for market milk. *J. Food Protect.*, **46**(1), 52.

Olley, J. (1976) Temperature indicators, temperature integrators, temperature function integrators and the food spoilage chain. *Aust. Natl. Comm. Joint Mtg. of Comms.*, C2, D1, D2, D3, & E1, Intl. Inst. of Refrig., Paris, p. l.

Patel, G.N. and Yang, N. (1983) Polydiacetylenes: an ideal system for teaching polymer science. *J. Chem. Educ.*, **60**(3), 181.

Patel, G.N. and Yee, K.C. (1980) Diacetylene time-temperature indicators. U.S. Patent 4,228,126.

Patel, G.N., Preziosi, A.F. and Baughman, R.H. (1977) Time-temperature history indicators. U.S. Patent 3,999,946.

Ross, E.W. (1998) Mathematical modeling of quality loss, in *Food Storage Stability* (eds. I.A. Taub and R.P. Singh), CRC Press, Boca Raton, FL, 331–352.

Saguy, I. and Karel, M. (1980) Modeling of quality deterioration during food processing and storage. *Food Tech.,* **34**(2), 78–85.

SAS Institute Inc. (1982) *SAS User's Guide: Statistics, 1982 Edition,* SAS Institute, Cary, NC.

Schoen, H.M. and Byrne, C.H. (1972) Defrost indicators: many designs have been patented yet there is no ideal indicator. *Food Tech.,* **26**(10), 46–50.

Schubert, H. (1977) Criteria for application of T-TI indicators to quality control of deep frozen products. *Science et Technique du Froid* IIF-IIR, **1**, 407–423.

Schwimmer, S., Ingraham, L.L. and Hughes, H.M. (1955) Temperature tolerance in frozen food processing: effective temperatures in thermal fluctuating systems. *Ind. Eng. Chem.*, **47**(6), 1149–1151.

Shellhammer, T.H. and Singh, R.P. (1991) Monitoring chemical and microbial changes of cottage cheese using a full-history time-temperature indicator. *J . Food Sci.*, **56**(2), 402–405 & 410.

Singh, R.P. (1976) Computer simulation of food quality during frozen food storage. *Int. Inst. Refrig. Bull. Supp.*, **1**, 197–204.

Singh, R.P. (1989) Computer-aided inventory management using time-temperature indicators, in *Food Properties and Computer-Aided Engineering of Food Processing Systems* (eds. R.P. Singh and A.G. Medina), Kluwer Academic Publishers, Amsterdam, the Netherlands.

Singh, R.P. and Heldman, D.R. (1976) Simulation of liquid food quality during storage. *Transactions of the ASAE* (American Society of Agricultural Engineers), **19**(1), 178–184.

Singh, R.P. and Heldman, D.R. (1993) *Introduction to Food Engineering*, Academic Press, San Diego, CA.

Singh, R.P. and Wells, J.H. (1985) Use of time-temperature indicators to monitor quality of frozen hamburger. *Food Tech.,* **39**(12), 42–50.

Singh, R.P. and Wells, J.H. (1987) Monitoring quality changes in stored frozen strawberries with time-temperature indicators. *Int. J. Refrig.,* **10**, 296–300.

Singh, R.P. and Wells, J.H. (1989) Time-temperature indicators in food inventory management, in *1989 Food & Beverage Technology International USA*, 195–198.

Singh, R.P., Barrett, E.L., Wells, J.H., Grisius, R.C. and Marum, W. (1986) *Critical evaluation of time-temperature indicators for monitoring quality changes in perishable and semi-perishable foods*. Report prepared for U.S. Army Natick Research & Development Center, Natick, MA. January. (Contract No. DAAK60-84-C-0076).

Singh, R.P., Heldman, D.R. and Kirk, J.R. (1975) Kinetic analysis of light-induced riboflavin loss in whole milk. *J. Food Sci.*, **40**, 164–167.

Singh, R.P., Wells, J.H., Dolan, K.D., Gonnet, E.J. and Munoz, A.M. (1984) *Critical evaluation of time-temperature indicators for monitoring quality changes in stored subsistence*. Report prepared for U.S. Army Natick Research Development Center, Natick, MA. September. (Contract No. DAAK60-83-C-0100).

Taub, I.A. and Singh, R.P. (1998) *Food Storage Stability*, CRC Press, Boca Raton, FL.

Taoukis, P.S. (1989) Time-temperature indicators as shelf life monitors of food products. *Ph.D. Thesis*. University of Minnesota, St. Paul, MN.

Taoukis, P.S. and Labuza, T.P. (1989a) Applicability of time-temperature indicators as food quality monitors under non-isothermal conditions. *J. Food Sci.*, **54**, 783.

Taoukis, P.S., and Labuza, T.P. (1989b) Reliability of time-temperature indicators as food quality monitors under non-isothermal conditions. *J. Food Sci.*, **54**, 789.

Taoukis, P.S., Fu, B. and Labuza, T.P. (1991a) Time temperature indicators. *Food Tech.*, **45**(10), 70–82.

Taoukis, P.S., Labuza, T.P. and Francis, R.C. (1991b) Time temperature indicators as food quality monitors. *Food Pack. Tech.*, ASTM STP 1113, ASTM, Philadelphia, PA 51.

Taoukis, P.S., Labuza, T.P. and Saguy, I.S. (1997) Kinetics of food deterioration and shelf-life prediction, in *Handbook of Food Engineering Practice* (eds E. Rotstein, R.P. Singh and K. Valentas), CRC Press, Boca Raton, FL, 361–403.

Taoukis, P.S., Reineccius, G.A. and Labuza, T.P. (1990) Application of time-temperature indicators to monitor quality of flavored products, in *Flavors and Off-Flavors '89* (ed G. Charalambous), Elsevier Applied Science Publishers, London and New York, p. 385.

University of California (1983) *BMDP Statistical Software* (ed W.J. Dixon), University of California Press, Berkeley, CA.

Van Arsdel, W.B., Coply, M.J. and Olson, R.L. (1969) *Quality and Stability of Frozen Foods.* Wiley-Interscience, New York, NY.

Wells, J.H. and Singh, R.P. (1988a) A kinetic approach to food quality prediction using full-history time-temperature indicators. *J. Food Sci.*, **53**(6), 1866–1871 & 1893.

Wells, J.H. and Singh, R.P. (1988b) Application of time-temperature indicators in monitoring changes in quality attributes of perishable and semiperishable foods. *J. Food Sci.*, **53**(1), 148–156.

Wells, J.H. and Singh, R.P. (1988c) Response characteristics of full-history time-temperature indicators for perishable food handling. *J. Food Proc. Preserv.*, **12**, 207–218.

Wells J.H. and Singh, R.P. (1989) A quality based inventory issue policy for perishable foods. *J. Food Proc. Preserv.*, **12**, 271–292.

Wells, J.H. and Singh, R.P. (1992) The application of time-temperature indicator technology to food quality monitoring and perishable inventory management, in *Mathematical Modelling of Food Processing Operations* (ed S. Thorne), Elsevier Applied Science, London, England.

Wells, J.H. and Singh, R.P. (1997) Temperature tolerance of foods during distribution, in *Handbook of Food Engineering Practice* (eds E. Rotstein, R.P. Singh and K. Valentas), CRC Press, Boca Raton, FL, 405–423.

Wells, J.H. and Singh, R.P. (1998) Quality management during storage and distribution, in *Food Storage Stability* (eds I.A. Taub and R.P. Singh), CRC Press, Boca Raton, FL, 369–386.

Wells, J.H., Singh, R.P. and Noble, A.C. (1987) A graphical interpretation of time-temperature related quality changes in frozen foods. *J. Food Sci.*, **52**(2), 435–444.

Wright, B.B. and Taub, I.A. (1998) Stored product quality: open dating and temperature monitoring, in *Food Storage Stability* (eds I.A. Taub and R.P. Singh), CRC Press, Boca Raton, FL, 353–368.

CHAPTER 2

The Methodology of Shelf-Life Determination

M.J. Ellis and C.M.D. Man

2.1 INTEGRATION OF SHELF-LIFE PROCEDURES INTO A TOTAL QUALITY SYSTEM

Shelf life is an important feature of all foods (Table 2–1): all those involved in the handling of foods should be aware of it, these may include growers, ingredient suppliers, manufacturers, wholesalers, retailers and the consumers. Shelf life of a food product may be defined as the time between the production and packaging of the product and the point at which it becomes unacceptable under defined environmental conditions. Storage and distribution are necessary links in the food chain. Quality (and safety) considerations dictate the conditions and maximum duration of these links in the chain although most food deteriorations take place gradually.

A total quality approach (Figure 2–1) must embrace all aspects of a food from its conception, through development and production to its consumption, and for a manufactured food product this will include:

- product design (including hazard analysis and risk assessment to ensure safety)
- specification and testing of ingredients and packaging materials
- manufacturing processes
- transport, storage, and retail display
- storage at home and consumption
- environmental impact.

Since the food must be safe and have an acceptable quality when consumed, the time for which this is maintained—the shelf life—is therefore an essential aspect of product design, the control of which is a requirement of good manufacturing practice (GMP) (IFST, 1998) as well as a requirement of the International Standard for Quality Management Systems, BS EN ISO 9000 (BSI, 1994).

Shelf-life determination of a new product often requires storage for significant periods, and includes samples from early development stages as well as initial production runs. Through the evaluation of stored samples, potential storage problems can be identified and either eliminated or controlled before the food goes into production. When in production, an ongoing quality assurance system is equally important and involves assessment of freshly made products, typically before the production has been released into distribution. Samples stored for up to the assigned shelf life should also be evaluated, so that any change in the storage performance can be noted, and appropriate actions taken if necessary. Furthermore, an efficient customer complaint management system will provide early warning signals concerning potential shelf-life problems. Finally, an up-to-date, operable, and effective product recall system is the last resort should the need to recall an unsafe or defective product arise. Thus an adequate shelf-life system will require appreciable resources, and these should be allowed for when planning the overall quality budget.

Since most aspects of shelf-life evaluation for new products and monitoring of standard pro-

Table 2–1 Examples of food products and their typical shelf lives

Food product	*Typical shelf life*	*Main limiting factors*
Bread	Up to 1 week at ambient	Stale flavor, texture change
Sauces, dressings	1–2 years at ambient	Stale/rancid flavor, color change
Pickles	2–3 years at ambient	Stale flavor, texture change
Chilled foods	Up to 4 months at 0–8°C	Microbial spoilage, flavor change
Frozen foods	1–1½ years in freezer cabinets	Color/texture change
Canned foods	1–1½ years at ambient	Tin pick-up (e.g., unlacquered cans)
Canned foods	2–4 years at ambient	Texture/flavor change (e.g., lacquered cans)

duction involve similar sample handling and analysis procedures to the rest of the quality system (perhaps with the added requirement of suitable storage facility), it is logical and normally most convenient to integrate shelf-life procedures into the total quality regime. As the manufacturer is likely to be responsible for most if not all product development activities, he will normally be responsible for both the shelf-life determination during product development and ongoing monitoring once regular production has begun. Even if development work has been subcontracted out, it is essential for the manufacturer's "due diligence" system to be

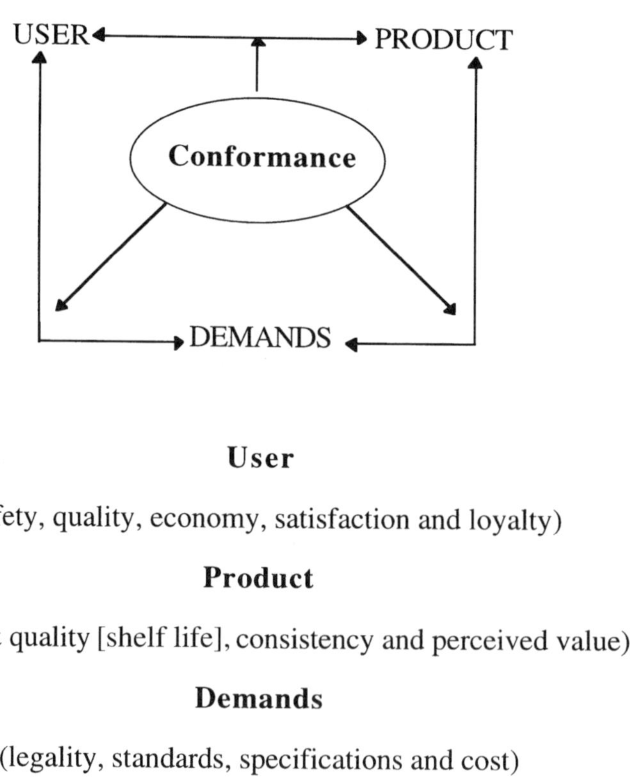

Figure 2–1 A total quality approach—the relationship between user, product and demands.

able to demonstrate the validity and reliability of any shelf life quoted.

Therefore, it is obvious that in order to carry out shelf-life work properly and professionally, there has to be in place within a food company, an effective safety and quality management system. Whether the latter is built upon GMP, BS EN ISO 9000, or any other principles is of secondary importance, so long as the system works, i.e., the system ensures safety and quality consistently. Essentially, there must be lasting management commitment which is translated into adequate provisions of appropriate facilities, resources, and up-to-date and relevant technical expertise. Without genuine management commitment, even the best system will not work. Figure 2–2 illustrates diagrammatically the importance of management commitment as the cornerstone of an effective quality management system.

2.2 FOOD LEGISLATION AND SHELF LIFE

Food legislation in many developed countries requires most prepacked foods to carry a date of "minimum durability." The inference is that procedures have been established (in the main by the manufacturers) for shelf lives to be evaluated. In the UK, for instance, "best before" date is the usual date mark required by current legislation to appear on the labeling of prepacked foods. The "best before" date, however, must be replaced by the "use by" date when the food is microbiologically highly perishable *and* therefore likely, after a short period, to contribute an immediate danger to human health. It is an offense to sell any food after its "use by" date.

The "best before" date is the date up to and including which the food "can reasonably be expected to retain its specific properties," providing it has been stored properly. Food may still be edible after that date, but its appearance and quality could suffer, for example crisp biscuits might become soggy (MAFF, 1991). Naturally, meaningful storage instructions are required alongside the "best before" date if the food is to be enjoyed at its best. While the term "specific properties" may not be well defined for many foods, the regulations, i.e., *The Food Labelling Regulations 1996* (HMSO, 1996), require that product names and descriptions are correct and it

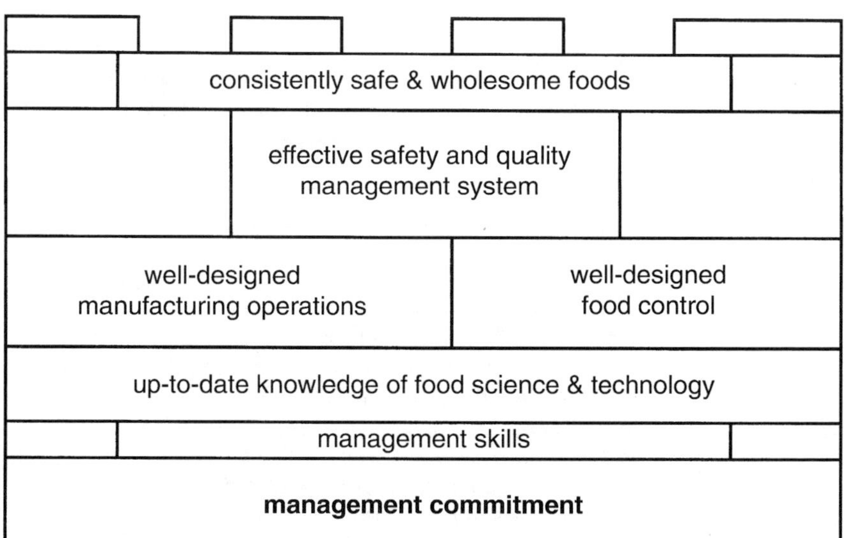

Figure 2–2 Safety and quality—the importance of management commitment. *Source:* Data from the Institute of Food Science & Technology, © 1995, London, United Kingdom.

can be expected that these will be considered misleading if the shelf life which is set allows marked changes in the properties of the food between being freshly made and the "best before" date. Examples of such changes may include the watery or oily separation which may occur during the shelf life of salad cream and mayonnaise. Foods in unlacquered cans may dissolve enough tin to infringe the UK *Tin in Food Regulations 1992* (HMSO, 1992). Such food may be confiscated as unfit for human consumption (HMSO, 1992). Vitamin claims may become invalid if loss occurs during storage. Cakes may not rise sufficiently during baking as a result of failure of the baking powder in the cake mix to effect the desired chemical aeration. Organoleptic properties of a beverage may be impaired due to either simple flavor loss or deteriorative chemical changes.

The seller is generally legally responsible for compliance when the food is sold to the consumer. However, in the case of manufactured foods a retailer would normally demand goods with sufficient remaining shelf life to sell and for the consumer to store for a reasonable period of time before use. An inappropriate shelf life set by a manufacturer could conceivably lead to prosecution since a retailer would be able to demonstrate due diligence in accepting the manufacturer's data.

It should also be recognized that if foods reach the marketplace in substandard condition the legal sanction imposed by the regulatory authority is likely to be trivial compared with the adverse publicity such action would attract, especially for a large or well-known firm. Publicity generated by product recalls, for example, due to yeast-contaminated fat free Thousand Island Dressing (*The Daily Telegraph,* 1991) and off-flavor Smarties (*The Globe and Mail,* 1993), respectively, is something that every manufacturer would do its best to avoid.

Examples of other shelf-life aspects that may need to be considered and are subject to legislative control in the UK are given in Table 2–2.

2.3 FORMS OF QUALITY DETERIORATION DURING STORAGE

Most foods change progressively during storage, and the changes in many cases render the food less attractive and palatable and eventually unfit for consumption. Figure 2–3 (IFST, 1993) summarizes the main deteriorative changes commonly encountered in foods.

Watery separation and "setting up" can occur in thickened foods in cans and jars while suspended particles tend to settle in thinner products. Oil can separate from emulsion products such as salad cream, giving a similarly unattractive appearance. Many food components change in the presence of oxygen. Thus the oil in emulsion products can develop "rancid" flavors. Many foods acquire a stale, sometimes "cardboard" like taste as they age. Natural pigments, e.g., lycopene, change color and in the case of

Table 2–2 Examples of some shelf-life aspects and their legislative control in the UK

Shelf-life aspects	*Examples of relevant legislation*
Tin contents	*Tin in Food Regulations* (1992)
Hygiene	*Food Safety (General Food Hygiene) Regulations* (1995)
Temperature control	*Food Safety (Temperature Control) Regulations* (1995)
Packaging migration	*Materials and Articles in Contact with Food Regulations* (1987)
	Plastic Materials and Articles in Contact with Food Regulations (1998)
Processing	*Food (Control of Irradiation) Regulations* (1990)
Use of preservatives	*Preservatives in Food Regulations* (1989)
Strict product liability	*Consumer Protection Act* (1987) Part 1 (Design, manufacture, etc.)

Source: Adapted with permission from D.J. Jukes, *Food Legislation of the UK—A Concise Guide, 4th Ed.,* © 1997, Butterworth-Heinemann, Oxford.

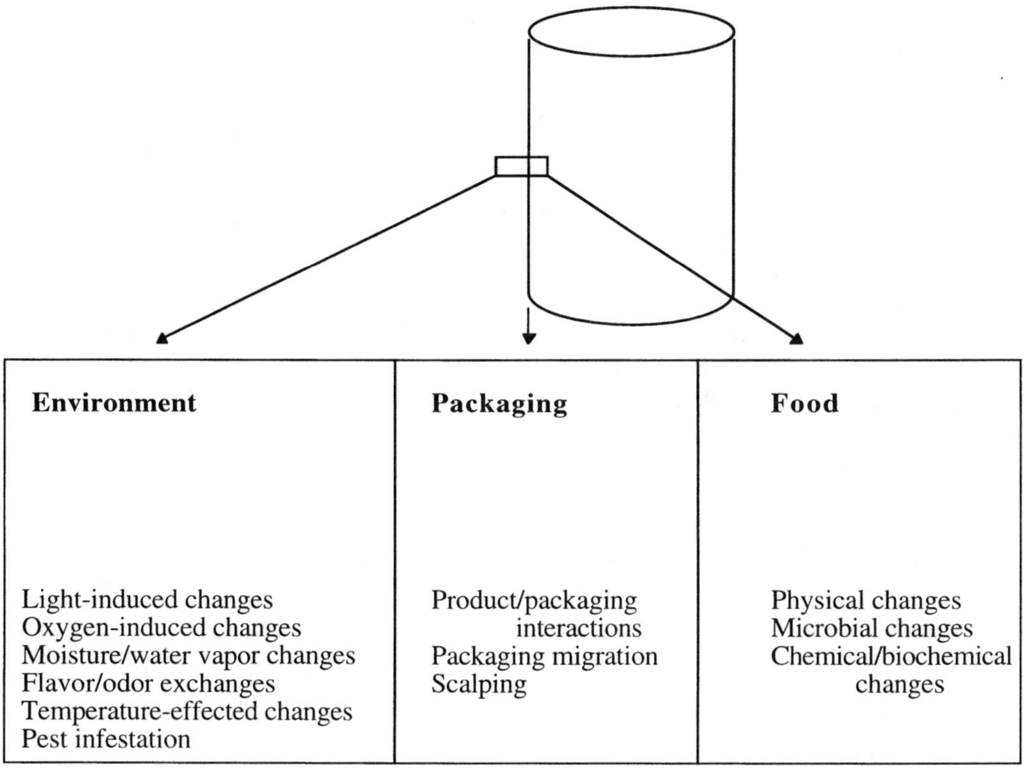

Figure 2–3 Food spoilage and deterioration—a basic model.

tomatoes from red to brown. Filling tomato products in unlacquered tinplate cans mitigates this color effect as the exposed tin scavenges the headspace oxygen during sterilization and storage, giving the food considerable protection from color and indeed flavor changes, but tin dissolves in the process. This can give a metallic taste and contravene the UK *Tin in Food Regulations* if the uptake of tin in the food exceeds 200 mg per kg within the assigned shelf life unless the latter is reduced to allow for this. Tin dissolution, however, can be uneven with some foods; localized attack can expose the steel base plate, which then reacts with the steel to generate hydrogen, creating internal pressure (swell), which would normally be interpreted as a sign of microbial growth. Continued attack can eventually perforate the can in extreme cases and allow contamination. Tin dissolution is increased by the presence of dissolved headspace oxygen, nitrate from water or ingredients, or residual sulfur dioxide (FAO, 1986; CCFRA, 1998, 1999). Although long-term effects are not expected from tin in the diet, high concentrations of tin in food may cause short-term stomach upset in some people (MAFF, 1998). Additionally, incidents of canned foods found to contain mean tin concentrations above the legal limit almost always lead to unwelcome publicity and costly product recalls (*The Daily Telegraph,* 1998a, 1998b, 1998c, 1999). Tin dissolution can be avoided, however, if the cans are fully lacquered internally, but oxygen-induced changes are then potentially greater.

Glass is inert to all foods so offers no oxygen scavenging effect, and color and flavor changes can occur in the surface layer of bottled foods due to reaction with oxygen present in the headspace or leaking in the event that the cap seal is not perfect. Sealed plastic packages (bottles and bowls) used for long-life foods can give more rapid oxidative changes as all plastics have some oxygen permeability.

During storage of frozen foods, "freezer burn" may appear as white surface areas caused by sublimation of water vapor from exposed portions. These areas are usually also toughened due to the dehydration. Some components such as vegetables can change color during frozen storage. Growth of ice crystals and other physical changes in sauces and dessert-type frozen foods can give an icy and granular mouthfeel, rather than the desired smooth texture. Ice cream can lose volume during frozen storage due to loss of the air incorporated during manufacture. Since ice cream is normally sold by volume, this loss may cause infringement of relevant weights and measures regulations.

In dried foods, oxygen-sensitive components may be more exposed than in the same foods when fresh, giving rise to rapid development of rancidity in oil-containing foods such as formation of "musty" or "hay" flavors in dried herbs. Loss of crispness can occur quickly if the food is hygroscopic unless protection is offered by the use of suitable packaging.

Deterioration in many fresh foods is largely due to growth of microorganisms, but wilting of vegetables is a result of water loss while staling of baked goods can be partly explained by changes in the structure of their starch component.

2.4 PROCEDURES FOR DIRECT DETERMINATION AND MONITORING OF SHELF LIFE

Direct shelf-life determination requires batches of samples to be taken at significant stages in the development or modification of a product. These samples should be examined during storage, usually under controlled environmental conditions, until their quality becomes unacceptable. The time when this occurs is the maximum product shelf life, and therefore the determination necessarily requires at least this time to complete. Section 2.5 outlines methods for shortening the process for long-life foods to avoid delays to product launches due to shelf-life testing.

Significant sampling stages within the program of shelf-life evaluation include (Figure 2–4):

- The successful experimental kitchen or pilot plant batch: While scale-up to production will normally highlight some differences, this stage may precede lengthy development activities such as market research or plant construction, allowing longer storage experience to be gained. At this stage it is possible to investigate formulation, process, or packaging changes to improve the shelf life without the costs of

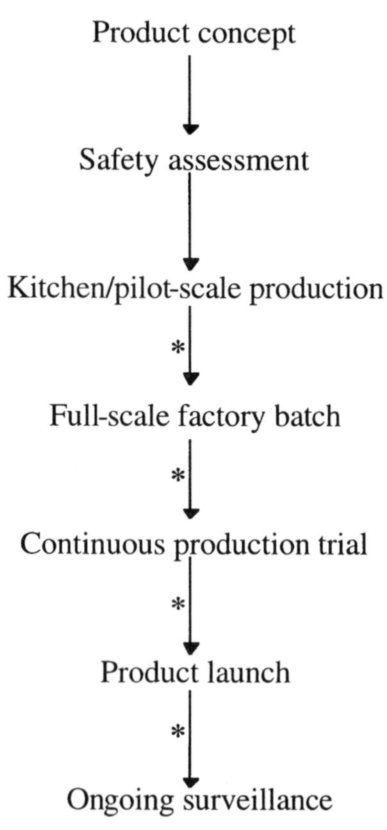

Figure 2–4 Evaluation of shelf life—significant sampling stages (*).

factory time and material quantities. Despite its complexity, the assurance of product safety and quality (i.e., achieving desired shelf life) and control of costs have remained the chief guiding principles of every scaling-up process.

- The successful full-scale factory batch: This is the most important sampling stage as the food should be substantially the same as subsequent production and examination of samples of the product will provide data for the setting of shelf-life and specification standards.
- The first continuous production trial: Examination of products should confirm the data from earlier samplings.
- Early runs of standard production not supervised by development personnel: This constitutes the transition between the development and production stages. Examination of samples from these runs forms a major part of the development review, allowing a first opportunity to adjust the shelf life should the need arise.

As part of an ongoing surveillance system, samples should be taken at suitable intervals for storage trial. The sampling interval should typically be 20% of the shelf life which will provide samples of about six different ages from fresh to the full shelf life, for examination at any time. For long-life products more frequent intervals may be useful to detect any changes in storage performance (e.g., every 2 months for a 2-year shelf life) while for shorter-life products a frequency longer than once-a-month is usually unnecessary.

Shelf-life samples should be subjected to conditions effectively simulating the normal storage and distribution conditions the food is likely to encounter. For ambient stable foods this usually means normal room temperature and relative humidity, but as laboratory and experimental kitchen temperatures are likely to be higher than that in the warehouse, a store cooled to and maintained at the mean warehouse temperature may be required. The following are a number of fixed (artificially created and controlled) conditions commonly used in the direct determination of shelf life:

1. frozen: –18°C or lower (relative humidity is usually near 100%)
2. refrigerated: 0 to +5°C, with a maximum of +8°C (relative humidity is usually very high)
3. temperate: 25°C, 75% relative humidity
4. tropical: 37°C, 95% relative humidity.

For foods packed in impermeable packaging, relative humidity of the storage environment is unlikely to be important in influencing shelf life. However, if shelf life is limited by moisture gain/loss (for instance, biscuits and cakes, respectively) or if the food is packed in moisture-sensitive oxygen barrier plastics such as ethylene vinyl alcohol copolymer (EVOH), control of the relative humidity becomes a major consideration in the shelf-life study.

If performance in tropical or subtropical conditions is required, similar storage conditions should be used as many chemical changes will be accelerated under these conditions. Similarly, if temperatures near or below freezing are envisaged, they should be used in testing as the conditions may destabilize thickeners such as modified starches.

Frozen foods are normally stored between –25 and –30°C for long periods by manufacturers and at –18°C during distribution, retail display, and in the home freezer. A realistic storage trial regime incorporating these temperatures should be employed, and, since frozen foods are frequently taken home after purchase in ambient or insulated containers only, a suitable period at an appropriate temperature should also be included. If storage instructions on the pack are to provide permissible periods in "star" (*) marked freezers or frozen food compartments of refrigerator, storage periods at –12°C (2*) and –6°C (1*) should also be included (UKAFFP, 1978).

Some foods such as jams, pickles, sauces, meat pastes, margarines, and low-fat spreads are normally used over a varying period of time (from days to weeks) after the pack has been opened. An "open" shelf-life storage test should be included for these products involving removal of portions for tasting or other examination at predetermined intervals and reclosing the pack over a suitable time span. Storage of the package between successive openings should be under recommended conditions (or likely conditions if misuse/abuse is expected). The latter usually means either room temperature or that of a domestic refrigerator.

Shelf-life examination of foods invariably involves, at some points, the assessment of product appearance, smell, texture, and flavor. Such assessments can be difficult as there may not be a suitable control sample with which to compare the stored samples, this being particularly so for new products. Freezing can often be a useful means of preserving reference/standard samples, but there is no universal method for keeping them. In many cases remaking of reference samples every time a comparison with stored samples is made may be the only option. It follows then the importance of proper handling, storage, and identification of reference samples cannot be overemphasised.

Quantitative measurements, for example, of color, texture, viscosity, amount of water, or oil separation, should be included if they either closely relate to organoleptic quality or can be used as reliable indicators of quality deterioration.

Given the assurance of food product safety, increasingly, carefully planned and properly conducted sensory evaluation involving the use of trained assessors is fast becoming an important part of every shelf-life study. Setting up an effective panel of assessors is a resource-intensive process, requiring expertise, facilities, training, patience, and commitment of all concerned (Kok, 1999). Just as critical chemical and microbiological parameters require standards against which they can be measured and monitored, from a sensory perspective, the key to arriving at a realistic and acceptable shelf life for a food product lies with the production of a precise sensory specification which identifies important organoleptic attributes and acceptable tolerances around them. One of the most important techniques that can be used to develop sensory specification is quantitative descriptive analysis (QDA) (Stone and Sidel, 1998), also known as conventional profiling, although other techniques can also be used. In general, appropriate sensory evaluation techniques enable questions such as "How long can a product be stored before its sensory properties noticeably change?" and "How long before the organoleptic quality changes make the product unacceptable?" to be answered, questions that are central to all shelf-life determinations.

In addition to sensory evaluation, other tests may be necessary. These may include tin content of products in unlacquered cans, vitamin content where a claim is made, or microbiological examination of fresh or chilled foods.

As previously mentioned, as a general rule, samples of product retained for shelf-life determination should be evaluated about six times during the assigned shelf life of the product. Thus short-life fresh or chilled foods may require daily examination until they become unacceptable, while long-life products, for example, canned foods, may need to be evaluated once every 6 months.

As a shelf-life scheme generates a substantial amount of information, control and documentation of this information are important, especially if a large number of different products are involved and there is a continuing program of innovation and development. A convenient basis of control is the allocation of a number to each different product and a serial number for each batch of samples tested during development and the commercial life of the product. For instance, a particular canned soup could be given the number 123 and thus 123/5 might be the fifth batch to be tested. Even if the product name is changed during or after development, the same shelf-life number can be retained for easy identification. Each batch of experimental or production samples can then be assigned an evaluation record identified by this number and kept as a file indexed by product shelf-life number, making the shelf-life history of a product readily accessible. Where a substantial number of different products are under development and being monitored on an ongoing basis, data control and, in particular, "cross indexing" to identify similar product or packaging changes over a period of time can still be cumbersome and in this case the use of a computer spreadsheet or database will probably be feasible and worthwhile even if the total record system is not computerized.

2.5 METHODOLOGY FOR ACCELERATED ESTIMATION OF SHELF LIFE

The system described in the previous section necessarily requires the full shelf life, possibly several years, to produce results. Development of new long-life foods will normally require results in a much shorter time to meet product launch schedules. More rapid alternative techniques are available, but they should always be supplemented by normal condition testings ("direct determination") for confirmation since it is hardly possible to predict the storage performance of a product under normal conditions with certainty from its behavior when "abused."

A number of accelerated techniques are in use. When using these it is a general rule that the more rapid the degradation induced, and thus the further from normal storage conditions, the less reliable the shelf-life estimate is likely to be. The potential problems and possible errors that can arise in the use of accelerated techniques have been described elsewhere (IFST, 1993; Robertson, 1993). There is no advantage in destabilizing a product which is perfectly stable during normal storage. The results obtained from accelerated techniques must be interpreted with caution as they are not applicable to all products. Using experience of similar products, the likely shelf life of a new or modified product can often be estimated before any storage tests are done. A conservative shelf life with a generous margin of safety should be given in this case so that the shelf life can be confirmed or modified as soon as accelerated and normal storage data have become available.

Raising the storage temperature will accelerate many aging processes and this is the basis of many of the accelerated methods. Storage at 30 to 33°C can give a 2- to 3-fold increase for many, in particular, flavor changes. Storage at 35 to 40°C can bring about a 4-fold speed-up of oil or water separation and facilitate tin dissolution in unlacquered cans. Storage at 55°C for a period of 4 to 6 weeks can show instability in pickle and sauce products. Products that remain stable after this storage will probably be stable for a long time at ambient temperatures; however, such a high temperature has been shown to cause acid hydrolysis of starches in normally stable foods.

Cycling the product between 0°C and room temperature will accelerate watery separation in starch-thickened foods. Absence of any separation after 30 cycles over 2 months normally suggests that the product will be stable for up to 2 years or more at ambient temperatures.

Where oxygen causes flavor or color changes in permeable (e.g., plastics) packaging, high oxygen atmospheres will speed up changes, as will a high relative humidity if EVOH is used as the oxygen barrier layer. Storage of the products in a nitrogen atmosphere should also be considered alongside air or enhanced oxygen conditions so that the effects of oxygen can be isolated from those due to other changes.

Controlled shaking (at 250 to 300 strokes per minute) to provide moderate product agitation for several hours will cause separation in unstable emulsion products. Shaking rates and package geometry must be closely controlled (or preferably standards should always be shaken alongside experimental samples) as these parameters greatly influence product movement and shearing.

Storage defects in frozen products can be accelerated by storage at higher than normal temperatures. Thus more rapid changes will occur at −18°C compared with the normal long-term storage temperature of below −25°C, and still faster changes will occur at −10°C. Certain forms of deterioration, such as ice crystal growth and freezer burn, will also be accelerated if the storage temperature is made to fluctuate while the food still remains frozen.

In recent years, developments in mathematical modeling, facilitated by the spectacular developments in personal computers, have meant that it is now possible to estimate shelf life of foods using computer-based models. Indeed, using computer models to predict microbial growth and non-thermal inactivation is now a reality (Anon., 1999). In the United States, the Department of Agriculture (USDA) has produced a predictive pathogen modeling program which is available free on the Internet (http://www.arserrc.gov/mfs/pathogen.htm). Models that can predict non-microbiological changes are also available. However, because the nature of these changes tends to be rather more product specific and there is a general need to ensure microbiological safety, models that are used to predict, for example, the development of rancidity or moisture migration in foods, are less publicized than microbiological ones.

2.6 THE IMPORTANCE OF SHELF LIFE TO QUALITY AND DISTRIBUTION

Production of foods of consistently acceptable quality with shelf lives adequate for their intended uses together with the correct communication to the consumer of their durability are important to the manufacturer, retailer, and consumer. The consideration which is most important is that the food must reach the consumer in good condition and retain its quality for the period expected, including a period after the package has been opened if normal use implies this. All other factors are subordinate: consumers will not continue to buy unsatisfactory products.

Modern manufacturing processes are often more efficient if long production runs are made. This will impose a constraint on shelf life if a number of different products are made on the same plant as inventory must be accumulated to cover the period between production runs. The quality of the food made before and after the required storage period must be similar, and both must be acceptable.

While consumer demand and rate of production will normally be matched to minimize storage and thus inventory costs, unforeseen fluctuations in demand can necessitate storage of products for longer than intended. For certain food products such as some bakery goods, frozen storage may be used successfully to extend the shelf life as a means of stockpiling prior to an expected period of increased sale. In addition, many foods are distributed nationally or internationally, so time is required for transport, handling, and transit storage at various stages. Careful considerations must also be given to the different environmental conditions prevalent in the different export markets in respect of adequate packaging protection to maintain shelf life. Today's large retailers will normally turn stocks over rapidly, small retailers will frequently be slower in selling and may well operate less controlled stock rotation systems.

Finally, the retailer will require time to sell the food to the consumer. Large retailers will usually have rapid stock turnover and thus an actual need for relatively short storage, while smaller shops with lower volumes may need a longer time to sell the food. Also, commercial considerations may dictate that a shorter shelf life will suffice so as to give the product a perceived fresher image and hence a commercial advantage. All will require a "best before" (or "use by") date sufficiently long after delivery for their needs. With their very large purchasing influence, major retailers may well be in a position to demand in their contracts a long residual shelf life on foods delivered to them.

Frequently there may be a conflict between quality and the logistic requirements of manufacture and delivery, so some compromise may have to be made. Quality (and of course safety) should always be given the greatest weight, and statutory quality standards must never be compromised. It may well be better in the long term not to produce and market a food product if quality and an adequate shelf life cannot be simultaneously realized.

2.7 CONCLUSIONS

For most foods, adequate shelf life can be achieved. Knowledge of the factors affecting shelf life and use of valid and appropriate methodology are vital to the delivery of an acceptable product to the consumer. Sufficient care and adequate resources should be devoted to shelf-life evaluation. A safe and quality product that consistently pleases its customer has its origin in good product design which includes properly conducted and documented shelf-life procedures and ongoing surveillance, evidence that can be used to prove that the manufacturer has considered his products and problems and that he does care about what he produces.

REFERENCES

Anon. (1999) *Food MicroModel Version 3.0*, Food MicroModel Ltd., Surrey, England.

BSI (1994) *BS EN ISO 9000 Quality Systems—Model for Quality Assurance in Design, Development, Production, Installation and Servicing*, British Standards Institution, London, England.

CCFRA (1998) *Investigation of sources of tin in canned foods*. Final Report to MAFF, Campden and Chorleywood Food Research Association, Chipping Campden, Gloucestershire, England.

CCFRA (1999) *Investigation of sources of tin in canned foods: Part Two*. Final Report to MAFF, Campden and Chorleywood Food Research Association, Chipping Campden, Gloucestershire, England.

The Daily Telegraph (1991) *Recall of Kraft Fat Free Thousand Island Dressing*. 9th September, London, England.

The Daily Telegraph (1998a) *Recall of Princes 20% Extra Free Chopped Tomatoes in Tomato Juice (480g)*. 21st November, London, England.

The Daily Telegraph (1998b) *Recall of Princes 227g Chopped Tomatoes in Tomato Juice*. 9th December, London, England.

The Daily Telegraph (1998c) *Recall of Napolina Chopped Tomatoes in Tomato Juice (400g)*. 9th December, London, England.

The Daily Telegraph (1999) *Recall of Tesco Italian Long Cut Spaghetti in Tomato Sauce (410g)*. 3rd February, London, England.

FAO (1986) *Guidelines for can manufacturers and food canners: prevention of metal contamination of canned foods*. FAO Food and Nutrition Paper 36, FAO, Rome, Italy.

The Globe and Mail (1993) *Smarties Recall Announced*. 21st August, Toronto, Canada.

HMSO (1992) *The Tin in Food Regulations 1992*, SI 1992 No. 496.

HMSO (1996) *The Food Labelling Regulations 1996*, SI 1996 No. 1499.

IFST (1993) *Shelf Life of Foods—Guidelines for Its Determination and Prediction*, Institute of Food Science and Technology, London, England.

IFST (1998) *Food and Drink—Good Manufacturing Practice: A Guide to Its Responsible Management*, 4th edn, Institute of Food Science and Technology, London, England.

Jukes, D.J. (1997) *Food Legislation of the UK—A Concise Guide*, 4th edn, Butterworth Heinemann, Oxford, England.

Kok, C. (1999) Development of sensory quality specifications for a ready-to-eat potato salad. M.Phil. Thesis. South Bank University, London, England.

MAFF (1991) *Best Before and Use By—A Guide to the Changes*, Food Sense, London, England.

MAFF (1998) *Cadmium, Mercury and Other Metals in Food*. Food Surveillance Paper No. 53. The Stationery Office, London, England.

Robertson, G.L. (1993) *Food Packaging*, Marcel Dekker, New York, NY.

Stone, H. and Sidel, J.L. (1998) Quantitative descriptive analysis: development, applications and the future. *Food Technol.*, **52**(8), 48-52.

UKAFFP (1978) *Code of Recommended Practice for the Handling of Quick Frozen Foods*, 2nd edn, United Kingdom Association of Frozen Food Producers, London, England.

Chapter 3

The Principles and Practice of Shelf-Life Prediction for Microorganisms

S. J. Walker

3.1 INTRODUCTION

Prior to determining the shelf life of a food, it is essential to determine which factor(s) actually limit the shelf life. Such factors may include chemical, physical, and biological changes which result in a sensory change in the food or in the food becoming unsafe. If the limiting factor is not correctly identified, then any subsequent studies will be flawed. Unfortunately, in many cases, the factors studied are those most easily measured rather than those which limit the shelf life. For the remainder of this chapter, only microbiological changes will be considered. With microbial growth, it is often useful to consider safety and spoilage separately although the controlling factors for both may be identical. Issues relating to safety must take precedence over those relating to spoilage.

Traditionally, food microbiologists have determined the microbiological shelf life of foods using "expert judgement," storage trials, or challenge tests. Unfortunately, "expert judgment" tends to be very subjective and with the greater onus on food manufacturers to demonstrate due diligence, this option has become less desirable. Both shelf-life trials and challenge tests, if done correctly, are very powerful weapons in the food microbiologist's armory and have served the food industry well (Brown, 1992). A challenge test is the laboratory simulation of what may happen to a food during its life and may involve inoculation with specific microorganisms and/or storage at abuse temperatures. Useful guidance on challenge testing is available (Anon., 1987). While the results from these are extremely valuable, challenge tests tend to be costly and time consuming. Furthermore, the results of a challenge test apply specifically to those conditions tested, and any change in a product's formulation or storage temperature may invalidate the results previously obtained. Consequently, there has been a great deal of interest in other ways to determine the shelf life of foods. Perhaps the most promising method is the use of predictive models for microbiology. These models allow the user to predict the fate of microorganisms when a food is stored under specific conditions of temperature, pH, gaseous atmosphere and so on.

The use of predictive models for foods is not new and the safe thermal processing of many canned goods is based on models for microbial inactivation developed more than 70 years ago (Esty and Meyer, 1922). In addition, there has been much interest in the development of models for microbial growth and non-thermal inactivation particularly in the last 15 years. Predictive models for microbial growth can be divided into two main types: those which predict the probability of an event occurring (probability models) and those which predict the change in microbial numbers (kinetic models). In some circumstances, probabilistic models are very powerful and have been used to predict the likelihood of *Clostridium botulinum* growing in cured meat products (Roberts and Jarvis, 1983). Kinetic models are often more useful as they not only indicate the ability to grow but also provide information about the rate and extent of growth that occurs in a given time period. The major advantage of predictive models is that they may be used to predict the fate of microorganisms in conditions not specifically tested during the construction of the model.

3.2 DEVELOPMENT OF PREDICTIVE MODELS

The development of predictive microbiological models for use in foods is neither simple nor inexpensive. A number of stages must be successfully completed if a model is to be of widespread use and applicable to foods.

3.2.1 Microorganism selection and inoculation

Most models have been developed following the inoculation of one or more strains into a test medium. The selection of the strain(s) is crucial. If it is atypical of the target species or group, then the resultant model will be flawed. Many workers use a cocktail of strains in an attempt to overcome this. The culture conditions of the strain prior to inoculation may also be important and affect subsequent growth (Walker et al., 1990). Finally, care is needed to ensure that the inoculation method does not alter the conditions of the test medium. For example, addition of a liquid inoculum to a dry foodstuff could alter the water activity and so affect subsequent microbial kinetics. In developing a model, all these factors need to be well defined and controlled.

3.2.2 Experimental design

As with all scientific research, the experimental data are crucial. An inappropriate design may result in the collection of vast amounts of unnecessary data or in data that cannot be successfully modeled. Furthermore, the factors modeled (e.g., temperature, pH, salt content, atmosphere) and their ranges must be appropriate to the intended use of the model. For example, there is little merit in a model for a psychrotrophic pathogen which applies only to high storage temperatures (25–35°C) if the major concern is with its growth at refrigeration temperatures.

3.2.3 Data collection

The accumulation of data on the growth or inactivation of microorganisms is undoubtedly the most costly stage of model development. Most work is done using microbiological media as they are of consistent composition and can be easily and reproducibly adjusted to the conditions required for study. The inherent variability of foods coupled with the microbial response tends to result in few studies directly in foods (Maxcy and Wallen, 1983), although simple models have been developed in largely homogeneous foods such as milk. Most data have been developed from studies using plate count techniques, although automated techniques such as optical density and conductimetry have been used. As workers develop models containing four or more factors, the requirement for automated data collection techniques will increase.

3.2.4 Modeling

The types of models that have been used vary and include linear and non-linear Arrhenius models, Bêlehrádek-type (e.g., Ratkowsky or square root) models, polynomial models, mechanistic models (all reviewed by McMeekin et al., 1993), combined growth-survival-inactivation models (Jones and Walker, 1993) and dynamic modeling approaches (Baranyi and Roberts, 1994). At present, it is inappropriate to recommend any single modeling technique. Rather, the technique should be suitable for the data used and should be economical. The modeling techniques currently employed are largely descriptive in nature. The development of mechanistic models, which are based on an understanding of how a microorganism behaves, will be one of the major advances in the future of modeling.

3.2.5 Model validation

Perhaps the most important stage associated with the development of models in microbiological media is their validation. One approach to model validation has been described by McClure et al. (1994). The validation should ensure that not only does the model describe accurately the data used to produce it, but it should also demonstrate the applicability of the model to foods. This latter stage is crucial if the users of models are to have confidence in the model's predictions. Food validation may be done using data from the published literature, industry-generated data, specific inoculated food studies (Figure 3–1) or by monitoring the development of the natural flora of a food. Pin et al. (1999) have considered how the validation process used may affect the errors associated with the predictions. Any deficiencies in a model to predict well for some, or all, foods should be evident to the model user.

3.2.6 Obtaining predictions

The final stage of model development is to ensure that the user can obtain useful predictions. Developments in computer hardware and software have permitted the output of models to be presented in a variety of ways. Initially, the vast majority of models available were for pathogenic bacteria. Part of the reason for this focus on pathogens is the high cost of developing models and major initiatives in the UK and USA were funded by national governments with a view to improving food safety. The UK government funded a major modeling programme to predict the growth, survival, and inactivation of a wide range of foodborne pathogens and is known as Food MicroModel. This is commercially available from the Leatherhead Food Research Association (Leatherhead, Surrey, UK). Similarly, the United States Department of Agriculture (USDA) funded the development of the Pathogen Modelling Program (PMP), which is available free-of-charge and can be obtained from the Internet (www.arserrc.gov/mfs/pathogen.htm).

In addition, systems are now available for predicting the growth of spoilage microorganisms.

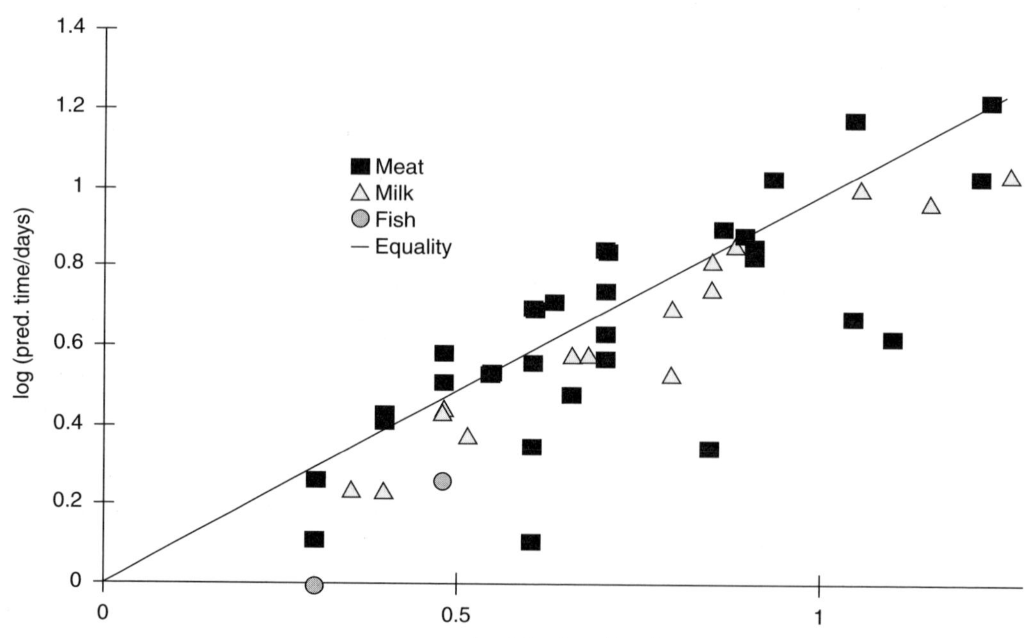

Figure 3–1 Comparison of model predictions and observed times for 1000-fold increase in numbers in foods inoculated with *Yersinia enterocolitica*.

Work in Tasmania has developed *Pseudomonas* predictor models applicable to milk and meat (McMeekin and Ross, 1996). In the UK, Forecast is a collection of models which can be used to assess the microbial spoilage rates or likely stability of foods. ERH CALC™ is a system which allows users to calculate the water activity and predict the minimum mold-free shelf life for flour confectionery products. Unlike other microbiological prediction systems, this model is integrated into a suite of other products for product development. Both Forecast and ERH CALC™ are available from Campden and Chorleywood Food Research Association (CCFRA, Chipping Campden, Gloucestershire, UK). More information about ERH CALC™ can be found in Chapter 9 (section 9.6).

3.3 USES OF MODELS

Once developed, models can be used in a variety of ways to help ensure the safety and quality of foods. Although the majority of applications has been developed for pathogenic bacteria, they may also be used for spoilage microorganisms provided appropriate, validated methods are available.

3.3.1 New product development

The design of a product formulation or reformulation is fundamental to the safety and quality of the food. Relatively small changes in the formulation may have a profound effect on microbial growth and may result in an unsafe product, or a product of poor quality. Increasingly, food manufacturers are under pressure to change the formulation of foods (e.g., reduction in salt levels or removal of preservatives), many of which have had a long and safe history. Such changes may potentially permit microbial growth in a food previously considered stable. Unfortunately, the microbiological safety or quality of a product formulation or reformulation is often one of the last parameters to be considered. If a product is subsequently shown to be microbiologically unstable, then the development costs to this point have been wasted. By using predictive models, the likely microbiological stability of a product formulation can be rapidly assessed. Models will enable the following example questions to be answered:

- What level of specific microorganisms will be present at different storage periods?
- What is the effect of reducing the salt content by 1%?
- What reduction in pH is required to prevent growth for 5 days?

Formulations that are intrinsically unsafe or unstable can be readily eliminated, so that subsequent development work is effectively targeted. Although this could be done using challenge tests, predictive models provide answers rapidly and cost-effectively. The effect of various product formulations on growth of a microorganism is shown in Figure 3–2.

Until predictive models have been widely accepted and have proven their applicability for food products, many food manufacturers may consider it prudent to do a microbiological challenge test on the final product to validate the model predictions. The nature and scope of this challenge test, however, will be considerably less than that required if predictive models had not been used.

3.3.2 Process design

The microbiological shelf life of many foods is dependent, not solely on formulation, but also on processes applied during manufacture (e.g., pasteurization). Often the product formulation cannot be such to prevent microbial growth and so processing is necessary. With the development of models for inactivation, the process can be designed to ensure that the target microorganism(s) are effectively eliminated.

Although the traditionally used *D* and *z* values have been widely used to determine safe processes, as with other types of predictive modeling, alternative approaches to modeling thermal inactivation, e.g., models involving non-linear kinetics, have been identified (Anderson et al., 1996; Bréand et al., 1998).

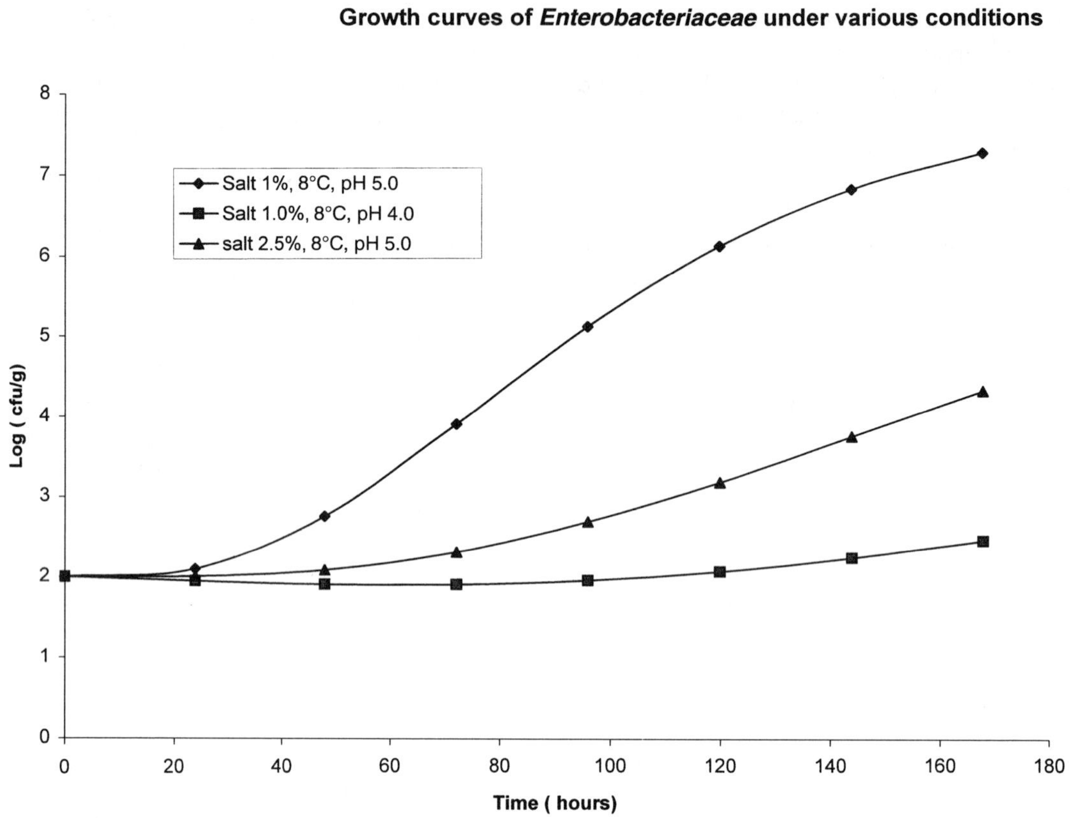

Figure 3–2 Effect of changes in product formulation on predicted growth of *Enterobacteriaceae.* Courtesy of Predictive Microbiology Software, Forecast, Campden & Chorleywood Food Research Association, Gloucestershire, England.

3.3.3 HACCP

Hazard Analysis Critical Control Points (HACCP) is a philosophy that has been applied to the production of food. This concept may be applied to both food safety and food quality. As a technique it has gained wide approval as a pro-active way to ensure food safety and quality. Seven principles have been identified in the application of a HACCP plan (Leaper, 1997):

1. Conduct a hazard analysis. Prepare a flow diagram of the steps in the process. Identify and list the hazards and specify the control measures.
2. Determine the critical control points. A decision tree can be used.
3. Establish critical limit(s) which must be met to ensure that each CCP is under control.
4. Establish a system to monitor control of the CCP by scheduled testing or observations.
5. Establish the corrective action to be taken when monitoring indicates that a particular CCP is not under control or is moving out of control.
6. Establish procedures for verification to confirm that HACCP is working effectively, which may include appropriate supplementary tests, together with a review.
7. Establish documentation concerning all procedures and records appropriate to these principles and their application.

When considering microbiological hazards, specific knowledge of factors affecting the growth, survival, and inactivation of microorganisms is required. Predictive models can be used to indicate if a hazard exists at a particular process step (principle 1). To determine if a process step is critical (principle 2), use of a decision tree is often recommended (see Chapter 5, Figure 5–6). Again valuable information to progress through the decision tree can be obtained from models. Once a process step has been identified as a critical control point (CCP), its control is essential. Predictive models can be used to identify the optimal target value for correct operation of this step and the tolerances that may be permitted (principle 3).

The application of models can therefore aid the application of HACCP to food manufacture by ensuring that good quality, systematic, and objective information is used in the decision-making process. A comprehensive and critical review of the HACCP system can be found in Chapter 5.

3.3.4 Risk assessment

Increasingly, government bodies are seeking to address the biological hazards associated with food safety using a risk analysis approach. This is a deliberate, structured, and formalized approach to understanding, and where necessary, reducing risk (Jouve et al., 1998). A key element is risk assessment: a scientific approach to estimating a risk and understanding the factors that influence it. Obviously, predictive models which estimate the growth or reduction in microbial numbers during processing and storage have an important role in producing quantitative data for this activity (Foegeding, 1997). Whiting and Buchanan (1997) demonstrate how models for thermal inactivation, growth, and non-thermal inactivation can contribute to a quantitative risk assessment for *Salmonella enteritidis* in pasteurized egg.

3.3.5 Time–temperature profiles

With many foods, particularly those stored chilled, the control of temperature during storage of a food is important in ensuring its safety and quality. During storage, the food will be subject to a fluctuating temperature profile. If this profile is known, predictive models can be used to determine the cumulative effects of this storage (Baranyi et al., 1995). Obviously the quality of the prediction is also dependent on the quality of the time–temperature profile that is collected. For example, if only the air temperature is monitored, the prediction of microbial growth in the food, which may not mimic the air temperature profile, will be erroneous.

A dynamic temperature function integration model was used jointly by regulators and processors in New Zealand, to develop justifiable criteria for the slaughter, processing, and storage of meats (Armitage, 1997).

3.3.6 Training and education

The development of predictive models in a user-friendly format can provide a useful training and educational tool. It provides a rapid and cost-effective mechanism whereby the benefits of doing a job correctly, or conversely the consequences of failing to do a job correctly, can be readily demonstrated. For example, the results of inadequate cooking, failure to use the correct product formulation, or improper storage of a food can be easily demonstrated and quantified.

As an educational tool, the use of predictive models in the basic training of food scientists and food technologists will allow them to appreciate more fully how factors (e.g., temperature, pH) act independently and together to affect the growth, survival, and inactivation of microorganisms.

3.4 LIMITATIONS OF MODELS

While predictive models for microbiology are very powerful, care is needed to ensure their correct use. Misuse of models may give the user a false sense of security about the microbiological safety or quality of a food.

First, it is important that the user understands which microorganism(s) is of concern. If this is incorrect then all subsequent predictions may be inappropriate. While most systems currently available are based on foodborne pathogens, the growth of food spoilage microorganisms often limits the product shelf life. Therefore, both must be considered when determining the shelf life.

Second, the factors (temperature, pH, etc.) used for the predictions must be carefully identified and measured. If there is a natural variation in these factors (e.g., batch-to-batch variation), this must be considered. Normally, the levels of the factors that are least inhibitory to microorganisms would be selected, otherwise the use of the prediction may result in some batches which may be of poor quality or even hazardous.

There has been concern with the errors associated with predictions from microbiological models. With validated models based on good quality data, experience has shown that the errors obtained are generally less than those associated with repeating microbiological challenge tests.

Many of the models available for use, particularly those for pathogens, have been developed in microbiological media and in the absence of a competing microflora. In addition, most models are limited to three, four, or five factors affecting the microbial kinetics, whereas other factors may be present in foods. Consequently, when such models are applied to some food situations, even validated models may give failsafe predictions, i.e., actual growth is slower than that predicted (Dalgaard and Jørgensen, 1998).

Finally, most models available indicate the likelihood or the extent of microbial growth or inactivation that will occur. They do not necessarily indicate that a food will be safe or do not indicate the shelf life of a food. Both these activities require judgment from an experienced microbiologist. In the absence of definitive data about the minimum infectious doses to cause disease for many pathogens and a poor understanding about the relationship between microbial numbers and food spoilage, the ability of predictive models to indicate the shelf life of a food will be limited.

3.5 THE FUTURE

The shelf life of a food is dependent on the raw material quality, the formulation of that food, the processes applied, the hygiene of production and storage, and the temperature of storage, distribution, and sale. This will be aided by the use of quality systems which incorporate well-trained staff. It is obvious that predictive models can be applied to many of the stages involved in the manufacture and sale of foods, which minimizes the risks posed by microorganisms. Models, therefore, do not replace the food microbiologist, but should result in the food microbiologist having better quality information on which to base judgments.

The development of validated predictive models for the microbial spoilage of food will be very important from a commercial perspective. Essential to this, however, is an understanding not only of microbial growth but also of microbial activity and how this relates to sensory changes in foods. When such understandings are available, it will then be possible to develop expert systems for the application of predictive models. Some developments in this area have started (Wijtzes et al., 1998).

Predictive food microbiology has emerged from within food microbiology to become a discipline in its own right with an important future. To date, much of the development of models has been progressed largely by food microbiologists. In the future, inclusion of other scientific disciplines, particularly microbial physiology, mathematics, statistics, computer programming, and knowledge engineering, will aid the next stages in the development of predictive models (some of which may become truly mechanistic), application software, and expert systems. Consequently, the power, flexibility, and application of predictive models will ensure their greater use by the food industry to provide good quality information in a rapid and cost-effective manner. This will in turn permit better decisions on microbiological safety and quality to be reached.

REFERENCES

Anderson, W.A., McClure, P.J., Baird-Parker, A.C. and Cole, M.B. (1996) The application of a log-logistic model to describe the thermal inactivation of *Clostridium botulinum* 213B at temperatures below 121.1°C. *J. Appl. Bacteriol.*, **80**, 283–290.

Anon. (1987) Microbiological Challenge Testing. *Campden Food and Drink Research Association Technical Manual No. 20.*

Armitage, N.H. (1997) Use of predictive microbiology in meat hygiene regulatory activity. *Int. J. Food Microbiol.*, **36**, 103–109.

Baranyi, J. and Roberts, T.A. (1994) A dynamic approach to predicting bacterial growth in food. *Int. J. Food Microbiol.*, **23**, 277–294.

Baranyi, J., Robinson, T.P., Kaloti, A. and Mackey, B.M. (1995) Predicting growth of *Brochothrix thermosphacta* at changing temperatures. *Int. J. Food Microbiol.*, **27**, 61–65.

Bréand, S., Fardel, G., Flandrois, J.P., Rosso, L. and Tomassone, R. (1998) Model of the influence of time and mild temperature on *Listeria monocytogenes* non-linear survival curves. *Int. J. Food Microbiol.*, **40**, 185–195.

Brown, H.M. (1992) Shelf-life determination and challenge testing, in *Chilled Foods: A Comprehensive Guide* (eds. C. Dennis and M. Stringer), Ellis Horwood, Chichester, England, 289–307.

Dalgaard, P. and Jørgensen, L.V. (1998) Predicted and observed growth of *Listeria monocytogenes* in seafood challenge tests and in naturally contaminated cold-smoked salmon. *Int. J. Food Microbiol.*, **40**, 105–115.

Esty, J.R. and Meyer, K.F. (1922) The heat resistance of spores of *Cl. botulinum* and allied anaerobes. *J. Infect. Dis.*, **31**, 650–63.

Foegeding, P.M. (1997) Driving predictive modelling on a risk assessment path for enhanced food safety. *Int. J. Food Microbiol.*, **36**, 87–95.

Jones, J.E. and Walker, S.J. (1993) Advances in modelling microbial growth. *J. Industrial Microbiol.*, **12**, 200–205.

Jouve, J.L., Stringer, M.F. and Baird-Parker, A.C. (1998) *Food Safety Management Tools,* International Life Sciences Institute Group, Brussels.

Leaper, S. (1997) HACCP: A practical guide (2nd edn). *CCFRA Technical Manual No. 38,* Chipping Campden, England.

Maxcy, R.B. and Wallen, S.E. (1983) Heterogeneity of samples as a problem in shelf-life prediction. *J. Food Protection,* **46**, 542–544.

McClure, P.J., Blackburn, C de W., Cole, M.B., Curtis, P.S., Jones, J.E., Legan, J.D., Ogden, I.D., Peck, M.W., Roberts, T.A., Sutherland, J.P. and Walker, S.J. (1994) Modelling the growth survival and death of microorganisms in foods. *Int. J. Food Microbiol.,* 265–275.

McMeekin, T.A. and Ross, T. (1996) Shelf-life prediction: status and future possibilities. *Int. J. Food Microbiol.*, **31**, 65–84.

McMeekin, T.A., Olley, J.N., Ross, T. and Ratkowsky, D.A. (1993) *Predictive Microbiology: Theory and Application,* Research Studies Press, Somerset, England.

Pin, C., Sutherland, J.P. and Baranyi, J. (1999) Validating predictive models of food spoilage organisms. *J. Applied Microbiol.,* **87**, 491–499.

Roberts, T.A. and Jarvis, B. (1983) Predictive modelling of food safety with particular reference to *Clostridium botulinum* in model cured meat systems, in *Food Microbiology: Advances and Prospects* (eds T.A. Roberts and F.A. Skinner), Academic Press, London, England, 85–95.

Walker, S.J., Archer, P. and Banks, J.G. (1990) Growth of *Listeria monocytogenes* at chill temperatures. *J. Applied Bacteriol.,* **68**, 157–162.

Whiting, R.C. and Buchanan, R.L. (1997) Development of a quantitative risk assessment model for *Salmonella enteritidis* in pasteurised liquid egg. *Int. J. Food Microbiol.,* **36**, 111–125.

Wijtzes, T., van't Riet, K., Huis in't Veld, J.H.J. and Zweitering, M.H. (1998) A decision support system for the prediction of microbial food safety and food quality. *Int. J. Food Microbiol.,* **42**, 79–90.

Chapter 4

Packaging and Food Quality

U. Stöllman, F. Johansson, and A. Leufvén

4.1 INTRODUCTION

In the developed world, the purchase of fresh farm produce is not a viable option. The time available for family shopping is too limited to allow the frequent visits to numerous retail outlets that such a system would require. The supply of exotic foodstuffs that consumers have learned to take for granted would also be impossible to maintain without some means of preservation and packaging. However, different products have different demands. Some products are sensitive to oxygen or moisture, others to microbial growth or light. Thus there is no general packaging solution and the type of packaging to be used must be determined individually for each product.

Packaging systems should not only ensure that the food is microbiologically sound, make it easy to handle, and convey pack information to the consumer, but also retain the desired sensory characteristics of the food. Such retention of sensory characteristics requires the package to act as an oxygen barrier, moisture barrier, aroma barrier, or light barrier depending on the sensitivity of the particular food and the prevailing environment. Thus packaging plays a vital role in the maintenance of the shelf life of foods and it is this aspect of packaging, and that of plastics packaging in particular, which will be dealt with in this chapter.

Primary packaging may be flexible, made from plastics film, paper, or thin laminate; semi-rigid, like thermoformed plastics, aluminium foil, laminate, and paperboard; or rigid, such as thick plastics, metal, and glass (Miltz, 1992). In many of today's food products, plastics are used in the primary packaging. These plastics may be used together with aluminium or paper in laminates. The main advantages of plastics are their great diversity and range of properties. Most plastics polymers have a density of around 1,000 kg/m^3, making them lightweight materials. They are also relatively cheap, and easily processed and shaped. The major technical drawback is that they do not provide an absolute barrier to gases and vapors and that they may interact with the packed food.

Environmental awareness has caused a great deal of attention to be focused on what is regarded as overpackaging. A long maximum storage time is not always perceived as a positive attribute. Nevertheless, an abundant supply of high quality food is demanded by the consuming public. To ensure this high quality in an environmentally friendly way is one of today's most challenging tasks for the packaging and food manufacturers.

In the following sections, the most commonly used plastics as well as some packaging systems including some recent developments will be reviewed. Finally, two examples (sections 4.4 and 4.5) are given to illustrate that packaging can have major influence on the storage quality and shelf life of foods.

4.2 PLASTICS PACKAGING MATERIALS FOR FOOD

4.2.1 Polyolefins

The most widely used polymers for food packaging applications are the polyolefins. Because of their characteristic heat sealing properties, polyolefins are often used as internal linings in flexible pouches or box-type containers for beverages (Shorten, 1982). The main properties of two polyethylenes (PEs), polypropylene (PP), and ionomers are outlined below.

4.2.1.1 Polyethylene (PE)

The molecular formula of PE is $(CH_2)n$ and the presence of branches prevents close packing of the polymer chains so that the density of the finished product is low. The presence of many branched chains also lowers the overall degree of crystallinity, which has a major effect on some of the properties of the polymer (Brydson, 1982). There are four main types of polyethylene available, which differ in structure and properties as well as with respect to the manufacturing processes that produce them. They are:

- high density polyethylene (HDPE)
- low density polyethylene (LDPE)
- medium density polyethylene (MDPE)
- linear low density polyethylene (LLDPE).

(a) LDPE. This variety accounts for the largest proportion of plastics for packaging. It is relatively inert chemically, moderately permeable to water vapor but highly permeable to oxygen. In general, the permeabilities to gases are high, with poor odor barrier characteristics. LDPE can be laminated, extrusion coated, or in some cases, coextruded. The main attractive features of PE, in addition to its low price, are good processability, toughness, flexibility, containing relatively few additives, and, in thin films of certain grades, transparency. LDPE has good heat sealability and is easily coated onto other materials such as paper and aluminium.

(b) HDPE. Compared with LDPE, HDPE is stiffer, harder, less transparent, and has a less waxy feel. It is more resistant to oils and greases, and has a higher softening point but lower impact strength and permeability to water vapor.

4.2.1.2 Polypropylene (PP)

PP is the next member in the series of polyolefins. It is possible to prepare PP in which the methyl (CH_3) groups are largely arranged on the same side of the polymer chain. The regular arrangement of the CH_3 groups produces a polymer with a higher crystallinity. This polymer is harder, has a higher softening point than HDPE, and greater resilience. Its permeability to water vapor and gases lies between those of LDPE and HDPE. Further uses of this polymer are in biaxially oriented films and bottles. The material is stretched in two directions at right angles, under suitable temperature conditions. The resulting material has improved clarity, impact strength, and barrier properties to water vapor and oxygen. Commercially available PPs have a density of 900–910 kg/M^3, one of the lowest among polymers.

4.2.1.3 Ionomers

Ionomers are defined as polyolefins copolymerized with a minor portion of an acid such as methacrylic acid, with some of the acid groups present in the form of a metal or quaternary ammonium salt. Ionomers are flexible, tough, and transparent and have excellent heat sealability. One of the trade names for ionomers is "Surlyn."

4.2.2 Polyvinyl chloride (PVC)

A number of plastics are available in which a hydrogen atom of ethylene has been substituted by a halogen or other group. These constitute the vinyl plastics. PVC is hard, stiff, and clear, with excellent resistance to moisture, low gas permeability, and high impact strength. Incorporation of plasticizers makes the resulting film softer and more flexible but lowers its tensile strength. PVC offers good resistance to oils and fats.

4.2.3 Polyvinylidene chloride (PVdC)

The monomer contains an additional chloride atom substituted into the ethylene molecule. The polymer is hard with a high degree of crystallinity. It produces a clear film with good strength properties and very low permeability to gases and water vapor. "Saran" is a common trade name for PVdC.

4.2.4 Ethylene vinyl alcohol (EVOH)

Ethylene vinyl alcohol (EVOH) is commonly used in packaging because of its good gas barrier properties. EVOH, trade name "EVAL," is an excellent gas barrier and commercially available in several grades with different ratios of ethylene/vinyl alcohol in the chain. The major limitation of EVOH is that its effectiveness as an oxygen barrier is severely reduced at high levels of humidity, due to disruption of the polymer–polymer hydrogen bonds. Also, because of the hydroxyl groups in the polymer backbone, EVOH tends to be extremely soluble in water.

4.2.5 Polystyrene (PS)

Styrene consists of an ethylene molecule in which one of the hydrogens has been replaced by a phenyl group. The bulky nature of the phenyl group prevents close packing of the macromolecular chains. Hence, PS is amorphous. The polymer is transparent and a good barrier to gases, except water vapor. It is brittle, but this may be overcome by copolymerization with butadiene and/or acrylonitrile (ABS and SAN, respectively). Another way of modifying PS is by making a blend of polystyrene and rubbery materials, i.e., a polystyrene SBR blend. This high impact polystyrene (HIPS) has greater impact strength than ordinary polystyrene but lower tensile strength as well as hardness, and a lower softening point.

4.2.6 Polyamides (nylons)

These comprise a series of polymers prepared by condensation of di-acids with di-amines or through self-condensation of certain amino acids. The amide group (CO–NH) is strongly polar and promotes strong hydrogen bonding between the CO and NH groups on adjacent polymer chains. Hence, nylons are strong, highly crystalline materials with high melting and softening points and low gas permeability. They absorb moisture depending on the relative proportions of hydrocarbons and amide groups in the polymer chain and may lose up to 50% of their strength when wet.

4.2.7 Polyesters

Polyester resins exist in many forms. The most widely used polyester is polyethylene terephthalate (PET). PET has good mechanical properties, excellent transparency, and relatively low permeability to gases. Biaxially oriented PET is often used in the production of bottles for carbonated beverages. Laminates with PET as one of the components are also very common in sterilizable pouches and boil-in-bag applications.

4.2.8 Polycarbonates (PC)

Polycarbonates are linear polyesters of carbonic acid with aromatic or aliphatic dihydroxy compounds. Most commercial polycarbonates are tough, stiff, hard, and transparent. Their disadvantages are high cost and the fact that they absorb moisture, with resulting deterioration of their mechanical properties and processability. An additional disadvantage, for packaging, is their poor gas barrier properties.

4.2.9 Lamination and coextrusion

Many new packaging materials or combinations of materials have been used in food packaging in recent years. The main driving force has been to increase the barrier properties of the package and thereby the protection of food products with more complex material structures. A single polymer is often unable to provide all the properties necessary to create suitable barrier characteristics. This has generated a need for multilayer structures and techniques such as lamination and coextrusion. Coextrusion is not simply a competitive alternative to lamination.

There are laminate properties that coextrusion is unable to meet, and non-thermoplastic materials such as paper and aluminium foil, often used in the lamination process, which obviously cannot be coextruded. In addition, printing ink, which can be protected if enclosed in a laminate, can only be applied on the surface of a coextruded film. There are also applications where the two technologies are combined, i.e., when a coextruded film is used in a laminate. For example, ionomers have excellent sealing characteristics, but are relatively expensive. PE/ionomer coextrusion can give the sealing properties needed in a more cost-effective way than an ionomer monofilm (Meadows, 1991). Such a structure may be produced by adhesive lamination of the materials or by coextrusion coating of the other material. The principal advantage of coextrusion over lamination is that the materials are extruded simultaneously, without a separate lamination process (Meadows, 1991).

In flexible packaging, laminates and metallized films have replaced single material structures. This has greatly improved the light and oxygen barrier properties of the materials, and extended the product shelf life. A drawback of the metallized laminates is the environmental load, even though only tiny amounts of metal are used. The long established trend towards using more laminates may be hit by recycling legislation when the industry will be controlled by requirements to make packaging products recyclable. Even though laminates exhibit many advantages such as flexibility in terms of stiffness, strength, barrier, and sealing properties, many structures may have to be simplified, and there may have to be an increased use of compatible base films so that the whole structure can be recycled via the same path (Meadows, 1991). However, the need for high performance barrier packaging to give shelf stability under a wide range of storage conditions continues to grow. Table 4–1 lists typical oxygen (OTR) and moisture vapor transmission rates (WVTR) for various plastics films for food uses (Turtle, 1987).

4.3 SOME PACKAGING SYSTEMS AND RECENT DEVELOPMENTS

4.3.1 Trends in food packaging

The rate of development of new packaging materials and packaging techniques in recent years has been remarkable. Many of the new packaging materials have been developed to complement developments in new preservation techniques, such as modified atmosphere packaging (MAP), or to meet increased demands for recyclability or other environmental considerations (Yokoyama, 1992).

Table 4–1 Oxygen and moisture vapor transmission rates for various plastics films

Film (25 μ)	*OTR (cm³/m²/24h/atm at 25°C and 50% RH)*	*WVTR (g/m²/24h at 25°C and 75% RH)*
PVdC	2	1
EVOH	0.5–10*	40
Nylon 6	50–150*	40
Polyester	80	8
PVC (unplasticized)	200	20
HDPE	1400	2
OPP	1600	1.5
LDPE	8000	5

*Humidity dependent.
Source: Turtle, 1987.

4.3.2 Silica-coated films

To achieve extra barrier properties, films have been coated with silica (SiO_2). Silica-coated plastics, also referred to as glass-coated, have several benefits: being clear barrier materials, they are microwavable, retortable, and recyclable without any step to separate components being required. So far, the SiO_2 coating has been deposited on PET containers, polyester, LDPE, and PP films. The film requires no pretreatment in order to obtain good adhesion of the SiO_2 coating to the film (Rice, 1992b). A 30 nm (300 Å) SiO_2-coated LDPE film has a 10 cm^3/m^2/day/atm. oxygen transmission rate. The oxygen transmission rates for coated OPP (biaxially oriented polypropylene) are said to be comparable to those of metallized OPP. Silica-coated PET films have a 1 cm^3/m^2/day/atm. oxygen transmission rate. Consequently, these coated films could offer barrier properties approaching those provided by multilayer plastics structures, foil laminates, or metallized films. Silica-coated plastics enable the use of a monolayer plastics film with an extremely thin SiO_2 layer.

Early problems reported in the development of these products were yellowing and fragility of the glass coatings. However, these problems seem to have been overcome (Rice, 1992a). There have also been reports of micro-cracks in the SiO_2 coatings. These cracks make tunnels through which oxygen may diffuse into the container (Yoshii, 1992). At this point, there seem to be three factors slowing down the commercial application of glass-coated packaging materials (Rice, 1992b):

- economy
- technical difficulties in ensuring good silica adhesion to certain substrates
- lack of converter and packer's experience of these new packaging materials.

In 1988 a multilayer microwavable pouch with a glass-coated polyester barrier layer was introduced in Japan. Today, several other applications are available. These are discussed by Rice (1992b).

4.3.3 Active packaging technologies

Active packaging systems involve an interaction between the packaging material itself, the internal atmosphere, and the food. The intention is to extend the shelf life of foods. New technologies which combine active packaging with modified atmosphere packaging (MAP) are now a commercial reality (Davies, 1995). Day gave a useful review of active packaging of foods (Day, 1998).

4.3.3.1 Packaging materials with preferential permeability

Selected permeable packaging materials can be regarded as highly functional barrier packaging materials. In general, the gas transmission rates of gas barrier resins such as PVdC and EVOH increase in the order of nitrogen, oxygen, and carbon dioxide, and their transmission ratios (e.g., CO_2/O_2) are constant for each barrier resin. Some foods require not only permeation of unfavorable gas generated from the inside but also maintenance of the oxygen barrier. Demands for packaging materials to meet these contradictory requirements have risen recently. An example of this is a packaging material which allows compounds from the smoking process to penetrate into the surface of, for instance, sausages under high temperature. After processing this material affords high barrier characteristics below ambient temperature, thus maintaining the quality of the product (Yoshii, 1990). Another example, this time from the dairy industry, is a packaging material with preferential permeability during the aging of cheese. This material allows permeation of the carbon dioxide formed during the aging process of Cheddar and Gouda cheeses, yet prevents oxygen ingress from the outside.

4.3.3.2 Oxygen scavenging

Another new trend in barrier packaging materials concerns chemical barriers or oxygen-absorbing materials. These materials can be effectively used together with modified atmosphere packaging techniques. These chemical barriers are used together with the so-called physical barriers, such as PVdC or EVOH. One example is the "OXBAR" system, with PET as the main component (96%), polymetaxylylene adipamide (MXD-6) as the oxidizable component (4%), and an organic cobalt salt added as the oxidizing catalyst at a 50–200 ppm level. This composition can be extruded to make blown bottles, which are said to be able to maintain the oxygen level inside the container at as low as 0% for two years (Yoshii, 1992). Barrier life time is a function of two main variables: scavenging capacity and rate of consumption. The former depends on the composition of the blend (MXD-6 and cobalt salt content) and the latter, the wall thickness of the container and the environmental conditions. Another example is an oxygen- absorbing plastic called "Long Life," which can extend the shelf life of oxygen-sensitive drinks. This material is said to be insoluble in water and of a structure similar to that of hemoglobin (Yoshii, 1992). The hydroxyl groups of the base polymer react with silica, which also combine with a metal to form a complex.

Another method for actively controlling the oxygen level in a food package is to use an enzyme reactor surface, which will react with certain substrates to scavenge incoming oxygen. A new active packaging system developed by PharmaCal Co. (USA) utilizes glucose oxidase and catalase enzymes for the removal of oxygen from liquid products (Yoshii, 1992).

There is one commercial application that uses a mixed iron powder–calcium hydroxide sachet in which both oxygen and carbon dioxide are scavenged (Labuza and Breene, 1989). It is being used to pack fresh ground coffee in flexible bags. This packaging system is said to more than triple the shelf life of the product. Other chemical scavenger systems include reactive dyes, ascorbic acid, and cathecol-based reducers.

4.3.3.3 Temperature-compensating films

A recently developed film called "Intelimer" has a temperature "switch" point at which the permeation of the film changes abruptly. The switching mechanism is accomplished by the long-chain fatty alcohol-based polymeric side chains patented by the Landec Corporation (Figure 4–1). Below a preset temperature, these side chains are crystalline and provide a relatively effective gas barrier. However, above this preset temperature, these side chains convert reversibly to an amorphous structure, which is up to 1,000 times more permeable to gas and may be tailored to the very large increase in respiration rates of fresh produce at temperatures above 5°C (Day, 1993). Because the change is physical and not chemical, this process is continuously reversible. The transition point can be moved by increasing or reducing the average number of carbon atoms in a chain. The switch point may be set anywhere between 0 and 45.5°C, and controlled within 2°C. The switch also changes the adhesiveness and stiffness of the film. Intelimer films are intended to be used for packaging highly respiring produce, but commercialization is still several years away (Anon., 1992).

Another temperature-compensating film for produce has recently been patented and relies on the same principle as a bimetallic strip, but uses two different plastics films instead of two metals. Temperature-sensitive apertures are cut into laminated plastics film in the form of small strips or "petals." At a selected temperature, the laminated film is flat and has minimum opening area. However, at higher temperatures the free area for gas exchange increases as the petals curl, and the permeability of the film can be engineered to increase proportionally to the rise in the respiration rate of the produce. The rate of curl is a function of the coefficient of expansion of the two separate films, and the extent of opening is controlled by the length and shape of the petals surrounding the aperture.

4.3.3.4 Moisture absorbers

Humidity will build up in the headspace of a food package either due to respiring produce or due to water vapor from moist foods. This in time will allow the growth and development of microorganisms. A desiccant film or sachet can be used to overcome this problem (Geeson et al., 1985). A Japanese company has developed a film which is made into a pillow with entrapped propylene glycol (Labuza and Breene, 1989). When placed in contact with the surface of meat or fish, it absorbs water and slows the growth of spoilage bacteria.

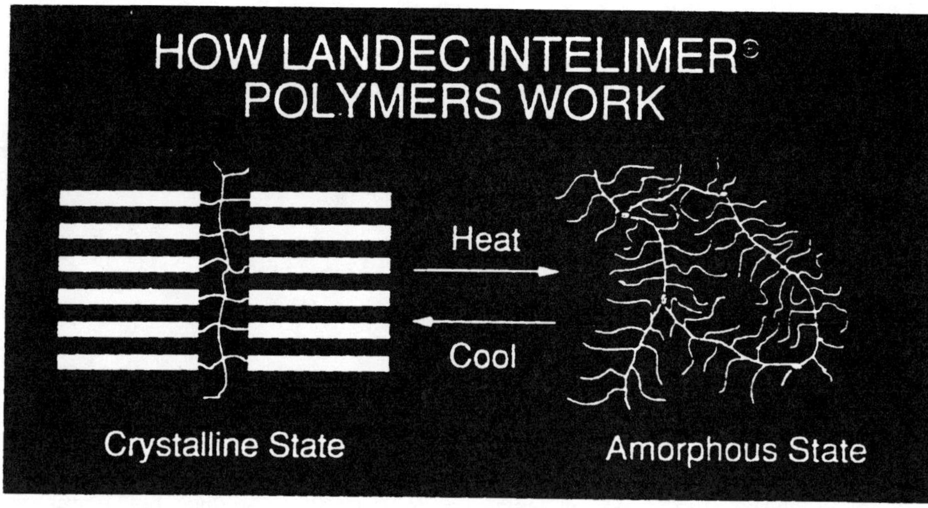

Figure 4–1 At a specified temperature the side chains of the Intelimer® film melt and convert to an amorphous structure that renders the film gas permeable (Anon., 1992).

4.3.4 Modified atmosphere packaging (MAP)

MAP is defined as the enclosing of food products in an atmosphere inside gas-barrier materials, in which the gaseous environment has been modified to slow down respiration rate and microbial growth, and reduce enzymic degradation, all with the intention of extending the shelf life (Parry, 1993). Such modifications usually result in a reduced oxygen content in the headspace atmosphere, but increased carbon dioxide and nitrogen levels.

Carbon dioxide is the most important constituent in the gas mixture. Several factors influence the antimicrobial effect of carbon dioxide, specifically the microbial load, gas concentration, temperature, and packaging film permeability. To be effective against aerobic spoilage microorganisms, a concentration of 20–60% carbon dioxide is required in the package headspace (Greengrass, 1993). One of the most important factors influencing the antimicrobial effect of carbon dioxide is the permeability of the packaging film (Gill, 1990). The success or failure of MAP for foods depends on the impermeability of the packaging materials to both oxygen and carbon dioxide, in order that the correct gas mixture in the package headspace can be maintained. In addition, films used in MAP packaging should have low water vapor transmission rates (WVTRs), so as to minimize changes in moisture content inside the pack.

Plastics commonly used for gas packaging of foods include polyester, PP, PVdC, EVOH, and PE. Since all of the desired characteristics of a packaging film, i.e., strength, impermeability, and heat sealability, are seldom found in one single material, separate materials are laminated to one another to produce films with the desired characteristics for gas packaging of foods. Examples of such laminated structures include nylon/PE, nylon/PVdC/PE, and nylon/EVOH/PE. Such composite structures have the desired characteristics for the gas packaging of food products: strength is provided by the outermost layer of nylon; gas and moisture vapor barrier is provided by EVOH and PVdC; and heat sealability by PE (Brody, 1989).

Development of optimum gas packaging materials is even more challenging for fruits and vegetables as these products respire. The ideal packaging material for gas packaging of these products must be able to keep low oxygen concentrations within the package headspace while preventing the build-up of high carbon dioxide concentrations. Packaging films commonly used to achieve such a balance include LDPE and PVC (Zagory and Kader, 1988). Packaging films with the correct permeability must be chosen to avoid depletion of oxygen and accumulation of carbon dioxide caused by respiring products. Recently developed highly permeable microperforated films such as P-Plus and Crop appear to be most suitable at present (Gill, 1990; Geeson et al., 1988). These films allow the product to breathe as it ripens inside the package.

A key feature of the MAP described below is the controlled permeation of O_2 and CO_2, as well as the scavenging of ethylene or controlled release of other compounds which could have a preservative effect on the food (Labuza and Breene, 1989). One way of controlling the oxygen level directly in a package is to use an oxygen scavenger system (Takahashi et al., 1984). The simplest scavenger is reduced iron. The amount of iron needed depends on the initial oxygen level in the headspace, the amount dissolved in the food, and the oxygen transmission rate of the film (Labuza and Breene, 1989). One gram of iron will react with 300 cm^3 of oxygen. The iron is separated from the food by being kept in a small highly oxygen-permeable pouch.

4.3.5 Aseptic/ultra-high temperature (UHT) packaging

Packaging materials for aseptic processing/packaging must be sterilizable, suitable for aseptic filling, and permit the maintenance of sterility of the product during storage and handling (Smith et al., 1990a). Glass containers offer similar advantages as metal containers, but have the additional disadvantages of fragility and high density. Plastics materials offer versatility, but cannot be used alone as they do not possess all of the required properties. Consequently, they must be laminated to other materials to give packages with the appropriate requirements.

At present, the Tetra Pak concept, using a paperboard-foil-plastics laminate, is widely used for aseptic packaging. This laminated structure consists of as many as six layers of materials: PP, Surlyn, foil, PE, paperboard, and PE as the inner food contact layer. Other combinations that can be used include laminations of EVOH, PE and PS; or a metallized polyester. Aluminium foil is the most commonly used barrier material. PP or PE, i.e., polyolefins, are the most common heat sealing and food contact materials used. Foils need to be protected from mechanical damage, and protection is usually provided by paperboard, which also gives rigidity. All of the composite materials described act as barriers to moisture, oxygen, light, and microbes, and have the strength and heat sealability needed for aseptic packaging.

A review of aseptic processing has been made by Lopez (1987). Historical perspectives on aseptic packaging technology and the use of plastics packaging materials are further described in other recent publications (Downes, 1989; Eidman, 1989; Fox, 1989; Kern, 1989; Rose, 1995).

4.3.6 Sous vide (vacuum cooking/packaging) technology

There has been a tremendous increase in recent years in the use of "sous vide" processing technology to extend shelf life and maintain the quality of fresh food (Smith et al., 1990b). This technology has been developed in response to consumer demands for ready-to-eat, microwavable, convenience foods with extended shelf life, that retain closer-to-fresh characteristics. The term sous vide means "under vacuum." In sous vide processing, foods are cooked in sealed, evacuated, heat-stable pouches or thermoformed trays in such a way that much of the natural flavor, aroma, and nutritional quality is retained by the product. This technique requires highly controlled packaging/cooking conditions. Packaging materials used in sous vide processing must withstand pasteurization and chill temperatures as well as protect and maintain product quality.

4.3.7 Edible films

Recently, an edible film of polysaccharides has been used in certain packaging applications

(LaBell, 1991). The polysaccharide film looks like a plastics film. The edible polysaccharide beads are seamless spheres of polysaccharide measuring 1–8 mm in diameter, which are dissolved by heat and moisture. Dry ingredients, such as spices and seasoning mixes, can be portion-packed in edible film pouches. The entire pouch can be added, for instance, to a soup, and it will dissolve quickly. However, the polysaccharide films are more permeable to gases than plastics films. In addition, their tensile strength is not as high as that of plastics films. This will create problems in automatic form/fill/seal machines.

In Japan, edible polysaccharide films are already widely used in the meat industry for packing of processed meats, such as ham and poultry products (LaBell, 1991). Processors can wrap the meat in edible film before smoking and steaming. The smoke flavor permeates the film during the smoking process and the film dissolves during steaming. The surface structure and texture of the finished ham are improved. Also, yields are better, as less moisture is lost during processing.

The shelf life of fresh fruit and vegetables can be extended by using an edible film. Plastics films can be used to provide a gas barrier to protect the products, but moisture from the packed material will collect inside the film. An edible packaging film will gradually absorb the moisture given off from the product, preventing it from accumulating inside the package. This would lead to reduced risk of mold growth. Eventually, the moisture absorbed may cause the film to dissolve.

4.4 THE INFLUENCE OF "SCALPING" ON FOOD

As plastics packaging is becoming more widely used for direct contact with foods, compatibility with the food must be considered. The sorption of flavor compounds, or scalping, is one of the most important compatibility problems. The problem of aroma absorption by plastics packages has been recognized for many years (Nielsen and Jägerstad, 1994). The tendency toward absorption varies with the type of plastics. For example, polyolefins can absorb certain flavor constituents selectively from products (Foster, 1987).

The quality of juices packed in aseptic cartons has been the subject of intensive research in recent years. Early work focused on volatile absorption from sources external to the package (Gilbert and Pegaz, 1969). Haydanek et al. (1970) and Gilbert et al. (1982) examined the volatile uptake by various packaging materials. PE and PP have been used widely in contact with food in packages due to their low costs, heat sealability, and relative lack of additives, but they are particularly susceptible to pervasion by organic vapors. The use of PE and PP in plastics packages has raised questions concerning sorption of intrinsic food flavors (Durr *et al.,* 1981).

4.4.1 The limonene controversy

Most juices that are filled aseptically are packed into laminated carton packages such as the Tetra Pak or Combibloc, in which the food contact material is PE (Mannheim et al., 1987). Mannheim found that the product shelf life of orange and grapefruit juices was significantly shorter in laminated cartons than in glass jars. Moshonas and Shaw (1989), Mannheim et al. (1988) and Sizer et al. (1988) have investigated how aseptic packaging can affect the quality of citrus products, with the emphasis on flavor loss. Hirose et al. (1988), Shimoda et al. (1984) and Durr and Schobinger (1981) have studied the sorption of citrus flavor components by packaging materials. Baner et al. (1991) found that the food contact layer, usually polyethylene, largely controlled the sorption behavior. This means that any improvements of flexible materials regarding aroma sorption must come from new food contact layers and not from new barrier layers.

Citrus flavor has often been used in model systems for flavor absorption studies (Durr and Schobinger, 1981; Marshall et al., 1985; Hirose et al., 1988; Kwapong and Hotchkiss, 1987). (+)-Limonene, the predominant volatile in citrus oils and essences, is often used as a probe compound in such studies. Compositionally, limonene is a non-polar, unsaturated terpene hydrocarbon. Many flavor compounds contain an ester, alcohol, and other polar functional groups. Essential oils from citrus fruit consist of terpenes, sesquiterpenes, oxygenated terpenoids, and non-volatile components. The oxygenated compounds are important contributors to the citrus flavor (Moshonas and Lund, 1969). Many research groups have reported that polar compounds are absorbed to a much lower extent in PE than limonene and other hydrocarbon volatiles. An understanding of the differences between various plastics materials as regards the sorption of essential oil constituents could help the processor reduce the flavor absorption problems.

4.4.2 The influence of pulp content on scalping

Shimoda et al. (1984) showed that in juice from satsuma mandarins, the limonene sorption by plastics is lower in the presence of juice pulp. Maeda et al. (1984) and Osajima (1983) reported that a low pulp content in orange juice resulted in higher absorption of aroma compounds. This is in agreement with the results of Gherardi (1982) and Granzer (1982). Fruit juice volatiles remain stable in the juice while they are absorbed by or suspended in the pulp (Yamada et al., 1992). However, the amount of pulp, such as cellulose powder, which must be added if it is to be used to control the sorption of volatiles by LDPE, is large.

4.4.3 Comparison of instrumental and sensory analysis

Mannheim et al. (1985) documented changes in aseptically packed citrus juices by sensory as well as chemical analysis. They related loss of orange juice flavor to the absorption of (+)-limonene into the polyolefin packaging material. However, in a comparison of the sensory quality of orange juice stored in polyethylene-lined paperboard cartons at different temperatures, Durr et al. (1981) reported the temperature to be the main variable affecting the flavor, even though orange oil expressed in terms of limonene was absorbed by the polyolefin food contact layer. They suggested that limonene absorption was an advantage, since limonene was known to be a precursor of off-flavor compounds. They also reported that desirable aroma volatiles were practically unabsorbed.

Although it is well documented that certain constituents of orange juice are absorbed by polymeric materials and that the absorption of flavor compounds changes the aroma profile (Mannheim et al., 1985), it has not been established unequivocally that these changes are caused by absorption of odor-active constituents of orange juice (Marin et al., 1992).

Most reported studies on flavor scalping in orange juice were performed with model solutions, not actual orange juice, or with orange juice which had not been commercially produced and packaged (Pieper et al., 1992). Many of the investigations, too, were performed without sensory evaluation. Pieper et al. (1992) showed that a sensory panel could not distinguish between orange juice stored in glass bottles and juice stored in laminated paperboard packages with LDPE as the food contact plastic, even though absorption of up to 50% of the (+)-limonene and small amounts of aldehydes and alcohols by the packaging materials was observed.

Marin et al. (1992) exposed plastics packaging materials, i.e., PE and Surlyn, to orange juice and measured the changes in the limonene content of the juice using a gas chromatograph equipped with a flame ionization detector (GC-FID). They determined the components which were most important to the aroma using gas chromatography–olfactometry (GCO). More than 70% of the limonene content was absorbed by the plastics in 24 hours at 25°C. However, results from the GCO analysis indicated that the limonene had only trace odor activity. Furthermore, the plastics did not substantially alter the odor-active components detected in orange juice. These reports emphasize the importance of using real food systems whenever possible, and supplementing the instrumental analysis with sensory evaluation.

4.4.4 Influence of scalping on barrier and mechanical properties

As has been shown in the reports outlined above, the scalping process may not alter the sensory quality of packed juices. However, absorption of flavor components by the package may alter the mechanical properties of the plastics, causing delamination, or modifying the barrier properties of the packaging materials. Miltz et al. (1990) showed that the absorption of (+)-limonene increased the oxygen permeability of the three plastics films used. If the efficiency by which a packaging material reduces the entry of oxygen into the package is impaired, foods sensitive to oxygen will deteriorate quickly. The result will be a shortened shelf life of the product.

4.4.5 Methods of suppressing scalping

Yamada et al. (1992) suggested methods of suppressing the sorption of volatiles by the food contact layer.

1. Sorption of volatiles could be reduced considerably by reducing the thickness of the heat seal layer. This could, however, cause the sorbed compounds to permeate the polyolefin layer and change the structure and properties of the barrier layer, or result in delamination of the laminate.
2. Sorption of volatiles could also be reduced by using a food contact layer that has already absorbed some fruit juice essence (i.e., presorption).

Treatments of the plastics surface and judicious selection of materials, such as use of plastics with crystalline polymers (Shimoda et al., 1988), use of polymer blends (Subramanian, 1990), fluorination of plastics surfaces (Hobbs et al., 1990), and inert coatings, e.g., silica, on the plastics surfaces (Charara, 1987), have also been proposed in the literature to circumvent the scalping of flavor compounds from food products.

4.5 FLAVOR TRANSFER PROBLEMS IN REFILLABLE PET BOTTLES

The debate about the environmental impact of packages and the demands for reuse and recycling of packages have led to the development of the refillable PET bottle system for soft drinks. All refillable systems involve certain problems with regard to product safety. Legislation and hygiene are essential aspects in this context. The washing procedure is a particularly critical step in the refilling of PET bottles, due to the limited heat resistance of the packaging material. To avoid shrinkage of the bottle the temperature must not exceed 58–59°C. It is crucial to the sensory quality of the packed product that aroma compounds from one product are not absorbed by the material and then transferred to the next product filled in the bottle. Numerous chemical substances can be absorbed in the PET material (Figure 4–2), and some of these compounds may impart off-flavors to subsequent fillings.

In order to assess the risk of flavor transfer between products when a returnable PET bottle is refilled with a different soft drink, sensory analyses are performed. First the bottle is filled with the test product. After storage the bottle is emptied, washed, and refilled with the second product on which the assessment of possible flavor transfer will be made. Different kinds of soft drinks such as citrus drink, cola, and mineral water have been used as the second fillings. For each of these products the intensity of any flavor transfer is evaluated separately. The test product will be approved for filling in returnable PET bottles only if it is approved in all the tests using the different soft drinks. Results so far showed that some products are unsuitable for the refilling system, especially those which have a very intense and strong flavor. For this reason some ciders and citrus fruit drinks have been rejected because some of their flavor has been detected in the second product filled in the bottle. Highly delicate products such as mineral water are to date only filled in new bottles.

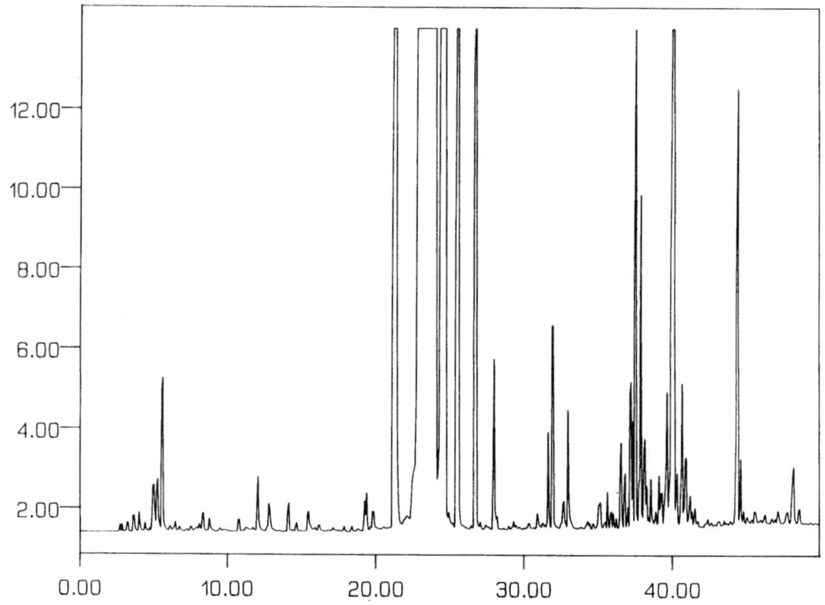

Figure 4–2 Gas chromatogram of a headspace sample of a PET bottle that has been filled with a citrus drink.

To supplement the sensory evaluation of the soft drinks, chemical analyses are needed to determine which flavor/aroma compounds are absorbed by the PET material and eventually migrate into the product filled subsequently. If this could be done, the next step would be to investigate their role as flavor components in the soft drinks and, if possible, replace or reduce them. In a production situation, it will be important to have equipment set up to identify foreign odors prior to refilling in order to avoid customer-introduced contamination.

4.6 CONCLUSIONS

The sensory quality of a packaged food is the result of a complex interaction between the food, the pack, and the environment. Substances from both the food and the environment can be absorbed/adsorbed in the packaging material. Such sorption from the food and the environment may not only change the taste of the food per se but also change the barrier and other physical characteristics of the packaging material. Changed barrier characteristics may result in a higher diffusion of oxygen into the package, resulting in the development of off-flavor due to oxidation of the food. The oxidation products may also, for better or worse, be absorbed into the packaging material. These interactions between aroma compounds, oxygen, and the pack can result in a dynamic and time-dependent change of the food quality.

A number of sophisticated technical solutions to preserve food for a long time are available today. Whether or not these solutions are economically and environmentally acceptable or even marketable has to be considered for each specific application. It is pointless to make a completely impermeable package if the consumer refuses to buy the product because it is considered to be artificial (it will "keep too long") or overpackaged (environmentally unsound) compared with other similar products on the market. In the future, it will be of increasing importance to a food manufacturer to use environmentally friendly packaging materials if he plans to stay in business. On the other hand, packaging in an environmentally friendly way cannot be allowed to comprise the fundamental requirement of having always to ensure the safety and quality of the food.

In addition to such obvious considerations as keeping the food microbiologically safe, a food manufacturer has to decide what his product should taste like, and the length of time it should remain acceptable. Once these decisions have been made, an economically and environmentally suitable packaging solution can be sought. To stay in business, food manufacturers and packaging suppliers must cooperate in order to face up to the demands of the product, the demands of the environment, and the expectation of the consumer.

Finally, it should always be remembered that the ultimate decision is made by the consumer in terms of whether to buy or not to buy.

REFERENCES

Anon. (1992) Temperature compensating films for produce. *Prepared Foods,* **95**, 82.

Baner, A.L., Kalyankar, V. and Shoun, L.H. (1991) Aroma sorption evaluation of aseptic packaging. *J. Food Sci.,* **56**(4), 1051.

Brody, A.L. (1989) Flavor interacts with packaging. *Prepared Foods,* pp. 128, 158.

Brydson, J.A. (1982) *Plastics Materials,* 4th edn, Butterworth Scientific, Oxford, England.

Charara, Z.N. (1987) Evaluation of orange flavor absorption into polymeric packaging materials, *M.S. Thesis,* University of Florida, Gainesville, FL.

Davies, A.R. (1995) Advances in modified-atmosphere packaging, in *New Methods of Food Preservation* (ed G.W. Gould), Blackie Academic and Professional, London, 304–320.

Day, B.P.F. (1993) *Modified atmosphere packaging—state of the art. SIK* MAP Seminar, April 27th, Gothenburg, Sweden.

Day, B.P.F. (1998) Active packaging of foods. *CCFRA New Technologies Bulletin No. 17,* pp. 1–12.

Downes, T.W. (1989) Food packaging in the IFT era: Five decades of unprecedented growth and change. *Food Technol.,* **43**(9), 228.

Durr, P. and Schobinger, U. (1981) The contribution of some volatiles to the sensory quality of apple and orange juice odor, in *Flavour '81* (ed J. Schreier), Walter de Gruyter and Co., Berlin, Germany, 179.

Durr, P., Schobinger, U. and Waldvogel, R. (1981) Aroma quality of orange juice after filling and storage in soft packages and glass bottles. *Alimenta,* **20**, 91.

Eidman, A.L. (1989) Advance in barrier plastics. *Food Technol.,* **43**(12), 91.

Foster, R.H. (1987) Ethylene vinyl alcohol copolymer resins for better solvent, aroma and flavor barrier. *Future-Pack '87, 5th International conference on packaging innovations,* Ryder Associates, Atlanta, GA.

Fox, R.A. (1989) Plastics packaging—The consumer preference of tomorrow. *Food Technol.,* **43**(12), 34.

Geeson, J., Everson, H. and Browne, M. (1988) Microperforated films for fresh produce. *Grower,* **109**(14), 31.

Geeson, J.D., Browne, K., Maddison, J., Sheppard, H. and Guraldi, F. (1985) Modified atmospheric packaging to extend shelf life of tomatoes. *J. Food Tech.,* **20**, 339.

Gherardi, D.S. (1982) Packaging and quality. *Proceedings of the International Congress of Fruit Juice Producers,* Munich, Germany.

Gilbert, S.G. and Pegaz, D. (1969) Find new way to measure gas permeability. *Package Engineering,* **Jan.**, 66.

Gilbert, S.G., Harzidimitriu, E., Lai, C. and Passy, N. (1982) Study of barrier properties of polymeric films to various organic aromatic vapors, in *Instrumental Analyses of Food* (eds G. Charalambous and G. Inglett), Vol. 1, Academic Press, New York, 405.

Gill, M. (1990) High permeability films for MAP of fresh produce. *International Conference on MAP.* Campden Food and Drink Research Association, Chipping Campden, Gloucestershire, England.

Granzer, R. (1982) *Proceedings of the International Congress of Fruit Juice Producers,* Munich, Germany.

Greengrass, J. (1993) Films for MAP of foods, in *Principles and Applications of Modified Atmosphere Packaging of Foods* (ed. R.T. Parry), Blackie Academic and Professional, Glasgow, Scotland, 63–100.

Haydanek, M.G., Woolford, G. and Baugh, L.B. (1970) Premiums and coupons as potential sources of objectionable flavor in cereal products. *J. Food Sci.,* **44**(3), 850.

Hirose, K., Harte, B., Giacin, J., Miltz, J. and Stine, C. (1988) Sorption of d-limonene by sealant films and effect on mechanical properties, in *Food and Packaging Interactions* (ed J. Hotchkiss), ACS Symposium Series 365, Washington, DC.

Hobbs, J.P., Anand, M. and Campion, B.A. (1990) Fluorinated high-density polyethylene barrier containers, in *Barrier Polymers and Structures* (ed W.J. Koros), ACS Symposium Series 423, Washington, DC.

Kern, C.L. Jr. (1989) High-performance polyester for food and beverage packaging. *Food Technol.,* **43**(12), 93.

Kwapong, O.Y. and Hotchkiss, J.H. (1987) Comparative sorption of aroma compounds by polyethylene and ionomer food contact plastics. *J. Food Sci.,* **52**(3), 761.

LaBell, F. (1991) Edible packaging. *Food Proc.,* **52**(12), 24.

Labuza, T.P. and Breene, W.M. (1989) Applications of "active packaging" for improvement of shelf-life and nutritional quality of fresh and extended shelf-life food. *J. Food Proc. Preserv.,* **13**(l), 31.

Lopez, A.A. (1987) *A Complete Course in Canning and Related Processing,* 12th edn, The Canning Trade Inc., Baltimore, MD.

Maeda, H., Takahashi, Y. and Ifuku, Y. (1984) Adsorption of citrus nastsudaidal Hayata juice with polyvinyl pyrrolidone, Nylon 66 and high porous polymer. *Nippon Shokuhin Kogyo Gakkaishi,* **26**, 1.

Mannheim, C.H., Miltz, J., Ben-Aryie, G. and Lavie, R. (1985) *Interaction Between Carton Packages and Aseptically Packed Citrus Juice.* Presented at the 45th Annual IFT National Meeting, Atlanta, GA, Abstract 131.

Mannheim, C.H., Miltz, J. and Letzer, A. (1987) Interaction between polyethylene laminated cartons and aseptically packed citrus juices. *J. Food Sci.,* **52**(3), 737.

Mannheim, C.H., Miltz, J. and Passy, N. (1988) Interaction between aseptically filled citrus products and laminated structures, in *Food and Packaging Interactions* (ed J. Hotchkiss), ACS Symposium Series 365, Washington, DC.

Marin, A.B., Acree, T.E., Hotchkiss, J.H. and Nagy, S. (1992) Gas chromatography-olfactometry of orange juice to assess the effects of plastic polymers on aroma character. *J. Agric. Food Chem.,* **40**, 650.

Marshall, M.R., Adams, J.P. and Williams, J.W. (1985) Flavor absorption by aseptic packaging material. *Proceedings ASEPTIPAK 85,* Princeton, NJ.

Meadows, P. (1991) Building barriers. *Packaging Today,* **October**, 59.

Miltz, J. (1992) Food packaging, in *Handbook of Food Engineering* (eds D.R. Heldman and D.B. Lund), Marcel Dekker, New York, NY.

Miltz, J., Mannheim, C.H. and Harte, B.R. (1990) Packaging of juices using polymeric barrier containers, in *Barrier Polymers and Structures* (ed W.J. Koros), ACS Symposium Series 423, Washington, DC.

Moshonas, M.G. and Lund, E.D. (1969) Aldehydes, ketones and esters in Valencia orange peel. *J. Food Sci.,* **34**(2), 502.

Moshonas, M.G. and Shaw, P.E. (1989) Flavor evaluation and volatile flavor constituents of stored aseptically packaged orange juice. *J. Food Sci.,* **54**.

Nielsen, T.S. and Jägerstad, M. (1994) Flavour scalping by food packaging. *Trends Food Sci Technol.,* **5**, 353–356.

Osajima, Y. (1983) *Study on the Influence of Plastic Films Against the Food Aroma,* Japan Food Industry Bulletin.

Parry, R.T. (ed.) (1993) *Principles and Applications of Modified Atmosphere Packaging of Foods,* Blackie Academic and Professional, Glasgow, Scotland.

Pieper, G., Borgudd, L., Ackermann, P. and Fellers, P. (1992) Absorption of aroma volatiles of orange juice into laminated carton packages did not affect sensory quality. *J. Food Sci.,* **57**(6), 1408.

Rice, J. (1992a) Silica-coated plastic packaging. *Food Proc. (Am.),* **53**(6), 78.

Rice, J. (1992b) Silica-coated films to go commercial in '93. *Food Proc. (Am.),* **53**(10), 82.

Rose, D. (1995) Advances and potential aseptic processing, in *New Methods of Food Preservation* (ed G.W. Gould), Blackie Academic and Professional, London, England, 283–303.

Shimoda, M., Ikegami, T. and Osajima, Y. (1988) Sorption of flavor compounds in aqueous solution into polyethylene film. *J. Food Sci. Agric.,* **42**, 157.

Shimoda, M., Nintanda, T., Kadota, N., Otha, H., Suetsuna, K. and Osajima, Y. (1984) Adsorption of satsuma mandarin juice aroma on plastic films. *J. Jpn. Soc. Food Sci. Technol.,* **31**, 67.

Shorten, D.W. (1982) Polyolefins for food packaging. *Food Chem.,* **8**, 109.

Sizer, C.E., Waugh, P.L., Edstam, S. and Ackermann, P. (1988) Maintaining flavor and nutrient quality of aseptic orange juice. *Food Technol.,* **42**, 429.

Smith, J.P., Ramaswamy, H.S. and Simpson, B.K. (1990a) Development in food packaging technology. Part 1: processing/cooking considerations. *Trends Food Sci. Technol.,* **Nov.**, 107.

Smith, J.P., Ramaswamy, H.S. and Simpson, B.K. (1990b) Developments in food packaging technology. Part 2: storage aspects. *Trends Food Sci. Technol.,* **Nov.**, 111.

Subramanian, P.M. (1990) Polymer blends—Morphology and solvent barriers, in *Barrier Polymers and Structures* (ed W.J. Koros), ACS Symposium Series 423, Washington, DC.

Takahashi, J., Yuda, S. and Yanagslawa, H. (1984) Oxygen scavenger systems. *Japan Food Packaging,* **28**, 40.

Turtle, I. (1987) The use of flexible plastics for food packaging. *Food Technology International Europe,* Sterling Publications, London, England, 289–296.

Yamada, K., Mita, K., Yoshida, K. and Ishitani, T. (1992) A study of the absorption of fruit juice volatiles by the sealant layer in flexible packaging containers (The effect of package on quality of fruit juice, part 4). *Packaging Technol. Sci.,* **5**, 41.

Yokoyama, T. (1992) Food packaging overview. *Packag. Japan,* **13**(68), 25.

Yoshii, J. (1990) Smokable casing Krehalon, "SMO Casing." *Packag. Japan,* **11**(55), 35.

Yoshii, J. (1992) Recent trends in food packaging development in consideration of environment. *Packag. Japan,* **13**(67), 74.

Zagory, D. and Kader, A.A. (1988) Modified atmosphere packaging of fresh produce. *Food Technol.,* **42**(9), 70.

CHAPTER 5

The Hazard Analysis Critical Control Point (HACCP) System

P.J. Smedley

5.1 INTRODUCTION

Shelf life of foods depends on a number of factors, as outlined in the two parts of this book. Intrinsic and extrinsic factors are the main ones. However, one area that needs to be carefully considered in any shelf-life evaluation is the hazards and risks associated with the product and its processing.

The Hazard Analysis Critical Control Points (HACCP) concept provides a well-tried and trusted system for risk assessments to be carried out. This chapter gives an overview of HACCP, its background, principles, and details of its introduction into a food manufacturing plant. The chapter is not intended as a definitive text on HACCP—there are several excellent texts available elsewhere—but to provide an overview for product developers who are charged with the establishment of shelf-life limits. Specific reference to shelf-life studies will not be made, but implicit in this chapter is the need for its principles to be taken into account during any shelf-life evaluation.

5.2 BACKGROUND

The food industry has undergone many changes over the past 50 years, including the introduction of highly automated, high-speed operations and innovations in packaging, formulations, and distribution systems. Large volumes of product can now be shipped almost immediately after production, via sophisticated distribution networks, delivering them ready for purchase by consumers a very short time later. This rapid transfer capability has focused attention on systems that can provide sufficient control over both the procurement of raw materials and the manufacture and distribution of these products, and thereby ensure the safety of the food during its shelf life.

With traditional quality control (QC) failing to meet the challenge, the focus has moved to quality assurance (QA), which in turn has served to revive and encourage interest in the HACCP system. Today, HACCP has come to be recognized by both industry and government authorities as a prevention-orientated, systematic, efficient, and cost-effective approach to the control of identified hazards when compared with traditional QC. Thus, while not without its drawbacks, HACCP in its many interpretations has emerged in the new millennium as the primary approach to ensuring the safety of the food supply.

5.3 CHRONOLOGY

The development of the HACCP system began in 1959 with the Pillsbury Company in the United States. It has continued to the present via a process of constant evolution, with many individuals and organizations having published guidelines for its application (ICMSF, 1988; NACMCF, 1989, 1992; Bryan, 1992; CFDRA, 1992; Codex, 1993; Mortimore and Wallace, 1998). The main evolutionary stages are detailed in Figure 5–1.

OVERVIEW OF HACCP CHRONOLOGY

1959	**Pillsbury Co.** required to produce the first foods to be used in space. HACCP concept developed in collaboration with **U.S. Army Natick Laboratories**—Modes of Failure. **National Aeronautics and Space Administration**—Zero Defects.
1971	First public exposure of the system at the first **National Conference on Food Protection** (USA).
1973	First comprehensive document on HACCP published by the **Pillsbury Co.**—Food Safety Through the Hazard Analysis Critical Control Point System.
1973	Use of HACCP mandated by Federal Regulations— **US FDA Thermally Processed Low Acid Canned Foods Packaged in Hermetically Sealed Containers Part 128B.**
1980	**Food Safety and Inspection Service** **Food and Drug Administration** **National Marine Fisheries Service** **US Army Natick Research and Development Center** entered into a contract with the National Academy of Sciences, which was to "provide guidance on the best manner in which to develop a uniform approach that could be used to establish microbiological criteria."
1985	**National Academy of Sciences** published its first report—*An Evaluation of the Role of Microbiological Criteria for Foods and Food Ingredients* (Green Book).
1987	**National Advisory Committee on Microbiological Criteria for Foods (NACMCF**) established initial recommendation, which is now its operational principle, to base all food safety inspection programs on the HACCP system.
1988	**International Commission on Microbiological Specifications for Foods (ICMSF)** published *Application of the HACCP System to Ensure Microbiological Safety and Quality.*
1989	**NACMCF** published first major document—*HACCP Principles for Food Production,* including principles, descriptions, and definitions.
1991	**Codex Alimentarius Commission WHO** published Draft Principles and Application of the HACCP System.
1992	**Campden Food and Drink Research Association (CFDRA)** published *HACCP: A Practical Guide,* Technical Manual No. 38.
1992	**NACMCF** adopted revised document, entitled *Hazard Analysis and Critical Control Point System.*
1993	**Codex Alimentarius Commission WHO** published Guidelines for the Application of the Hazard Analysis Critical Control Point System.

Figure 5–1 Evolutionary stages of HACCP.

The need to manufacture product for a particularly demanding environment often provides the impetus for the development of new quality systems (e.g., the DEFSTANs of the British Ministry of Defence which subsequently gave rise to BS EN ISO 9000) and the HACCP system is no exception, being a by-product of the American space program.

The requirement of the Pillsbury Company to produce food for consumption during space missions brought with it the need for extremely effective quality assurance, due to the potentially catastrophic consequences of an illness or injury caused by contamination with any bacterial or viral pathogens, toxins, chemicals, or physical hazards (Bauman, 1990).

Faced with this challenge, the Pillsbury Company found the traditional, "reactive," methods of quality control involving end product testing (Figure 5–2) to be wholly inadequate, demand-

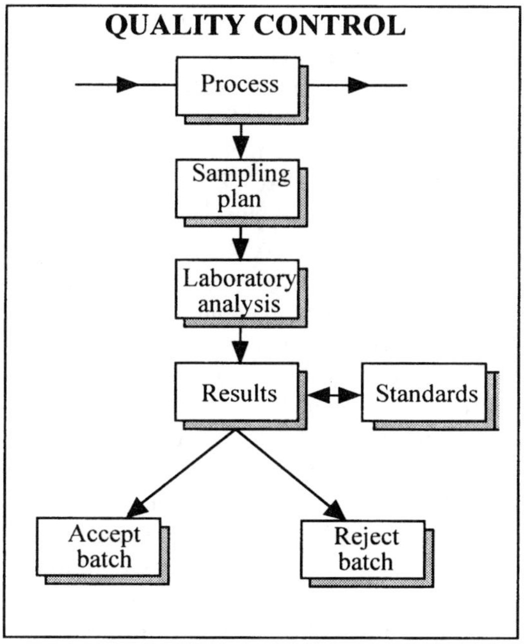

Figure 5–2 Traditional quality control.

Table 5–1 Drawbacks of traditional QC

The traditional way of ensuring product safety by end product testing has a number of drawbacks:

1. The control is reactive in that remedial actions are taken after the problem has occurred.
2. Considerable expertise can be needed to interpret the results of the tests.
3. Testing can be very slow.
4. The cost of the product sample and its analysis can be high.
5. The operation is controlled by a scientist at a laboratory often remote from the point of manufacture.
6. Only a limited number of samples can be assessed.
7. Not all potential hazards are taken into account.
8. Generally, only a section of the workforce, primarily technical personnel, are directly responsible for product safety.

ing large quantities of product for testing without providing the assurance of the quality demanded (Bauman, 1990; Pierson and Corlett, 1992). Today, the disadvantages of such systems are widely recognized (Majewski, 1992; Mitchell, 1994) (Table 5–1).

It was concluded that the only way forward was a preventative approach, in which control was exercised over raw materials, process, environment, personnel, storage, and distribution, as early in the system as possible.

Working in cooperation with the National Aeronautics and Space Administration (NASA), the Natick Laboratories of the US Army and the US Air Force Space Laboratory Project Group, the Pillsbury Company examined concepts embodied in previous "hardware" safety assurance systems and found the Modes of Failure system (used by the US Army), after some modifications, to be the best model for a hazard analysis system for the food industry (Pierson and Corlett, 1992). Every ingredient used had to be examined at each stage of its processing as it moved through the food chain, to determine what might happen to it and what could be expected in the way of problems when it appeared at the plant. Also analyzed were the manufacturing process and the environment to determine which areas could contribute to a safety hazard.

The HACCP system developed was designed to be both proactive and preventative (permitting prescribed remedial action to be taken quickly before problems occur), thereby allowing a high degree of product safety assurance and, when implemented properly, an end to the need for finished product testing (except for monitoring purposes). The system is based on two main concepts:

1. the identification of hazards, encompassing the entire process from raw materials to consumer use
2. the elimination and/or control of identified hazards via an integrated system of controls, centered on the concept of critical control points (CCPs), i.e., points in a system where loss of control may result in an unacceptable health risk and on which resources should thus be concentrated.

This new system of control was first exposed to the public in 1971 during the first US National Conference on Food Protection (US Dept. HEW, 1972). Subsequent cooperative work between Pillsbury and the US Food and Drug Administration (FDA) led to the publication of the first comprehensive document on HACCP (Pillsbury Co., 1973), and the first inclusion of its principles in national legislation in the form of the acidified and low-acid canned food regulation (US FDA, 1973). However it was not to be seriously considered by the food industry for broad application for another 15 years.

Figure 5–1 shows an overview of the early evolution of the HACCP, which culminated in the adoption of the Guidelines for the Application of the Hazard Analysis Critical Control Point System (WHO, 1993) by the Codex Alimentarius Commission (established in the 1960s by the Food and Agricultural Organization [FAO] and the World Health Organization [WHO] to develop internationally agreed codes), which not only provided an international consensus on the basic principles of HACCP but also demonstrated that HACCP has become the internationally accepted approach for ensuring the safety of food.

5.4 HACCP PRINCIPLES

The HACCP concept is a systematic approach to hazard identification and control. CFDRA (1992), NACMCF (1992), and Codex (WHO, 1993) have all proposed a common approach, allowing for an international consensus on the basic principles and how they should be implemented.

In fact, none of the steps involved in HACCP are unique, and often several of the criteria in the HACCP system are already utilized in many food operations, especially if written specifications and a good quality control system are available. The primary difference is that, rather than having isolated quality control procedures at various points of the process, they are linked together as a system. The various controls should interrelate and interlock so there is no way the system can go out of control, which would result in a hazard, without the information being picked up through the monitoring systems (Bauman, 1990).

However, it is also important to recognize the scope of the system. HACCP is not a substitute for the application of the principles of hygiene found in Codes of Good Manufacturing Practice (GMP). Though it is technically possible to build these into an HACCP system, this would overly complicate its application. Similarly, HACCP is only likely to be successful where the principles of good hygienic practice are already in place. If the structure, equipment, and cleaning standards are inappropriate, for example, it will be extremely difficult to develop an effective HACCP system.

The current HACCP system consists of seven principles which outline how to establish, implement, and maintain an HACCP plan for the operation under study (Figure 5–3). These are normally augmented by a further six preliminary stages (NACMCF, 1992; CFDRA, 1992; WHO, 1993) which serve to lay the groundwork for the HACCP study process (Figure 5–4).

5.4.1 Preliminary stages in an HACCP study

5.4.1.1 Assemble an HACCP team

The team should consist of individuals who have specific knowledge and expertise appropriate to the product and process. Such a team should be multidisciplinary to provide a broad base of knowledge, and where necessary outside expertise should be called upon. However, an HACCP plan should not be developed entirely by outside resources as it will likely be erroneous, incomplete, and lack the support of those charged with its implementation.

5.4.1.2 Describe the food and the method of its distribution

This consists of a full description of the food and its distribution and should include consideration of the potential for abuse during distribution and by consumers.

5.4.1.3 Identify the intended use and the consumers

The intended consumers may be the general public or a particular segment of the population, potentially at greater risk (e.g., infants). Such in-

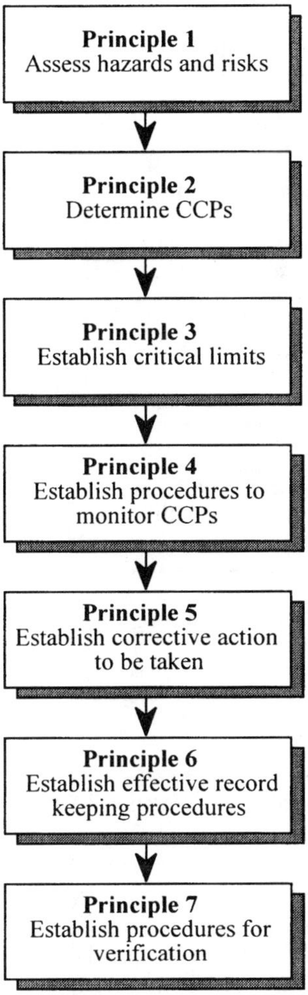

Figure 5–3 HACCP principles. Principles 6 and 7 are often used interchangeably.

formation, along with the product's intended use (e.g., cooked by the consumer or ready-to-eat), is an important factor for consideration when hazards are identified and assessed.

5.4.1.4 Develop a flow diagram which describes the process

The flow diagram should provide a clear and simple description of the steps involved in the process, this being the basic reference point for the rest of the study.

5.4.1.5 Verify flow diagram

The process should be inspected to ensure the validity of the flow diagram before the study is commenced.

5.4.2 HACCP principles

The following definitions of the HACCP principles are taken from the Codex Guidelines (WHO, 1993). Most have seen variations of interpretation during the evolution of the system and detail is given where contentious issues are still relevant.

5.4.2.1 Principle 1

> Identify the potential hazard(s) associated with food production at all stages, from growth, processing, manufacture, and distribution, until the point of consumption. Assess the likelihood of occurrence of the hazard(s) and identify the preventative measures for their control.

The HACCP team should conduct a hazard analysis and identify the steps in the process where hazards of potential significance can occur. For inclusion in the list, the hazards must be of such a nature that their prevention, elimination, or reduction to acceptable levels is essential to the production of a safe food. Hazards which are of a low risk and are not likely to occur would not require further consideration.

Hazard analysis is the most important single stage of the HACCP study, placing the greatest demands on available information and expertise. Failure to correctly identify the hazards presented by a process can lead to an ineffective HACCP system. It is therefore of no surprise that it is the scope and execution of this stage which have been the most contentious issue within the HACCP system and the root of many problems in its application.

First, the issue of whether to limit the HACCP system to safety-related concerns or to include quality parameters for non-safety regulations (e.g., indicators of economic fraud) is one that has consistently divided food safety professionals and

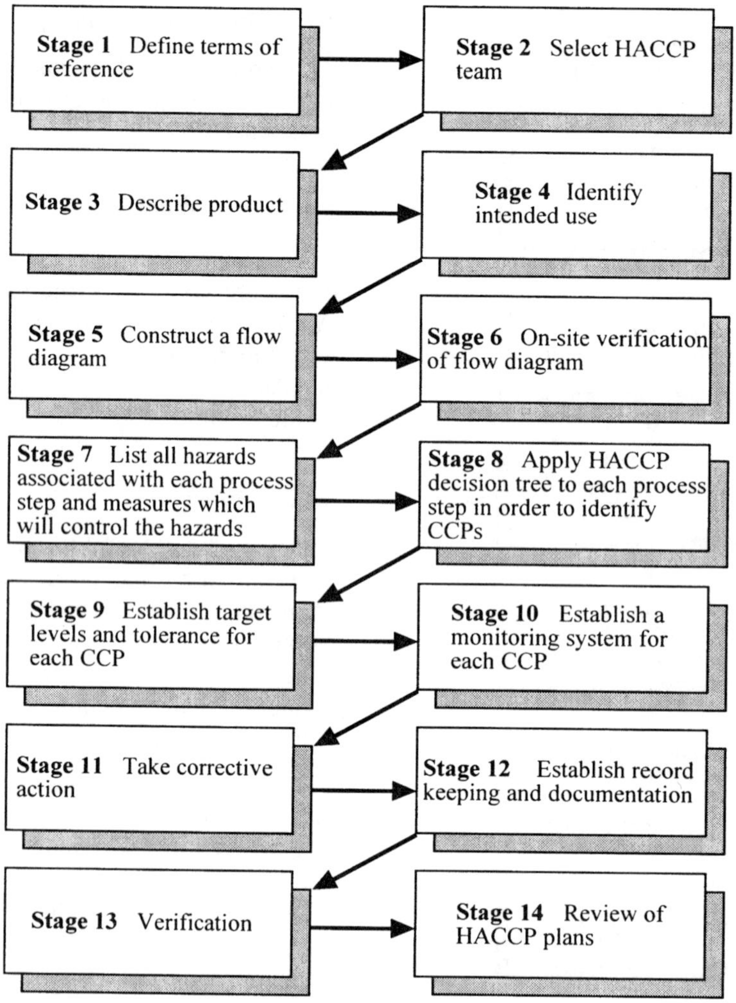

Figure 5–4 Stages for an HACCP Study. Courtesy of Campden & Chorleywood Food & Research Association Group, 1992, Gloucestershire, United Kingdom.

regulators. For instance, during the 1970s many manufacturers made attempts to implement HACCP but failed to achieve any quantifiable result. A reason put forward was that these programs often combined quality and regulatory programs with HACCP, diluting out the focus on safety. Such programs typically had many more critical control points (CCPs) than were needed to ensure the production of safe foods and became too cumbersome to sustain (Nat. Food Processors Assoc., 1993). More recently, the ICMSF considered both safety and quality in its discussions (ICMSF, 1988), whereas the NACMCF specifically recommended that formal HACCP programs be restricted to safety concerns (NACMCF, 1992). The latter recommendation has been tentatively adopted by the US Food Safety and Inspection Service in its planned implementation of an HACCP approach to meat and poultry inspection. In contrast, the HACCP program being considered by the US National Marine Fisheries Service (Garrett and Hudak-Roos, 1990) incorporated a definition of hazards that included economic fraud issues, such as species identification and net weight. While there is no inherent scientific reason why a system approach similar to HACCP could not be used to deal with quality-related issues, due to the danger of diluting efforts to control hazards associated with public health it is generally recommended that, if a program for quality or non-safety regulatory factors is desired, it should be established and administered separately from a safety-oriented HACCP program (Buchanan, 1990).

The actual definition of a hazard has also evolved along with the system, moving away from the original microbiological emphasis and inclusion of quality concerns to also encompass physical and chemical hazards with a safety scope only.

Another major issue of contention is that of risk assessment. The belief has long been held that "the analysis of hazards must be quantitative if it is to be meaningful" (ICMSF, 1988). This requires the assessment of two factors in respect of any identified hazard, namely risk and severity, where risk is an estimate of the likely occurrence of a hazard, and severity is the seriousness of the hazard. To this end there have been a number of schemes proposed which attempt to assign a value to a hazard reflecting its importance. The following are examples:

- ranking of severity and risk combined to give an overall assessment (CFDRA, 1989)
- the NACMCF system which groups products and raw materials into Risk Categories on the basis of the number of Risk Characteristics which they possess (NACMCF, 1989; Corlett and Stier, 1991).

However, there was no clear link between the hazard category of the product and the identification of the critical steps in the process. This shortcoming became more apparent as a greater proportion of the industry was educated in HACCP applications (Adams, 1994).

Thus, in line with the removal of risk-related terms from the definition of a hazard, the current international approach does not advocate categorization of hazards on the basis of risk categories, but instead requires control of any hazard which can reasonably be expected to occur. While there is no uniform acceptance for the discarding of structured risk assessment, the current system still requires some assessment of risk based on a considerably more pragmatic approach. The NACMCF has suggested a list of questions which can aid such an assessment (NACMCF, 1992).

5.4.2.2 Principle 2

> Determine the points/procedures/operational steps that can be controlled to eliminate the hazard(s) or minimize its likelihood of occurrence—critical control point (CCP). A step means any stage in food production and/or manufacture, including raw materials, their receipt and/or production, harvesting, transport, formulation, processing, storage, etc.

All hazards identified in the hazard analysis should be considered and assigned CCPs, thereby targeting resources on those points in the process critical to safety.

Since the introduction of HACCP, the definition of CCPs has changed markedly. As originally stated, the critical control point is a concept adopted by the panel to describe the location(s) or point(s) in a food processing operation at which failure to prevent contamination can be detected by laboratory tests with maximum assurance and efficiency (American Public Health Association, 1972). This original narrow definition for CCPs was modified by the ICMSF to "a location or a process which, if not correctly controlled, could lead to contamination with unacceptable growth" (WHO/ICMSF, 1982) and then further, to "a location, practice, procedure or process at which control can be exercised over one or more factors which, if controlled, could minimise or prevent the hazard" (ICMSF, 1988). This was then further modified to the current definition of a CCP, which clearly states that a CCP exists only where control can be exercised: "A point, step or procedure at which control can be applied and a food safety hazard can be prevented, eliminated or reduced to acceptable levels" (WHO, 1993). While the definition of CCPs has been clarified, their correct identification remains a stumbling block for many users of HACCP.

There are several points which have been put forward as important when determining the CCPs within a process:

- No minimum or maximum number of CCPs must be identified.
- The identification of CCPs is a factory- and product-specific process.
- The unnecessary duplication of CCPs should be avoided.
- Controls should not be introduced where they should not exist or are not necessary.
- Where serious doubt occurs, an expert should be consulted.
- Above all, common sense should be used (Kirby, 1994).

It is generally believed that HACCP is a tool and, provided the procedure is followed correctly, there is no reason why two different teams should necessarily always come up with exactly the same CCPs for the same process.

However, a common problem of some unsuccessful attempts to institute HACCP has been the identification of too many points in the process considered to be CCPs, where control must be exercised and critical limits must be met in order to produce a safe product. Excessive and inappropriate CCPs are too cumbersome for many manufacturing operations and can actually result in a less safe product by reducing emphasis on control points which truly are critical (National Food Processors Association, 1993). To avoid this problem, a clear differentiation needs to be made between critical control points and the control points which form an integral part of a food processor's comprehensive product control system. To this end, the NACMCF defined a control point as "any point in a specific food system where loss of control does not lead to an unacceptable health risk" (NACMCF, 1992). The disadvantage of this definition is of course its reintroduction of the risk assessment concept into the HACCP process. A complementary decision tree has also been proposed (Figure 5–5).

In general, control points (CPs) are non-safety points related to product quality or regulatory compliance and while important (and requiring control), they should not be regarded as part of a HACCP food safety management program. However, if a CP is repeatedly violated or several related CPs are simultaneously violated, the situation may warrant placing the affected product on hold.

A further problem related to principle 2, which is still to be fully resolved, is the decision process whereby CCPs are assigned.

The current solution is provided by the Codex Committee and the NACMCF which together have devised a logic sequence decision tree as an aid to determining which control steps are to be defined as CCPs (Figure 5–6). The adoption of the decision tree was a direct result of the increasing application of HACCP principles in the USA and the growing confusion regarding the appropriate designation of CCPs. Application of the decision tree to each step should determine whether the step is a CCP for the identified hazard. An explanation of the questions is given in Table 5–2. However, not only does its use often pose problems but the end results (in some situations) have also been called into question (Tompkin, 1994).

It is these first two principles which represent the most important stages of an HACCP system and about which there has been the greatest debate. The remaining principles relate mostly to the operation of the HACCP system and have remained relatively less controversial in their application. They are thus included here only briefly, for the sake of completeness.

5.4.2.3 Principle 3

Establish critical limit(s) which must be met to ensure a CCP is under control.

Each CCP will have one or more preventative measures that must be properly exercised to ensure prevention of the hazard(s) or reduction of associated risk(s) to acceptable levels. Critical limits must be specified for each preventative measure, where a critical limit is "a value which separates acceptability from unacceptability."

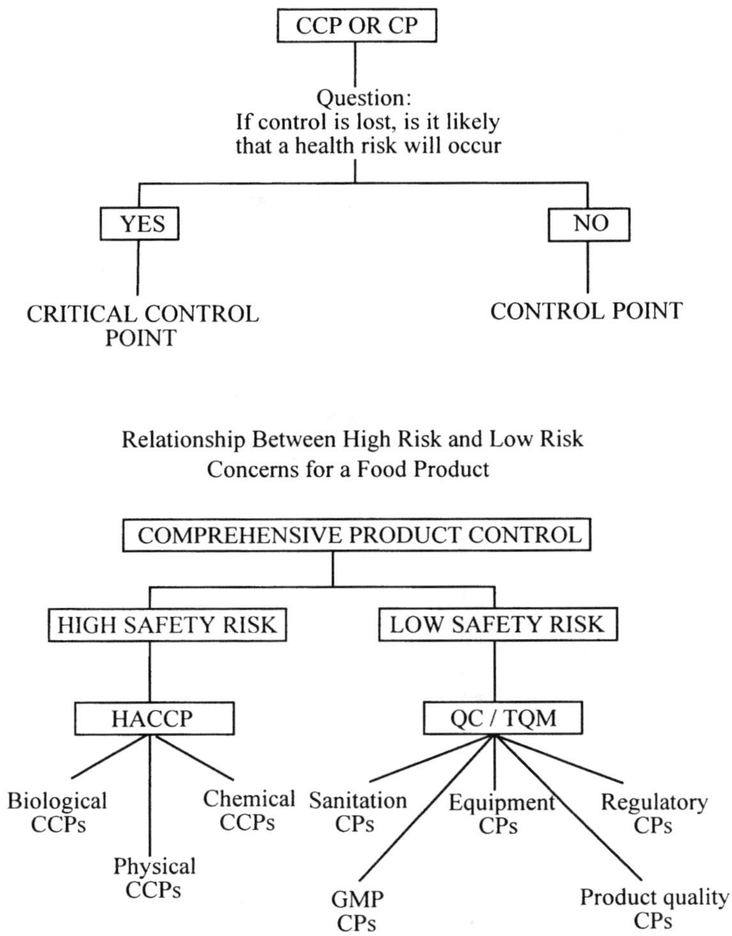

Figure 5–5 CP/CCP differentiation. *Source:* Reprinted with permission from M.D. Pierson and D.A. Corlett, *HACCP Principles and Applications, 1/E,* © 1992, Plenum Publishers.

Parameters often used include temperature, time, moisture level, pH, and a_w. Control can be complicated when product safety is affected by multiple critical factors that are known to interact. In this case, critical limits for each factor have to be determined and specified.

5.4.2.4 Principle 4

Establish a system to monitor control of the CCP by scheduled testing or observations.

Monitoring is the scheduled measurement or observation of a CCP relative to its critical limits. The monitoring procedures must be able to detect loss of control at a CCP and ideally provide such information in time for corrective action to be taken to regain control of the process before there is need to reject the product. Continuous monitoring is the ideal.

5.4.2.5 Principle 5

Establish the corrective action to be taken when monitoring indicates that a particular CCP is not under control.

Specific corrective actions must be developed for each CCP in the HACCP system in order to deal with deviations when they occur. The ac-

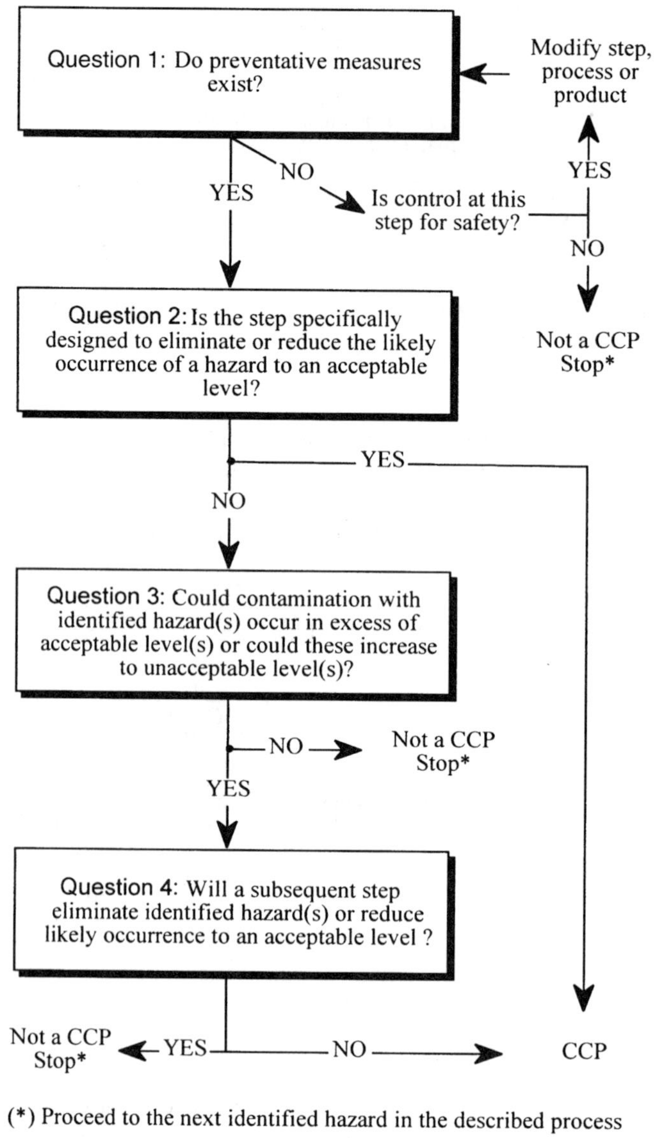

Figure 5–6 CCP decision tree. *Source:* Adapted with permission from *Codex Alimentarius Guidelines for the Application of the Hazard Analysis Critical Control Point System,* Vol. 1, Supp. 1, © 1993, World Health Organization.

tions must ensure that the CCP will be brought under control.

5.4.2.6 Principle 6

Establish procedures for verification which include supplementary tests and procedures to confirm that the HACCP system is working effectively.

Monitoring and auditing methods, procedures, and tests, including random sampling and analysis, must be used for verification.

5.4.2.7 Principle 7

Establish documentation concerning all procedures and records appropriate to these principles and their application.

Table 5–2 Explanation of Decision Tree Questions

Q1	Do preventative measures exist? If preventative measures cannot be introduced, then by definition the step cannot be a CCP because it is not possible to effect control at this step. If the answer is yes, then carry on to Q2.
Q2	Is the step specifically designed to eliminate or reduce the likely occurrence of a hazard to an acceptable level? If the answer is yes, then the step can be considered to be a CCP. This question allows flexibility, which would otherwise be denied by Q4. For example, steps, such as evisceration or skinning in a slaughtering operation, can be considered to be CCPs in addition to those that would be determined by consumer handling of the raw meat. If the answer is no, then carry on to Q3.
Q3	Does contamination with identified hazards occur in excess of acceptable levels or could these increase to unacceptable levels? This question forces us to consider those steps that could permit the contamination or the growth of microorganisms, such as storage or handling of the food, between processing stages. If the answer is no, then the step cannot be a CCP because there is nothing that needs to be controlled. If the answer is yes, then carry on to Q4.
Q4	Will a subsequent step eliminate identified hazard(s) or reduce their likely occurrence to an acceptable level? If the answer is yes, then the step is not a CCP and the whole decision tree is repeated for the next hazard or step in the operation. This question has a very important function to play when determining CCPs: it allows the presence of a hazard at a step, if that hazard will subsequently be eliminated or reduced to an acceptable likelihood of occurrence either as part of the operation or by some action of the end user. Otherwise every step in an operation might be critical, leading to too many CCPs for an effective, practical control system to be drawn up. Again the keys to success with this question are flexibility and common sense. This question is designed to work in tandem with Q2. For example, the presence of low levels of *Salmonella* in the meat ingredient for a ready-to-eat cooked meat prior to the cooking stage may be a concern but it might not necessarily be critical. On the other hand, its presence in a fresh meat product, that will leave the factory raw and will be cooked by the end user, if not a CCP under the terms of Q4, can be considered a CCP under the terms of Q2.

Source: Data from B. Mitchell, How to HACCP, *British Food,* Vol. 94, No. 1, © 1992, MCB University Press.

Documentation of HACCP procedures at all steps should be included and assembled into a manual.

5.5 HACCP AND LEGISLATION

As has been previously stated, the first instance of the inclusion of HACCP principles in legislation took place in the US, with the HACCP approach to the control of microbiological hazards in low-acid canned foods being mandated by the Food and Drug Administration (US FDA, 1973).

The success of these regulations is attributed to a number of factors:

1. Industry and government worked cooperatively to identify and develop monitoring procedures for critical control points. The procedures included details of inspection frequency with which the tests should be performed, the keeping of records, and the actions to be taken when monitoring results indicated lack of compliance, i.e., failure at a critical control point.
2. The FDA required that operators of retorts, aseptic processing and packaging systems, product formulation systems, and container closure inspectors be under the supervision of a person who had satisfac-

torily completed an FDA-approved training course.

3. FDA inspectors were trained in the principles of the HACCP system. As a result, FDA inspectors became knowledgeable of the critical control points in the production of these products. At the time of plant inspections, major emphasis was placed upon review of monitoring documentation.
4. The use of the HACCP system was mandated by federal regulations.
5. Microbiologically, the manufacture of low-acid canned foods is a well-established "control situation." Only two general CCPs have to be controlled to ensure the safety of the product, i.e., the adequacy of the thermal treatment and the integrity of the container. The parameters involved can be monitored by rapid physical measurements (Felix, 1987).

Further HACCP-based legislation was not forthcoming for many years due to the absence of such conducive factors in other sectors of the food industry. However, as HACCP has gained international recognition, a growing number of countries have recognized the advantages of including the system in their legislative approaches to food safety. In the US, further regulations have been introduced that include mandatory HACCP studies: seafood safety regulations (US FDA, 1995); The pathogen-reducing HACCP systems applied to the meat and poultry regulations (US FSIS, 1996); 10 generic HACCP models (US FSIS, 1999); proposed HACCP for fruit and vegetable juices (US FDA, 1998).

In its 23rd session, the Codex Alimentarius Commission (July 1999) recommended that "Governments should be encouraged to integrate risk analysis into their legislation." In the UK, the issue of inclusion of HACCP in food legislation was first raised by the report to the Government on the Microbiological Safety of Food chaired by Sir Mark Richmond (Richmond, 1990). Part 1 of the Committee's report was released prior to the passing of the Food Safety Act (1990), in which it was recommended that "the adoption of the principles of HACCP at all stages of food production will greatly enhance food safety. . . . all food processes should be designed on HACCP principles."

While the Food Safety Act (1990) did not explicitly include HACCP, its due diligence provision (Section 21), whereby a person charged with an offense will have the possible defense that he "prove that he took all reasonable precautions and exercised all due diligence to avoid the commission of the offence by himself or a person under his control," gave strong impetus for its use.

The first part of the defense, taking reasonable precautions, requires a system to ensure things will not go wrong, and HACCP is just such a system, anticipating hazards and identifying points where control can be exercised. Also the estimates of risk made within HACCP are based on the best available knowledge and thereby comply with Article 6 of the EC liability directive (85/374/EEC) and the product safety directive, which requires the producer to "adopt measures to ensure suitable monitoring of the safety of the products, taking account of their particular characteristics, with a view to being properly informed of the risks which these products might present," and Article 7(e) of the liability directive, which states, "The producer shall not be liable if he proves that the state of scientific and technical knowledge at the time when he put the product into circulation was not such as to enable the existence of the defect to be discovered." Thus a business applying an appropriate HACCP system is much more likely to have an acceptable defense before the EC as well as UK law.

The principles of HACCP have also been introduced into the European food legislation, implicitly in several vertical directives and explicitly in the horizontal food hygiene directive.

The vertical directives, which lay down the hygienic conditions required for the production and marketing of specific food commodities (e.g., DIR 91/493 on fishery products, DIR 92/5 on meat-based products, and DIR 92/46 on milk and milk-based products), contain a general provision to require persons responsible for an establishment to carry out their own checks based on the following principles (Jouve, 1994):

- identification of critical points in their establishment on the basis of the manufacturing process used
- establishment and implementation of methods for monitoring and checking such critical points
- taking samples for analysis in an approved laboratory by the competent authority for the purpose of checking cleaning and disinfection methods and for the purpose of checking compliance with the standards established by the directive
- keeping a written record of the preceding points with a view to submitting them to the competent authority.

The horizontal directive on the Hygiene of Foodstuffs (DIR 93/43/EEC) is a framework directive which sets out the general rules of hygiene for foodstuffs and the procedures for verification of compliance with these rules. Article 3 establishes a general requirement for hygiene at all stages of the food chain and a further requirement that "food business operators shall identify any step in their activities which is critical to ensuring food safety and ensure that adequate safety procedures are identified, implemented, maintained and reviewed on the basis of the following principles, used to develop the system of HACCP." Reference to HACCP is explicit only in terms of its principles, meaning that the HACCP system as such is not a formal requirement of the directive. The requirement is for a system approach which has to be developed under each food business operator's responsibility, integrating hazard analysis, risk assessment, and other quality assurance techniques, to provide evidence that relevant safety hazards have been properly identified and assessed and that appropriate procedures for safety have been identified, implemented, monitored, and reviewed. Different tools can therefore be used to comply with this requirement of the directive, provided they are based on the same principles as HACCP.

Implementation of the Hygiene of Foodstuffs Directive in the UK has taken the form of The Food Safety (General Food Hygiene) Regulations 1995. Here, Regulation 6 places a duty on food businesses to identify and control potential food hazards based on the principles set out in Schedule 2: "The proprietor of a food business shall identify any step in the activities of the food business which is critical to ensuring food safety and ensure that adequate safety procedures are identified, implemented, maintained and reviewed on the basis of the following principles—

1. analysis of the potential food hazards in a food business operation;
2. identification of the points in those operations where food hazards may occur;
3. deciding which of the points identified are critical to ensuring food safety ("critical points");
4. identification and implementation of effective control and monitoring procedures at those critical points;
5. review of the analysis of food hazards, the critical control points and the control and monitoring procedures periodically and whenever the food operations change."*

To facilitate the application of these principles a number of legally recognized Industry Guides to Food Hygiene Practice have been published in concert with various industry sectors (Industry Guide: Retail/Catering/Baking, 1997). Included within these guides are generic assessment tables for the identification of hazards and possible control measures.

Further examples of the inclusion of HACCP in legislation can be found in Canada where the Health Protection Branch under the Food and Drugs Act has proposed Good Manufacturing Regulations for Foods that include the requirement for manufacturing controls following HACCP principles. Also, as of February 1, 1992, the Canadian fish processing industry was the first industry worldwide to be regulated

---

**Source: Food Safety (General Food Hygiene) Regulations,* Regulation 6, 1995, is reproduced under the terms of Crown Copyright Policy Guidance issued by HMSO, London, United Kingdom.

through an HACCP-based program called the Quality Management Programme.

Thus, along with the resurgence of interest in HACCP in particular and QA in general, HACCP is increasingly finding its way into legislation worldwide. This can only serve to increase its use and therefore its potential to provide a universal system for food safety that could both facilitate the international movement of foods and provide a high level of safety assurance. FDA Deputy Commissioner, Michael Taylor, summarized the FDA's vision with respect to HACCP: "We envision HACCP becoming in the future years the central element of the harmonized food regulatory standards on which international trade in food will depend" (Taylor, 1993).

Without doubt, the future of food legislation thus lies not only in the inclusion of HACCP principles but more importantly the use of HACCP as a basis for the formulation of such legislation.

5.6 HACCP AND SHELF LIFE

With the central tenet of the HACCP approach being the control of hazards throughout the entire food production process, up to and including the point of consumption, its application by a manufacturer goes by definition far beyond the factory gate. While most factors related to safety with reference to the retail and consumption of products are outside the direct influence of the manufacturer, two of the most important, i.e., the definition of accurate label information and a safe shelf life, are not and should form part of a full HACCP study.

To have the greatest effect, HACCP principles should ideally be applied from the outset of a product's design and development process, thereby enabling required controls to be built into the process and product and thus the implementation of QA.

Two of the major ex-factory considerations in product design are the intended/anticipated circumstances of use and the safe shelf life. As previously identified, the stages of the HACCP process directly following the selection of a HACCP team are the description of the food and the method of its distribution, followed by the identification of its intended use. A major part of a product description will be its shelf life, the initial basis for which will often be a customer's requirements. While the actual achievable shelf life will be based on a combination of safety and/

Table 5–3 Technical data required for a HACCP study

- Epidemiological data on microbial pathogens
  - Incidence of foodborne illness
  - Surveillance programs
- Raw material, intermediate, and final product data
  - Formulation
  - pH
  - a_w
  - Presence of preservatives
  - Packaging materials and conditions
  - Product structure
  - Processing conditions
  - Storage and distribution conditions
  - Shelf life
  - Consumer use instructions, pack labeling
  - Consumer target groups
- Processing data
  - Number and sequence of all processing stages including storage
  - Range of product time–temperature conditions at each process stage
  - Rework handling operations
  - High/low risk separation systems in use
  - Flow conditions (for liquids)
  - Presence of void spaces in processing equipment
  - Efficacy of cleaning and disinfection
  - Possibilities for cross-contamination
- Microbial data
  - Likely presence of microbiological hazards in raw materials (see also epidemiological data)
  - Growth rates of microbial hazards in food products
  - Death rates of microbial hazards under a range of processing conditions

Note: This table is a list of the types of technical data required for an HACCP study. Not all data will be required for all studies.

Source: Reprinted from *Food Control,* January, Mayes, Technical Data Required for a HACCP Study, Table 1, p. 17, Copyright © 1992, with permission from Elsevier Science.

or organoleptic considerations, in terms of HACCP it is the safe shelf life which is the important factor. Thus, within these terms of reference it is the introduction, survival, and multiplication of pathogenic microorganisms (or their toxins) which are of concern. A number of considerations will have an influence in this respect and will thus require consideration within the HACCP study (see also Table 5–3):

1. ***The type of raw materials used***—What is the potential initial microbiological loading, and what storage/handling conditions are on site to prevent multiplication/cross-contamination?
2. ***The intrinsic nature of the product***—Will the composition (e.g., acid/salt/a_w/pH) retard or prevent pathogen growth or the introduction of a preservative be permissable (desirable)?
3. ***Processing conditions and technology***—Are pathogens likely to be introduced from the manufacturing environment, or reduced by antimicrobial treatments (e.g., pasteurization, irradiation)?
4. ***Product packaging***—What are the barrier properties and, in the case of several modern materials, the antimicrobial properties of the packaging used?
5. ***Conditions during storage and distribution***—In particular, what is the chill chain?
6. ***Consumer use (abuse)***—What are the requirements for both storage and final preparation conditions?

In terms of safety, these considerations will lead to the classification of the product as contaminated, likely to become contaminated with, or supporting the growth of pathogenic microorganisms. The shelf life can then be determined on the grounds of product quality alone. However, where the nature of the product results in the unavoidable risk of pathogen presence/growth, the period of time until the product becomes unsafe will need to be ascertained and the shelf life set to control this hazard. To aid in this process, mathematical microorganism growth models (e.g., MAFF Food MicroModel) are now available as an adjunct to standard challenge testing methodology.

Therefore the definition of a product's shelf life lies at the heart of safe product design and thus the basis of the hazard analysis procedure as well as the subsequent in-process controls.

REFERENCES

Adams, C.E. (1994) HACCP as applied in the USA. *Food Control,* **3**.

American Public Health Association (1972) *Proceedings of the 1971 National Conference on Food Protection.* US Government Printing Office, Washington, DC.

Bauman, H. (1990) HACCP: concept, development and application. *Food Technol.,* **48**(5), 156–158.

Bryan, F.L. (1988) Hazard analysis critical control point: What the system is and what it is not. *J. Environ. Health,* **50**, 400.

Bryan, F.L. (1992) *Hazard Analysis Critical Control Point Evaluations. A Guide to Identifying Hazards and Assessing Risks Associated with Food Preparation and Storage.* WHO, Geneva, Switzerland.

Buchanan, R.L. (1990) HACCP: A re-emerging approach to food safety. *Trends Food Sci. Technol.,* **1**(5), 104–106.

CFDRA (1989) *Hazard Analysis Critical Control Points (HACCP)—A Practical Guide,* Technical Manual No. 19, Campden Food and Drink Research Association, Chipping Campden, England.

CFDRA (1992) *Hazard Analysis Critical Control Points (HACCP)—A Practical Guide,* Technical Manual No. 38, Campden Food and Drink Research Association, Chipping Campden, England.

Codex Alimentarius Commission (1999) *23rd Session Summary Report.*

Corlett, D.A. and Stier, R.F. (1991) Risk assessment within the HACCP system. *Food Control,* **2**(4), 71–72.

Felix, C.W. (1987) *Food Protection Technology: Proceedings of the '86 Conference for Food Protection,* Lewis Publishers, Chelsea, MI.

Garrett, S.E. and Hudak-Roos, M. (1990) Use of HACCP for seafood surveillance and certification. *Food Technol.,* **48**(5).

ICMSF (1988) *HACCP in Microbiological Safety and Quality,* Blackwell Scientific Publications, Oxford, England.

ICMSF (1998) *Application of the HACCP System to Ensure Microbiological Safety and Quality,* Book 4, Micro-organisms in Foods, Blackwell Scientific Publications, Oxford, England.

Industry Guide (1997) *Industry Guide to Good Hygiene Practice: Retail Guide/Catering Guide/Baking Guide,* Chadwick House Group Ltd., London, England.

Jouve, J.L. (1994) HACCP as applied in the EEC. *Food Control,* **5**(3), 181–186.

Kirby, R. (1994) HACCP in practice. *Food Control,* **5**(4).

Majewski, M.C. (1992) Food safety: The HACCP approach to hazard control. *Communicable Disease Report,* **2**(9), 105–108.

Mayes, T. (1992) Simple users' guide to the Hazard Analysis Critical Control Point concept for the control of food microbiological safety. *Food Control,* **1,** 14–19.

Mayes, T. (1994) HACCP Training. *Food Control,* **5**(3).

Micro and Food Safety Committee of the National Food Processors Association (1992) HACCP and total quality management—winning concepts for the 90s: A review. *J. Food Protection,* **55**(6).

Micro and Food Safety Committee of the National Food Processors Association (1993) Implementation of HACCP in a Food Processing Plant. *J. Food Protection,* **56**(6).

Mitchell, B. (1994) How to HACCP. *British Food J.* **1**.

Mortimore, S., and Wallace, C. (1998) *HACCP A Practical Approach,* Aspen Publishers, Gaithersburg, MD.

NACMCF (1989) *HACCP Principles for Food Production,* USDA Food Safety and Inspection Services, US Government Printing Office, Washington, DC.

NACMCF (1992) *Hazard Analysis and Critical Control Point System,* USDA Food Safety and Inspection Service, US Government Printing Office, Washington, DC.

National Academy of Sciences (NAS) (1985) *An Evaluation of the Role of Microbiological Criteria for Foods and Food Ingredients,* NAS National Research Council, National Academy Press, Washington, DC.

National Food Processors Association (1993) Implementation of HAACP in a food processing plant. *J. Food Protect.,* **56**(6), 548–554.

Pierson, M.D. and Corlett, D.A. (1992) *HACCP Principles and Applications,* Van Nostrand Reinhold, New York, NY.

Pillsbury Company (1973) *Food Safety Through the Hazard Analysis Critical Control Point System,* The Pillsbury Company, Minneapolis, MN.

Richmond, M. (1990) The Microbiological Safety of Food: Report of the Committee of Enquiry, Part 1 HMSO. London: HMSO.

Richmond, M. (1991) The Microbiological Safety of Food: Report of the Committee on Enquiry, Part 2 HMSO.

Rine, D.C. (1988) *Improving Software Productivity: An Expert Database Approach,* IEEE Expert, Summer.

Taylor, M. (1993) FDA's Plans for Food Safety and HACCP—Institutionalizing a Philosophy of Prevention. International Life Sciences Institute Symposium on Foodborne Microbial Pathogens, August, Atlanta, Georgia.

Tompkin, R.B. (1990) The use of HACCP in the production of meat and poultry products. *J. Food Protection,* **53**(9).

Tompkin, R.B. (1994) HACCP in the meat and poultry industry. *Food Control,* **5**(3).

US Department of Health, Education & Welfare (1972) *Proceedings of the 1971 Conference on Food Protection,* US Government Printing Office, Washington, DC.

US Food Safety and Inspection Service (US FSIS) (1996) Pathogen reduction: Hazard Analysis and Critical Control Point (HACCP) systems—final rule, *Federal Register,* **61**(144).

US FSIS (1999) *Guidebook for the Preparation of HACCP Plans and Generic HACCP Models (HACCP 1 to 15),* US Government Printing Office, Washington, DC.

US Food and Drug Administration (US FDA) (1973) Acidified Foods and Low Acid Foods in Hermetically Sealed Contrainers, 21 CFR, Parts 113 and 114.

US FDA (1995) *Procedures for the Safe and Sanitary Processing and Importing of Fish and Fishery Products.* December.

US FDA (1998) Hazard Analysis and Critical Control Point (HACCP); Procedures for the Safe and Sanitary Processing and Importing of Juice. Proposed Rule. 21 CFR, Part 120, *Federal Register,* **63**(79).

US National Advisory Committee on Microbiological Criteria for Foods (1991) Recommendations of the NACMCF: I HACCP principles, II meat and poultry, III seafood. *Food Control,* **2**(10).

WHO (1993) *Codex Alimentarius Guidelines for the Application of the Hazard Analysis Critical Control Point System,* Vol. 1, Supp. 1.

WHO/ICMSF (1982) *Report of the WHO/ICMSF Meeting on Hazard Analysis Critical Control Point in Food Hygiene,* Document VPH/82.37, WHO, Geneva.

Chapter 6

Preservation Technology and Shelf Life

R.J. Winger

6.1 INTRODUCTION

Time–temperature tolerance (TTT) of foods was first defined in frozen foods, with a series of articles initiated by van Arsdel (1957) and summarized by van Arsdel et al. (1969). TTT and the reaction kinetics associated with prediction of shelf life assume the milieu of reaction systems are identifiable, predictable, and reproducible from one sample of food to the next. Unfortunately this is far from the truth. In fact, foods are derived from biological systems, which themselves are inherently variable and notoriously unpredictable. Each food product contains an extraordinary array of physical, chemical, biochemical, and biological reaction systems, such that it is almost impossible to catalog, identify, or understand this variety in any one food product. One of the features of food systems is the complex interrelationship among the numerous concomitant reactions that dictate the food's distinct structure, physical and chemical characteristics, and the aesthetic values defining its unique organoleptic characteristics.

All food products progressively, inexorably decline in quality from harvest. Shelf life is the period of time at the end of which the "quality" of a given food product is perceived as significantly, unacceptably different from the expected "fresh" quality. This may be assessed in terms of microbiological numbers, chemical or physical changes in foods, or one of an array of sensory tests of varying complexity. The reason for the end of shelf life may be due to one of many changes occurring simultaneously in a given sample of food. It is not realistic to define all chemical components and the myriad reactions that occur in real food systems. Thus a concise and meaningful definition for shelf life of all foods is unachievable.

6.2 THE THREE Ps—PRODUCT, PROCESSING, AND PACKAGING

The TTT concept can be mathematically modeled for any given set of conditions, for a given batch, or individual sample of food. This model can offer powerful options, such as the prediction of shelf life of a similar batch of food from its "fresh" characteristics, or following an accelerated shelf-life testing procedure. However, the impact of temperature on shelf life can be dramatically overridden by other factors within the control of food manufacturers. This ability to manipulate the food (product), processing conditions, and packaging has been coined as the product, processing, and packaging (PPP) concept (Dalhoff and Jul, 1965; Jul, 1984). In every instance, it is essential to define PPP parameters for a food in order to adequately assign a suitable TTT model. In addition, it is logical that the Hazard Analysis and Critical Control Point (HACCP) system, total quality management, and an holistic technological approach are mandatory requirements in the production of a consistent, safe, economic, and acceptable food product which meets a desirable commercial shelf life.

Pragmatically, the temperatures foods are subjected to throughout the distribution chain are variable and inconsistent. This coupled with the impact of PPP variables on food quality means that the prediction of the end of shelf life for a given food is a problematic event. Over a period of time there will be a statistical distribution of shelf-life failure for a particular food product. For most natural biological systems this would resemble a normal statistical distribution. The manufacturer needs to decide which part of this biological distribution would be used to define the shelf life: the point of first observed failure, or a point closer to the middle of the statistical curve. This decision is often determined by the type of food and the mechanism of shelf-life termination. For example, if pathogenic microorganism growth was the determinant of shelf life, one would choose the safe option of the shortest failure time.

There are two more important Ps which every food technologist must consider as part of food product development: people and price. The former—the consumer—is the ultimate target of the food being produced. The latter relates to the cost of production of the foodstuff and its relativity to the purchase price and quality expectations of the consumer.

6.3 TWO ADDITIONAL Ps: PEOPLE AND PRICE

6.3.1 People

Food design without consideration of the people consuming it is a nonviable commercial option. Food must meet the current (and fickle) expectations of the consumer, particularly in terms of organoleptic acceptability, convenience, variety, safety, nutritional and health needs, value for money, and the aesthetics of presentation (particularly packaging, portion control, color, etc.). Shelf life must also be considered in terms of the consumers and their response. The practical end to shelf life occurs when consumers reject the food.

The consumers' ability to perceive the end of shelf life varies considerably. A small proportion of consumers are often very sensitive to subtle changes in one or more quality attributes in a particular food product. At the other extreme some consumers will be surprisingly insensitive to those same quality changes. Each consumer is unique. Cultural and social patterns, traditional eating habits, and genetic inheritance have a profound influence on a consumer's sensitivity and acceptance of foods. Thus the ability of consumers in a given market to detect a quality change in a specific food will follow a statistical distribution. In addition, the consumers' reaction to these quality changes (in terms of like/dislike) is complex and follows a non-symmetrical statistical distribution. Because of the unpredictable nature of the consumer response to storage-induced changes in food quality, sensory tests are usually mandatory, these being a benchmarking method for shelf-life testing.

Two examples stem from the author's personal experiences with the sensory testing of meat. While selecting panelists for a sheep meat flavor profile study, minced sheep meat was mixed with minced beef and triangle tests were performed. The best New Zealander could detect 15% sheep meat in beef (compared to pure beef), with around 60 of the 200 people tested being able to detect ≤25% sheep meat. A small group of Southeast Asians could unfailingly detect 5% sheep meat in 95% beef.

The second example relates to rancidity detection. For trials on shelf life of frozen meat, panelists were selected who were sensitive to rancid flavor. Although brilliant on rancidity, these panelists were insensitive to most other meat flavors.

There are two levels of statistical variability which need to be considered when defining shelf life: the natural food product variability, and the ability of the consuming population to detect a significant, undesirable difference in the food. Shelf life is sometimes defined by trained, very sensitive panelists (coined as the "high quality life"). Sometimes shelf life is defined in terms of the "average" consumer and has been coined the "practical shelf life" (Jul, 1984). In effect shelf life may be defined in terms of the number of consumers' complaints that are commercially acceptable to the marketing team.

6.3.2 Price

Price is the commercial driver of all company operations and covers all matters from the cost of production and profitability to all parts of the growing, manufacturing, and distribution chain. Manufactured food products are constantly being reviewed by technical teams to find ways to reduce costs. These may include new ingredients, or new technologies. Reducing costs is an important technical role in industry. For food products this often translates into ingredient substitution, which in turn can impact on shelf life.

Long-term company viability is directly related to new product development. Food product life cycles are becoming shorter, competition is strengthening, and new products can be copied and extended by other companies often faster than the innovative company can in introducing the new product. Assurance of shelf life becomes an essential component of product development. The tools associated with modeling this shelf life and predicting the outcome of product changes are paramount in our information technology portfolio. It is essential to understand the impact of PPP factors, particularly in relation to the ability to modify price, in order to utilize these models.

6.4 PRODUCT

The natural biological variability of real food systems is enormous. Freshly harvested crops exhibit variations in maturity, composition related to dietary or fertilizer inputs, genetic purity, and many quality features important to consumers (such as blemishes, size, shape, color, etc.). This natural variation may be as high as ±50%.

The first steps in food manufacture usually involve selection of individual food units from among this variable stock. Revolutionary technologies, such as genetic engineering, offer novel, rapid techniques for introducing new variability. Raw materials may be specifically selected or tailored to meet clearly defined quality characteristics. Ironically any change in the genetic template of a plant or animal, no matter how well targeted and controlled, could (theoretically) modify the safety, shelf life, or reaction milieu of the system.

Literature is replete with examples of natural biological variability and its impact on food quality. However, it is possible to convert natural food products into new custom-designed, highly consistent raw materials commonly grouped as "food additives" or industrial ingredients. The use of food additives as part of the food manufacturing process is a critical option which is defined as part of the "product" component of PPP.

Food additives may be used in the form of a wax coating on fresh horticultural products to minimize water loss and reduce respiration rates. They may be used to control reactions, such as antioxidants for delaying the onset of rancidity. They may be used to modify structures and produce new foods which cannot exist in the natural state, such a starches. In doing so, this can change the shelf life and predictable behavior of the formulated food, based upon knowledge of the natural raw materials being used. This is merely a glimpse of the use of additives in food manufacture. Their impact will be discussed later in this book, and the reader is referred to some of the many excellent texts in food chemistry and ingredient technology that exist (Angelo, 1992; Birch and Campbell-Platt, 1994; Charalambous, 1993; Fennema, 1996; Potter and Hotchkiss, 1995; Russell and Gould, 1991).

6.5 PROCESS

Since time immemorial, humans have learned to process perishable food products to prolong shelf life and ensure the safety of the foods from microbiological proliferation. Processing ranges from simple selection and grading operations, through highly sophisticated technologies, such as extrusion and aseptic liquid processing, that have been introduced only in the latter stages of the 20th century.

Processing can result in shelf-life extensions that are unparalleled by any other means. For example, fresh milk has a shelf life of 1–2 days. Pasteurization followed by refrigeration increases this to 1–2 weeks. UHT processing results in milk storage at room temperature for 6 months. Removal of water (cheese, butter) can result in shelf lives measured in months and years. Concentrating and canning operations result in a milk product which is stable for several years. It becomes apparent from this example that mathematical models looking at shelf life from a TTT approach alone will be dramatically in error when attempting to predict the maximum life from the characteristics of the original fresh food product.

A variety of processing methods have been used or are being developed for purposes of food preservation. These include drying and related processes (removing water), refrigeration (chilling and freezing), thermal processing (pasteurization, blanching, cooking, sterilization), radiation (irradiation using gamma rays, light pulses, ultraviolet), high hydrostatic pressure (including extrusion), separation processes (filtration, membranes, centrifugation), ultrasound, pulsed electric fields, oscillating magnetic fields and recombination processes (e.g., homogenization). In addition, many industrial process lines may incorporate several preservation processes.

These processes may be used prior to packaging. However, with the advent of minimally processed foods and modified atmosphere packaging, there is a growing use of key technologies post-packaging. Apart from the traditional thermal preservation techniques (such as canning), innovative heating methods are being used. These include microwaves, pulsed electric fields, dielectric heating and ohmic heating. Light pulses (particularly ultraviolet), irradiation, pulsed X-rays, and magnetic fields provide different options for preservation within packages. The effectiveness of these techniques depends on the type of food being preserved (Barbosa-Canovas et al., 1998).

6.5.1 Drying

The oldest technology, drying, ameliorates food stability through the removal of water. This impacts on food stability in two ways. First, microorganisms need a minimum level of moisture to grow. The availability of water in foods has been defined as water activity (a_w), and is a fundamental physical chemistry property of foods. a_w is realized in food in terms of fugacity of water. This, in turn, depends upon the temperature, atmospheric pressure, and the number of moles (and type) of solutes present. Although this concept is mathematically defined for dilute solutions, it is poorly predictable in food systems because of the complex and highly concentrated solute environment present. These issues have been discussed in greater detail elsewhere (Fennema, 1996; Rockland and Beuchat, 1987).

The second impact water has on foods is its action as a plasticizer of solutes, particularly high molecular weight materials. This property is called "molecular mobility," though it is often termed "glass transition." Although identified for foods in the 1940s (see Fennema, 1996), recent interest was rekindled in the early 1980s (Franks, 1982; Levine and Slade, 1986). This concept has a far-reaching impact on a wide range of food stability issues. Any diffusion-based reaction system in foods, of which there are many, can be examined using this concept. Here water acts to first hydrate and then lubricate the solutes. The impact is on the rheology of the food. The most stable molecular form is the crystal (white sugar, sucrose, is an example). Upon heating, crystals will melt and become a liquid. If this liquid is cooled slowly, it will recrystallize. However, if the melt is cooled rapidly, it will solidify and remain as an amorphous, non-crystalline solid. This is the basis of hard candy confectionery. These amorphous "solids" are called "glasses" and are in fact highly viscous solutions. The viscosity of water at 20°C is 1 cP. The relative viscosity of a glass at 20°C is of the order of 10^{14} cP. Diffusion coefficients (molecular mobility) change from about 10^{-5} cm^2/sec in water at 20°C to values measured in terms of a few cm^2/year in a glass. The concepts of molecular mobility have made a significant contribution to explaining many apparently anomalous phenomena observed in the processing and storage of foodstuffs. This concept is still being defined and better understood.

6.5.2 Thermal preservation (heat processing)

Heat processing modifies food safety and shelf life through the impact of high, lethal temperatures on microorganisms and natural biological components in foods (e.g., enzymes) (Zeuthen et al., 1990b).

Pasteurization and blanching are partial heat treatments designed to destroy pathogenic vegetative microorganisms and plant-based enzymes, respectively. Provided the processed foods are properly packaged, the shelf lives of these foods are dramatically increased in comparison to the natural product. These mild heat treatments cause minimal changes in the food product and are therefore often used for minimally processed foods.

Sterilization using heat is designed to destroy all bacteria, including spores. The effectiveness of this process on shelf life is achieved only if reinfection is avoided, which means effective packaging systems must be employed. Heat also modifies other components of the food and this is often an undesirable side-reaction. Heat-sensitive nutrients, enzymes, and proteins and textural and organoleptic characteristics are all modified from the original raw products. The relative impact of the thermal process on microbial destruction, nutrient loss, and physical, chemical, biochemical, or organoleptic changes is directly related to the time–temperature combination provided the food. Fortunately for food technologists, microbial destruction is very sensitive to increasing process temperature whereas most non-microbial changes are relatively insensitive. This means, for example, that a sterilization process for milk at 100°C (approximately 10 minutes) will completely destroy all B vitamins, while an equivalent microbial kill at 135°C (approximately 2 seconds) will result in only 30% destruction of most B vitamins. The organoleptic quality of the milks from each process is also dramatically different.

Thermal processing may be in-package (e.g., canning, retort pouches) where the food product is first hermetically sealed in a suitable container and then thermally processed. A suitable time–temperature combination can be mathematically modeled but requires experimental heat penetration data for the specific food/container combination being used. This in-package thermal processing tends to cause detrimental effects to other quality characteristics of the food, such as organoleptic and nutritional attributes. This is particularly true of sterilization processes, though novel techniques such as "sous vide" are highly effective for minimally processed pasteurized foods (Rees and Bettison, 1991).

Alternatively, thermal processing can be in-flow. Liquid foods are pumped through a heat exchanger where the pasteurization or sterilization process is achieved. Either direct (e.g., steam injection) or indirect (e.g., heat exchangers of several types, ohmic or dielectric) heating methods may be used. The food must then be packaged without reinfection with microorganisms. This involves specialized equipment, aseptic processing and handling, and sophisticated packaging systems. These in-flow systems can allow very short heat treatments and all components of the food products can be efficiently heated and cooled prior to packaging. In-flow systems allow high-temperature–short-time (HTST) or ultra-high temperature (UHT) processing, with a concomitant improvement in non-microbial food quality attributes (Holdsworth, 1993; Singh and Nelson, 1992; Reuter, 1993).

Aseptic processing is an important developing technology. In essence, foods are sterilized outside the package, then maintained in a sterile state (asepsis) until hermetically sealed into pre-sterilized packaging. Sterilization is achieved normally using thermal processing, but this is not a mandatory requirement. Any sterilization treatment can be used. Singh and Nelson (1992) provide a useful review of aseptic processing principles and practices.

For effective thermal processing, all parts of the food must receive the heating process. While this is easy in liquid systems, solids require the geometric center of the biggest particle to receive the ideal heating treatment. As a result, all other regions of the food (the outside edges of the solid particle and the liquid media) are overheated, often to the detriment of food quality.

There are several methods of applying heat. Apart from the traditional techniques involving conduction and convection of heat to the food, novel rapid methods have been used. These include microwaves and dielectric (ohmic) heating. These are radiation processes which specifically excite the water molecules in foods, thereby causing them to vibrate rapidly. The friction generated from this excited molecular movement of the water molecules provides heat which cooks the food. Microwaves, for example, use radiofrequencies typically of either 915 MHz or 2450 MHz, which specifically relate to the vibrational frequencies of water. Because these are radiation sources, heat is generated throughout the food (rather than being conducted through the food from the outside) and the rate of food heating at the geometric center of large food particles is relatively high. Solid food particles can be heated quickly, and this reduces the time it takes to impart a lethal dose of heat through the food. Microwaves and dielectrics have been important in processing high-quality liquid food products containing solid particles.

6.5.3 Refrigeration

Refrigeration involves the reduction of temperature, which in turn results in a change in kinetic properties of foods. This is the basis of TTT and the foundation of the kinetic models discussed in this book.

It is important to note that a phase change from water to ice, which occurs on freezing foods, can result in unpredictable results in relation to reaction rates. There is invariably a discontinuity in the Arrhenius curve of a specific food system at this point of phase change. In addition, some foods exhibit an accelerated rate of reaction at high sub-freezing temperatures. This may be due to concentration of reactants, as freezing results in the crystalization of water as pure ice. It may also be due to the disruption of cellular structures, thus allowing previously separated chemicals to react.

Interestingly, frozen foods tend to follow similar patterns of reaction rate kinetics to that seen in dry foods. The water activity of any food containing ice is directly proportional to temperature (and not food composition). The reaction rates are usually slower than those of dry foods because of the difference in temperature (Erickson and Hung, 1998; Jeremiah, 1995; Jul, 1984; Zeuthen et al., 1990a). Frozen foods exhibit molecular mobility characteristics. The shelf life of frozen peas at –24°C is about 6 months before oxidation is noticeable. At –26°C the shelf life is in excess of 2 years because the peas are below their glass transition temperature (Reid, 1999).

6.5.4 Radiation

Irradiation utilizes high-energy particles (gamma irradiation) to randomly interact with biological molecules in food and microorganisms. Normally products are irradiated after packaging. One use of irradiation is to destroy microorganisms, but sterilization is only achieved realistically in dehydrated food products.

Although well researched, this processing technology is met with a degree of skepticism and mistrust by consumers and thus has a somewhat checkered history. It is particularly effective as a sterilant of dry foods, for pasteurizing some specific fresh products (e.g., poultry), or for controlling some biological conditions which limit food shelf life (such as sprouting in potatoes) (Thorne, 1992).

New radiation technologies involving pulsed radiation are being evaluated for future use. These include submerged arc discharge, pulsed X-rays, and pulsed ultraviolet (UV) rays. These tend to be most effective on liquid foods, as penetration of the radiation into solid matrices can be limited (particularly with UV). The efficiency of these pulsed radiations tends to improve with increased peak intensity, but the techniques are critically influenced by the composition of the food, including dissolved gases (Rosenthal, 1992).

6.5.5 High hydrostatic pressures

Thermal processing above 100°C involves the use of increased hydrostatic pressures. These pressures, however, are small compared to those used for either critical fluid extraction or high hydrostatic pressure processing of foods. Critical fluid extraction involves pressures of 10–45 MPa (compared to 0.1 MPa atmospheric pressure at sea level). The most commonly used supercritical fluid for food applications is carbon dioxide, which becomes supercritical around 10 MPa at 40–60°C. This has been used to decaffeinate coffee beans or extract oleoresins from spices (Rizvi, 1994).

To have a significant destructive impact on microorganisms or undesirable enzymic activity, much higher pressures and moderate increases in temperatures (to 60–80°C) are required. Sophisticated pressure vessels capable of withstanding 700–900 MPa pressures are available for both experimental and limited commercial use. Considerable work is still required to understand the constraints of this technology, particularly with regards to issues on safety. Certain microorganisms can survive these extreme pressures. Until the safety issues involved with selective destruction of microorganisms are fully understood, these high hydrostatic pressure processing technologies are unlikely to gain commercial acceptance as a sterilizing technology. However, they may offer some innovative techniques for modifying food structures (Isaacs, 1998).

6.5.6 Separation processes

Food systems are often difficult to manipulate or sterilize because of the multiple components present. For example, pure sugar solutions and pure amino acid solutions can be sterilized separately using heat, with limited impact on the properties of the solutes. However, if they are combined before sterilization, as might be the case in many foods, multiple reactions will occur during heating (e.g., browning reactions) which will create significant and obvious quality changes in the food. The ability to separate components from foods offers an important processing step to either concentrate or remove reactants, desirable or undesirable components, or to produce specialized ingredients which can be reused in other food systems. In addition, sensitive liquid streams can be separated from and sterilized separately to particulate components before being recombined aseptically and packaged.

These separation processes are numerous and varied. They include supercritical fluid extraction, precipitation and filtration, a range of membrane processing procedures, freeze-concentration, centrifugation, and many similar technologies. Separation processes may include pre-processing activities, such as sorting, peeling, or trimming.

Membrane processes specifically involve liquids passing over membranes which may be selectively permeable to small organic molecules or may contain micropores (a few micrometer or nanometer pore diameters) allowing all small molecules (including water) to pass through. The processes may involve distillation or ion exchange. The combination and variety of membrane processes are significant (Singh and Rizvi, 1995).

There are numerous opportunities for significant added-value processing using these techniques. Separation can often be achieved at low temperatures, which means there is limited damage to biological molecules. These techniques have been widely used to produce pharmaceuticals (separating the active ingredient from a fermentation broth) and are commonly used to extract highly valued macromolecules from very dilute liquid streams. For example, membrane processes are used to extract immune proteins from milk, where millions of liters of milk must be processed to obtain a few grams of the protein.

6.5.7 Ultrasound, pulsed electric fields, and pulsed magnetic fields

Still in their infancy of development, these techniques offer exciting promise, particularly for liquid foods. The principles behind these technologies rest in their ability to excite water or solute molecules inside food, causing them to rapidly heat or modify their organoleptic properties. Rapidly alternating electrical or magnetic fields are used to reorient molecules, creating the friction needed to achieve heating. The use of processing equipment is one of the key competitive advantages offered a manufacturer in today's modern food industry. One of the greatest challenges facing food product developers is the application of their knowledge to fit the equipment available within their manufacturing company (Povey and Mason, 1998).

6.6 PACKAGING

Whereas processing can provide competitive advantages because it is secreted behind closed doors, packaging is an obvious and essential element of the food preservation chain. Apart from some items of fresh produce, all foods are packaged to some extent. For some food products (e.g., canned, bottled, aseptic processed), hermetically sealed packaging is a fundamental and necessary part of the product survival. For other foods, packaging may be used to ensure a degree of protection from contamination, to define a unit size, or merely to contain the food in a convenient form to be handled. However, packaging of foods invariably has a significant impact on the shelf life of foods—in both a negative and positive manner (Robertson, 1993).

Some packaging is an impermeable barrier, maintaining the food in its own unique environment (e.g., glass, tin and aluminium cans). Others, such as flexible packaging films, are variably permeant to the external environment—particularly gases and moisture. Some packaging is edible (e.g., chocolate coating on confectionery or biscuits, wax on apples). Most packaging reacts with the food system itself, adding another level of complexity to the assessment of shelf life. The ability of light to penetrate the package may also be an important consideration, particularly if the food is sensitive to photooxidation.

Packaging allows the manufacturer to impose a barrier between the food and the external environment. In doing so, the immediate environment around the food can be altered significantly and maintained in an altered condition for considerable lengths of time. This opportunity has led to the introduction of modified atmosphere packaging (MAP). In turn this has facilitated the revolutionary moves toward minimally processed foods with commercially pragmatic shelf lives that are far longer than those achievable with the unpackaged original fresh foods (Farber and Dodds, 1996; Parry, 1993).

MAP allows a new set of shelf-stable products to be produced with minimal processing and few, if any, added chemical preservatives. MAP incorporates an atmosphere in the package which is different from normal atmosphere (78% nitrogen, 21% oxygen, 0.03% carbon dioxide). The technique extends from simple vacuum packaging to more sophisticated gas flushed packaging. MAP is effective *only* if the package is correctly sealed and appropriate packaging laminates are used to minimize gas transfer across the packaging film. Once the package is sealed, the atmospheric composition inside is determined by the biological and chemical processes occurring within the food and the rate of gas transfer across the packaging film itself.

Recommended gas atmospheres vary depending upon the food and the desired result. For microbial stability, high levels (>60%) of CO_2 to control spoilage organisms are used in conjunction with a low amount (1–2%) of oxygen to help control obligate anaerobes (such as *Clostridium botulinum*). Where fat oxidation is a problem, the package may be flushed to remove all traces of oxygen. However, pragmatic success in this instance is dependent upon other means of removing oxygen (such as oxygen scavengers, or the reliance on microorganisms or the food biochemistry to consume the last traces of oxygen remaining in the sealed package). As CO_2 is highly soluble in water, it tends to be absorbed into moist food systems. This results in a significant change in the headspace volume inside the package. In such systems, it is common to include large volumes of nitrogen as an inert and essentially insoluble gas. Nitrogen, however, does not have the same microbial stasis effect of CO_2. Nitrogen is a dangerous alternative if pathogenic bacterial growth is a possibility as it has no significant effect on many of these bacteria.

MAP alone, all other conditions being held constant, will often treble the shelf life of a perishable food product. Chilled chicken shelf life is extended from about 6 to 18 days by flushing the packages with 100% CO_2 before sealing. Roasted coffee can be stored for as long as 18 months after being flushed with CO_2 and N_2, then vacuumed packaged—up from about 3 days without packaging (Blakistone, 1998).

A rapidly emerging technology is the use of "active" packaging. Special gas or moisture absorbers are included inside the package or incorporated directly into the packaging film itself. This is commercialized already for oxygen and water scavengers. There are also products available for ethanol release, carbon dioxide release, or ethylene removal (Farber and Dodds, 1995; Blakistone, 1998). These materials can dramatically increase shelf stability and prolong shelf life under inclement climatic conditions (e.g., the tropics) (Rooney, 1996).

One major concern of the use of MAP is food safety. The technique has its main advantage in the production and storage of minimally processed food products containing few (if any) additives. MAP is very effective at controlling spoilage organisms which commonly limit the shelf life of normally packaged chilled foods. However, MAP is not so effective at controlling pathogenic microorganisms. This latter group do not produce visible signs of growth and can be present in large numbers before the food is overtly spoiled. In fact, the delay in growth of spoilage organisms allows pathogens an ideal opportunity to grow with relatively little competition. This is true for *Aeromonas hydrophila,* nonproteolytic *Clostridium botulinum, Listeria monocytogenes,* and *Yersinia enterocolitica.* MAP packaged, minimally processed foods rely heavily on other protective mechanisms (e.g., refrigeration) and high standards of manufacturing practice for their effectiveness in controlling pathogenic microbial growth. The propensity for abuse of this complex set of control systems has limited the widespread commercial use of MAP packaging.

6.7 COMBINED PPP EFFECTS

In practice, food technologists tend to use all three PPP factors simultaneously. The satisfactory control of many of the biological, physicochemical, or biochemical changes that occur in foods is rarely achieved using only one control measure.

Shelf-life extension of perishable food products essentially targets the control of microorganisms. To achieve this control, any one of a variety of food conditions can be manipulated. These might include increasing the concentration of solutes such as salt or sugar, the modification of gas atmosphere in the package, mild heat treatments, or reducing water content, for example. The effective use of any one of these control measures on their own is usually impractical. Such use of one control measure may result in organoleptic characteristics that are unacceptable to the consumer, or selective control of spoilage microorganisms and the uncontrolled growth of some pathogens.

An array of food products have been produced which incorporate several of these options at the same time. Although each option may be at a suboptimal level, in combination they are extremely effective at controlling all microbial growth in the food. This concept has been called "hurdle technology" (Leistner, 1978; Leistner and Gorris, 1995). Hurdle technology, although defined for microbial aspects of shelf-life extension, is applicable across all forms of food preservation.

Fresh red meat without any form of preservation treatment needs to be cooked and consumed within 24 hours to avoid food poisoning. Refrigeration increases this shelf life to about 1 week. Packaging plus refrigeration can extend this to 3–4 weeks. MAP plus refrigeration can extend this to 12 weeks. However, fermented sausages can be stored at room temperature for several months. This is achieved through a combination of relatively high acidity (pH about 4.5), low water activity (a_w about 0.9), added salt (about 5%), and a mild heat or smoke treatment during processing. Packaging is even unnecessary. None of those PPP factors alone will ensure the microbial stability of this product. They are effective only in combination. If any one of those PPP factors is not achieved, the product will fail to meet its predicted shelf life.

Control of oxidation in potato crisps involves careful selection and constant cleaning of the cooking fat, use of high solids potato raw material, antioxidants in the flavoring and special packaging (which may include modified atmospheres).

Combination processes can play a vital role in prolonging the storage life of raw milk. This stores less than 1 day without any preservation treatment. Refrigeration can increase this to about 3 days. Pasteurization plus refrigeration (and packaging) increases this to 1 week. Pasteurization, addition of a starter culture (to make acid) and an enzyme (rennet) to facilitate protein precipitation, separating the majority of the water (as whey), then adding salt, pressing out more water, and holding under refrigeration (that is making cheese) can increase the milk shelf life to several years. Simply removing the water (to make dried milk powder) will result in a shelf life of months or years, depending upon the packaging used and the milk composition (particularly fat and vitamin content). Bottled milk will develop noticeable off-flavors if exposed to sunlight for 20 minutes because of light reactions with riboflavin, which in turn oxidizes the fat.

6.8 A TYPICAL FOOD PROBLEM

A biscuit manufacturer had been selling a water cracker product successfully in New Zealand. This packet of 45 cracker biscuits had a net weight of 125g (actual biscuit weight 130 ± 3 g) and a use-by date of 6 months from date of manufacture. In this instance, the use-by date had been set by the practical distribution chain (all product had been sold within that time frame) and did *not* represent the shelf life due to product failure. The distribution system involved 14 days in the company premises (20°C, 75% relative humidity), 3 days in transport (up to 40°C and 50% relative humidity) and 163 days in wholesale storage, retail display, and consumption (average 20°C, 75% relative humidity). The cardboard, square carton had the following dimensions: 60 mm square by 200 mm long. Inside the carton, the crackers were tightly wrapped in a laminate comprising 30-micron cellophane and 6-micron polyvinylidine chloride (PVdC, Saran). The diameter of the circular crackers (and hence the cellophane/PVdC package) was 57 mm with a total package length of 193 mm.

The company wished to export this product to Asia. They wanted the same shelf life plus 6 weeks (the sea shipping time). The environmental conditions in Asia are tropical (typically 35°C and 85% relative humidity). The logistics staff indicated there would be up to 30 days required from manufacture to on-board the ship (at 20°C and 75% relative humidity), with the remaining 193 days in the marketplace under tropical conditions.

Following sensory testing of these crackers in their existing packaging at 20°C and 75% relative humidity, three apparently different deteriorative reactions were identified: a "funny flavor" after 210 days, loss of crispness after 240 days and a rancid flavor after 300 days. Clearly, the first detectable change was an undesirable funny flavor which was first noticed 30 days after the currently defined use-by date. This was the "cause of failure" under New Zealand conditions. This emphasizes the difference between use-by date and shelf life. They may be mutually exclusive. Note also that the consumers would probably experience neither loss of crispness nor rancidity with these crackers, as the product would be unacceptable for funny flavor reasons before the other changes were obvious.

Unless the reason for the observed change is known, it is important to identify key factors (apart from temperature) influencing these results. One important parameter is moisture content. Freshly made, unpackaged biscuits were placed in desiccators equilibrated at 20°C, each with a different relative humidity, provided by various saturated salt solutions. Using a sensory panel, it was found that crackers stored at or below a_w 0.35 were excellent after 4 weeks' storage. However, at an a_w of 0.4 they had developed the funny flavor (but were still crisp) following 5 days' storage. Crispness was maintained for several weeks at an a_w of 0.40. However, it was noticeably lost after 7 days' storage when the a_w was 0.45. Rancidity was first noted at both a_w 0.40 and 0.45 after 30 days. A further test was run at 40°C, using the same saturated salt solutions. It was found that the corresponding data were 4 days, 2 days, and 17 days, respectively, for funny flavor, crispness, and rancidity. Note that crispness became the first noticeable change in quality and thus the shelf-life limiting change.

From the equations presented in Chapter 1, the rate of moisture transfer into the package can be estimated. It is clear from the kinetic changes occurring in the unwrapped crackers at an a_w of 0.4 that the funny flavor occurred very quickly (less than 1 week) when the moisture content was sufficiently high. Clearly, if this change takes 210 days inside the package, the package was a reasonably good barrier to moisture. Although not scientifically accurate, it does not make a difference to the answer if we assume that the critical moisture for this funny flavor occurs when the a_w inside the package reaches a value of 0.4. The initial moisture content of the crackers was 0.05 g water/g dry matter (a_w 0.20); the critical moisture content was 0.08 g water/g dry matter (a_w 0.40) and in the New Zealand conditions the equilibrium moisture content (75% RH at 20°C) was 0.13 g water/g dry matter.

Key parameters related to the biscuits include weight of dry biscuits (130/(1 + 0.05) g = 123.8 g). The package surface area was 0.04 m^2. At 20°C, from steam tables, the saturated moisture content of the air was 17.53 mm Hg (and at 75% relative humidity this becomes 13.15 mm Hg). Packaging thickness is 36×10^{-6} m. Substituting these figures into the equation provides a solution for the permeance of the packaging film for a shelf life of 210 days. This permeance value was 0.0738 g water $day^{-1} \cdot m^{-2} \cdot mm\ Hg^{-1}$.

If all the cracker characteristics remained unchanged and the package was shipped to the tropics, the moisture content in the air changes because of the temperature. From the steam tables this value at 40°C and 85% RH is 47.03 mm Hg. This change alone reduces the shelf life of the crackers from 210 days to 59 days (assuming the permeance of the film is unchanged). It is also clear that this atmospheric change would also reduce the shelf life in respect of crispness and rancidity to substantially less than the required 7½ months. Thus to provide a suitable shelf life all three problems will need to be fixed.

Having identified the issues, possible solutions can be assessed. The most obvious change should be to reduce the permeance of the packaging. This is not always possible in an economic or aesthetic manner. Some of the best moisture barriers are opaque and/or expensive. Doubling packaging thickness will approximately double shelf life (but this still does not meet the minimum shelf life criteria, and there are limits to the pragmatic thickness of a packaging film). Another option may be to wrap the biscuits in two layers of packaging (provided there is an air space between the two layers). This provides a buffer air space between the crackers and the external atmosphere which (mathematically) more than doubles the shelf life when using two layers of the same film. An added option is to include a moisture scavenger between the two layers of film. It may also be possible to change the atmospheric moisture content by using refrigeration (or air conditioning).

The relationship between moisture content and shelf life is logarithmic. The initial moisture content could be reduced—a doubling in shelf life can be achieved by about 30% reduction in initial moisture content. Invariably, however, crackers have been cooked to meet desirable consumer traits such as color. Further drying during production (e.g., prolonging cooking) will probably result in deleterious quality. In other words, this option was not feasible in practice.

The critical moisture content can also be moved toward a higher a_w. For example, to control rancidity, shelf-life extension might involve the addition of an antioxidant. A suitable antioxidant, in fact, will extend the shelf life with regards to oxidation to achieve the necessary shelf life. Modification of the formula may also result in changes for the funny flavor or crispness, though the solution for each may be deleterious to the other. Once again, small changes in the critical moisture content (product formulation, processing) may have a profound impact on shelf life.

The practical solution for this problem rested in a combination of better packaging (reducing the permeance substantially and adding a moisture scavenger), the use of an antioxidant, and some modifications in the formulation to compensate for the funny flavor changes that occurred during storage. It was possible to achieve the new marketing requirements by modifying the PPP parameters alone.

This example shows the value of the mathematical models for kinetic changes in food systems and how they can be used to provide useful guidance to an industrial situation. It also provides a clear indication of the complexity of food systems, where there is invariably a series of reactions which occur simultaneously, at different rates, and often mutually independent. The solutions to this problem clearly indicate the importance of PPP variables in food formulation and production and that these PPP factors can (and usually do) exceed the impact of temperature control (TTT) alone.

6.9 EMERGING OPPORTUNITIES

The rate of change of raw materials is staggering. Genetic technologies offer revolutionary tools to design and engineer specific new foods and ingredients. Rape-seed has been transformed using traditional technologies from a toxic, lubricating oil-producing seed to a high-yielding food crop which produces an oil (canola) that is nutritionally ideal for humans. This took 10 to 20 years to commercialize, whereas modern techniques can shorten this to 5 years.

One key to reducing the variability in food production is to use purified, specific ingredients. For example, natural spices are relatively unhygienic, and are a major source of microbial contamination in foods. Gas sterilization of these spices is now illegal in most countries around the world. In some countries irradiation—the only other way to sterilize natural spices—is illegal or met with suspicion. Spices are also highly variable in efficacy from batch to batch. The development of oleoresins—standardized, sterile liquid extracts of the active spice ingredient—has provided a valuable ingredient that can be accurately metered into foods, without the side effects of the natural product.

Industrial ingredients are developing rapidly. These ingredients are produced often from natural products (e.g., oleoresins, starches, gums) or through biotechnology (e.g., citric acid). Products such as milk are being progressively dissected and modified into a range of subcomponents which have specific functionality in foods. These ingredients have stringent specifications and are of guaranteed quality. Production recipes can be considered a mixture of chemicals of known and guaranteed quality, rather than an artisan cookbook. This is particularly important for final consumer product consistency and quality.

The new technologies being researched today fall into two categories: those that provide new methods for undertaking old preservation processes, and those that are totally new preservation technologies. The former, such as new rapid methods for heating foods, are being commercially exploited with new and improved product characteristics. The latter group, in the main, have still considerable barriers to cross with regards to their safety, before being approved for commercial use. Some, such as high-pressure techniques, may have potential for creating totally novel food systems. These have already been seen with technologies such as extrusion cooking. However, the preservation method being employed is unlikely to be high pressure alone. For example, in extrusion cooking the product often reaches temperatures as high as 180–200°C and will likely be in a dry state (a_w below 0.5) during storage.

6.10 CONCLUSIONS

Preservation technologies—the so-called product-process-packaging effects—have a profound impact on the shelf life and stability of foods. Whereas the kinetic impact of temperature on reaction rates can be modeled and sophisticatedly predicted, the impact of PPP is unpredictable. Food systems can be significantly modified to result in major extensions to their shelf life. These modifications in many foods are well documented and offer the food technologist an exciting set of variables which allow their effective design of new and innovative solutions to complex storage problems.

In striking contrast, this means that models associating temperature to food storage life must be defined strictly for a specific food and its unique formulation, processing method, and packaging system. There are growing sets of data which cover a wide range of these PPP conditions, but it is dangerous to define a predictive model by extrapolating from these data to new food products. The final and inalienable requirement for any predictive model for shelf life of a given food system is the experimental shelf-life test. This test should be a routine part of every quality control laboratory protocol, for every product produced by a company, in order to obtain a pragmatic and experimentally proven set of data. These results can be compiled to include accelerated test data, so new products, line extensions, or modifications in formulation can be readily and quickly assessed prior to market launch.

Shelf-life determination and modification require a fundamental understanding of the food system in question: its chemistry and related properties, physical attributes, manufacturing options and opportunities, and the packaging options. These need to be integrated with the needs of the consumer, within the financial and economic constraints of the company and market. It is this complex set of interactions, and the ability that exists to modify them, that is discussed in specific product chapters in the remainder of this book.

REFERENCES

Angelo, A.J., St. (1992) *Lipid Oxidation in Food,* American Chemical Society, Washington, DC.

Barbosa-Canovas, G.V., Pothakamury, U.R., Palou, E. and Swanson, B.G. (1998) *Nonthermal Preservation of Foods,* Marcel Dekker Inc., New York, NY.

Birch, G.G. and Campbell-Platt, G. (eds) (1994) *Food Safety. The Challenge Ahead,* Intercept Ltd., United Kingdom.

Blakistone, B.A., ed. (1998) *Principles and Applications of Modified Atmosphere Packaging of Foods,* 2nd edn, Blackie Academic and Professional, Glasgow, Scotland.

Charalambous, G. (1993) *Shelf Life Studies of Foods and Beverages. Chemical, Biological, Physical and Nutritional Aspects,* Elsevier, Amsterdam, The Netherlands.

Dalhoff, E. and Jul, M. (1965) Factors affecting the keeping quality of frozen foods, in *Progress in Refrigeration Science and Technology,* Pergamon Press, New York, NY, 1: 57–65.

Erickson, M.C. and Hung, Y.C. (eds) (1998) *Quality in Frozen Food,* Chapman and Hall, London, England.

Farber, J.M. and Dodds, K.L. (eds) (1995) *Principles of Modified-Atmosphere and Sous Vide Product Packaging,* Technomic Publications, Lancaster, PA.

Farber, J.M. and Dodds, K.L. (eds) (1996) *Principles of Modified Atmosphere and Sous Vide Product Packaging,* Technomic Publishing Co. Inc., Lancaster, PA.

Fennema, O.R. (ed) (1996) *Food Chemistry, 3rd edn,* Marcel Dekker, Inc., New York, NY.

Franks, E. (1982) The properties of aqueous solutions at subzero temperatures, in *Water—A Comprehensive Treatise,* Vol. 7 (ed) E. Franks, Plenum Press, New York, NY, 215–338.

Holdsworth, S.D. (1993) *Aseptic Processing and Packaging of Food Products,* Elsevier Science Publishers Ltd., London, England.

Isaacs, N.S. (ed) (1998) *High Pressure Food Science, Bioscience and Chemistry,* Royal Society of Chemistry, Turpin Distribution Services Ltd., United Kingdom.

Jeremiah, L.E. (1995) *Freezing Effects on Food Quality,* Marcel Dekker, Inc., New York, NY.

Jul, M. (1984) *The Quality of Frozen Foods,* Academic Press, London, England.

Leistner, L. (1978) Hurdle effect and energy saving, in *Food Quality and Nutrition* (ed W.K. Downey), Elsevier Science Publishers, London, England, 553–557.

Leistner, L. and Gorris, L.G.M. (eds) (1995) *Food Preservation by Combined Processes.* Final Report. FLAIR Concerted Action Number 7. European Commission, Luxembourg.

Levine, H. and Slade, L. (1986) A polymer physico-chemical approach to the study of commercial starch hydrolysis products, *Carbohydrate Polymers,* **6**, 213–244.

Parry, R.T. (1993) *Principles and Applications of Modified Atmosphere Packaging of Food,* Blackie Academic and Professional Ltd., Glasgow, Scotland.

Potter, N.N. and Hotchkiss, J.H. (1995) *Food Science,* Chapman and Hall, London, England.

Povey, M.J.W. and Mason, T.J. (eds) (1998) *Ultrasound in Food Processing,* Blackie Academic and Professional Ltd., Glasgow, Scotland.

Rees, J.A.G. and Bettison, J. (eds) (1991) *Processing and Packaging of Heat Preserved Foods,* Blackie Academic and Professional, Ltd., Glasgow, Scotland.

Reid, D. (1999) Personal communication.

Reuter, H. (ed) (1993) *Aseptic Processing of Foods,* Technomic Publishing Co., Basel, Switzerland.

Rizvi, S.S.H. (ed) (1994) *Supercritical Fluid Processing of Foods and Biomaterials,* Blackie Academic and Professional Publishers, Glasgow, Scotland.

Robertson, G.L. (ed) (1993) *Food Packaging. Principles and Practice,* Marcel Dekker Inc., New York, NY.

Rockland, L.B. and Beuchat, L.R. (eds) (1987) *Water Activity: Theory and Application to Food,* Marcel Dekker, Inc. New York, NY.

Rooney, M.L. (1996) *Active Food Packaging,* Blackie Academic and Professional Ltd., Glasgow, Scotland.

Rosenthal, I. (1992) *Electromagnetic Radiations in Food Science,* Springer Verlag, Heidelberg, Germany.

Russell, N.J. and Gould, G.W. (eds) (1991) *Food Preservatives,* Blackie Academic and Professional, Glasgow, Scotland.

Singh, R.K. and Nelson, P.E. (eds) (1992) *Advances in Aseptic Processing Technologies,* Elsevier Science Publishers Ltd., London, England.

Singh, R.K. and Rizvi, S.S.H. (eds) (1995) *Bioseparation Processes in Foods,* Marcel Dekker Inc., New York, NY.

Thorne, S. (ed) (1992) *Food Irradiation,* Elsevier Science Publishers Ltd., London, England.

van Ardsel, W.B. (1957) The time temperature tolerance of frozen food, I. Introduction: The problem and the attack. *Food Technol.,* **11**:28–33.

van Arsdel, W.B., Copley, M.J. and Olson, R.L. (1969) *Quality and Stability of Frozen Foods. Time-Temperature Tolerance and Its Significance,* Wiley-Interscience, New York, NY.

Zeuthen, P., Cheftel, J.C., Eriksso, C., Gormley, T.R., Linko, P. and Paulus, K. (1990a) *Processing and Quality of Foods. Vol. 3, Chilled Foods: The Revolution in Freshness,* Elsevier Applied Science, London, England, and New York, NY.

Zeuthen, P., Cheftel, J.C., Eriksso, C., Gormley, T.R., Linko, P. and Paulus, K. (1990b) *Processing and Quality of Foods. Vol. 1, High Temperature/Short time (HTST) Processing: Guarantee for High Quality Food with Long Shelf Life,* Elsevier Applied Science, London, England, and New York, NY.

Part II

The Practice

CHAPTER 7

Chilled Yogurt and Other Dairy Desserts

M. Lewis and R.H. Dale

7.1 INTRODUCTION

For many years the production of short-life perishable food products was organized and carried out on a regional or local level. This enabled products to be sold within a very short period following manufacture. However, since World War II a major revolution has taken place within the food industries of Europe, the USA, and Japan. This has occurred in the chilled, short-life dairy products sector in particular.

Modern marketing methods and techniques have created ever-increasing mass markets with an apparently insatiable appetite for new and novel products. The drive for increasing the efficiency of production plants in order to reduce costs and increase profits has led to a centralization of production units. Major improvements in the ability of manufacturers and retailers to transport refrigerated products over long distances coupled with the lowering of trade barriers within the European Union (EU) have created export opportunities. All this now means that dairy desserts once considered perishable within very short periods are required for consumption many days after their manufacture. This tremendous change within the market both in volume and requirement has created immense technical problems and has led to a quiet revolution in the field of food technology.

Ensuring the safety of these products has of course been paramount, but to underwrite the commercial success it has been equally important to ensure that other product parameters remain good. Indeed, the sensory, visual, and physical characteristics should exhibit the same quality of freshness at the end of the product's life as at the beginning.

This requirement for freshness (often associated by the consumer with healthiness and wholesomeness) together with the safety requirements have taken place in an environment where the removal of artificial preservatives and other additives was also demanded. The challenge created for the food and dairy technologists has been enormous.

In order for the public to be confident of the quality of these products, governments introduced mandatory date coding. Initially this was adopted by the food industry in the form of "sell-by" dates. However, recent changes in UK and EU legislation require "best before" dates for nonperishable products and a more definitive "use-by" date for microbiologically perishable products such as dairy desserts. This legislation also requires manufacturers to be able to justify the codes which are applied to the products if requested to do so by the enforcement authorities. Therefore, it has been necessary for the food technologist, not only to extend the life of the products, but also to improve the methods by which the shelf life is determined.

This chapter will review many aspects of shelf life associated with dairy desserts, not in complicated scientific terms but in the practical language of the dairy industry. It will attempt to categorize dairy desserts into broad groupings and discuss key factors affecting shelf life from raw materials through to the consumer.

7.2 PRODUCTS, PRODUCT GROUPS, AND CHARACTERISTICS

Over the last 15 years, the sale of yogurt, fromage frais, and other desserts has shown a steady growth, benefiting from market segmentation and the healthy eating trend. More recently, there has been an increasing trend for the more indulgent products containing higher fat levels, exotic fruits and the use of "Bio" cultures.

The different varieties and approximate percentages of sales by volume of yogurts, fromage frais, and traditional desserts in recent years are shown in Figures 7–1, 7–2, and 7–3.

7.2.1 Product groups

For shelf-life evaluation the products are classified into two main groups:

1. cultured products
2. desserts (unfermented/blends).

The manufacture of cultured products involves the controlled use of lactic acid producing microorganisms for prescribed time and temperature combinations to achieve a final product with the desired characteristics of viscosity, texture, mouthfeel, flavor, and acidity.

The desserts are either simple products such as custards and jellies or complex products including trifles, mousses, and cheesecakes which are a mixture of textures, taste, and acidity levels. General classifications of chilled yogurts, soft cheeses, and desserts are given in Figures 7–4, 7–5, and 7–6.

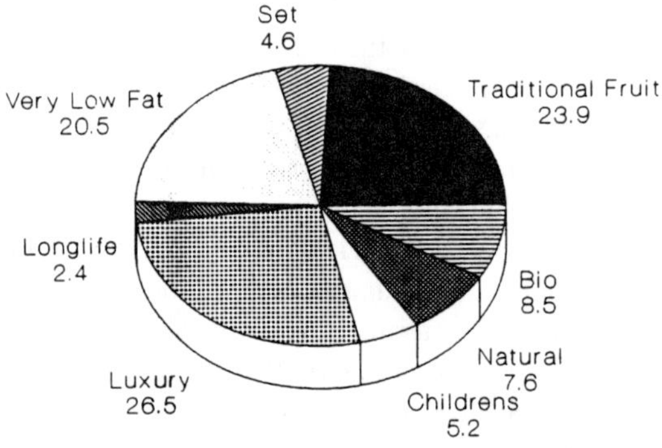

Figure 7–1 Yogurt—percentage of total sales.

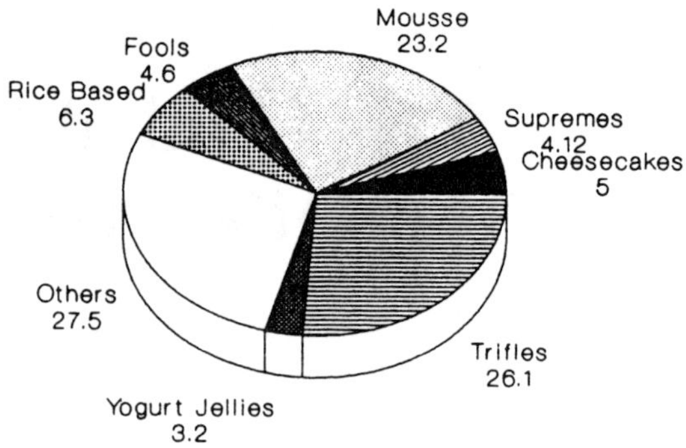

Figure 7–2 Traditional desserts—percentage of total sales.

7.2.2 Product characteristics

7.2.2.1 Yogurts

(a) Stirred yogurt. The texture should be smooth with a glossy appearance, thick, and almost pourable, but must not be grainy or nodulated. The color of natural yogurt should be creamy white. If fruited, it should represent the color of the fruit without being too bright. The typical yogurt flavor—slightly acidic, clean, and fresh with the acetaldehyde note—should be distinguishable and complementary to the fruit flavor.

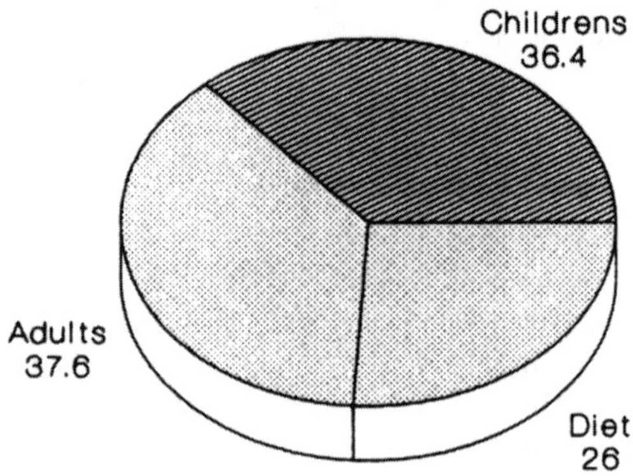

Figure 7–3 Fromage frais—percentage of total sales.

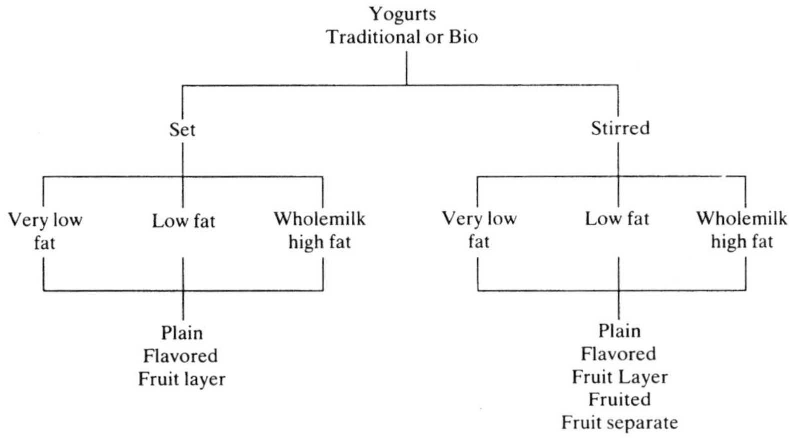

Figure 7–4 General classifications of chilled yogurts.

(b) Bio yogurts. The Bio yogurts should be milder and creamier in taste than the traditional yogurts, but of similar color and texture.

(c) Set yogurts. Set yogurts should have a glossy surface appearance without excessive whey. The texture is smooth and almost junket-like, giving a clean cut when spooned.

7.2.2.2 Fromage frais

Natural low fat fromage frais is milky white in color has a smooth texture with a spreadable consistency. It is clean and mildly acidic in flavor with no appearance of syneresis.

7.2.2.3 Desserts

(a) Layered. Generally the layered products will have a spectrum of textures. Taste will range from sweet to acidic, strong to bland and the colors and flavors should complement the individual components. Cake and biscuit layers should retain their characteristic crunchy or open texture and not become wet and soggy.

(b) Custards. The texture should be smooth and thick with a glossy surface, but not gelatinous or grainy.

(c) Jellies. The gels should not be too brittle, but firm with "melt-in-the-mouth" characteristics and a good fruity flavor.

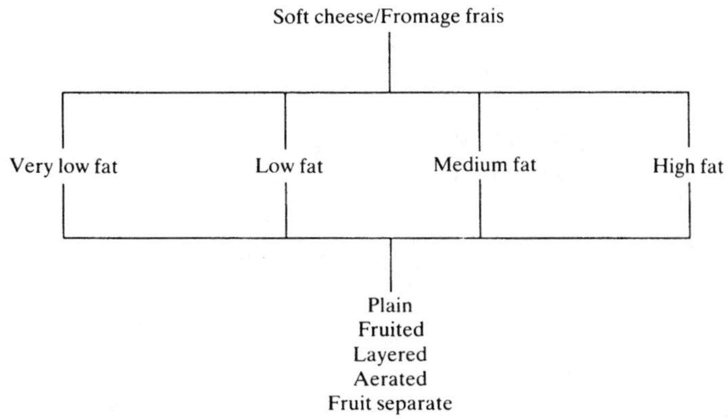

Figure 7–5 General classifications of chilled soft cheeses.

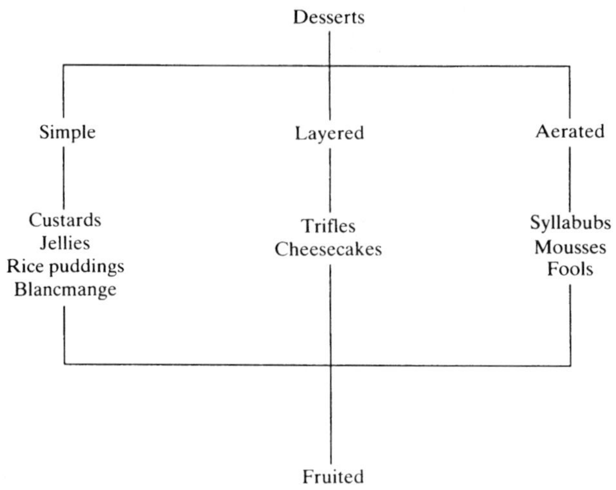

Figure 7–6 General classifications of chilled desserts.

(d) Mousses. Mousses may be whipped for varying periods to obtain either high or low overruns depending on the final textural characteristics required. In all cases a mousse should have uniform air cells. It should retain its texture and overrun without collapsing during its shelf life.

7.3 FACTORS AFFECTING SHELF LIFE

Commercially the shelf life of a product may be defined as the number of days after production that the product can be consumed while still remaining safe, retaining its quality appeal and meeting customer expectations. In other words, it should remain microbiologically safe and organoleptically acceptable within its stated shelf life.

To evaluate the shelf life of any product it is first necessary to identify which characteristics of the ingredients, the process, and the storage conditions are responsible for, or will have an influence on, its shelf life.

The life of chilled dairy desserts is influenced by:

- raw materials
- product formulation
- processing parameters
- implementation of Good Manufacturing Practice
- filling and packaging
- storage and distribution
- consumer usage and handling.

Measurement, monitoring, and control of the above are of paramount importance in the evaluation process. The point at which the product becomes unacceptable either organoleptically or microbiologically, whichever manifests itself first, is the ultimate end point from which the life can be established.

7.3.1 Raw materials

The raw materials most likely to be used for the production of yogurt, fromage frais, and desserts are given in Table 7–1. All the listed ingredients introduce microorganisms and chemicals which may affect the quality of the final product. Partnerships with approved suppliers and agreed specifications will help the dessert manufacturer to consistently supply a product which remains acceptable within the determined life. Unwarranted changes in source and supply of these ingredients will result in a variation of factors which influence shelf life and are not to be recommended.

Table 7–1 Raw materials commonly used in the manufacture of chilled yogurt and other dairy desserts

Yogurt	*Soft cheese/ fromage frais*	*Desserts*
Milk	Milk	Milk
Cream	Cream	Cake/sponge/biscuits
Sugar	Sugar	Stabilizer: gelatine, starch, pectin, gums
Cultures	Cultures	Sugar
Milk concentrates	Fruit/fruit conserves	Fruit/fruit conserves
Milk powders	Flavors	Cream
Fruit/fruit conserves	Colors	Flavors
Stabilizers		Colors
Flavors		Decoration: cherries, nuts, vermicelli
Colors		

7.3.1.1 Milk

This ingredient, which is common to all the products, is variable in all the main constituents—protein, lactose, fat—and in the microbiological flora.

The variation occurs between different breeds of cattle and different geographical regions, and is seasonal. The control of the type and level of microorganisms, together with the monitoring of the protein content, is essential as these significantly affect the texture and stability of the final product.

Microorganisms can gain access to the milk from the cow, personnel, or equipment used during milking or during transportation to and storage at the processing dairy. To keep the microbiological flora to a minimum it is essential to cool the milk to below 5°C directly after milking. This temperature should be maintained until the milk is processed at the manufacturing site.

Recent trends have been toward larger dairy units having milk collection every other day or even twice a week. This has encouraged dairy farmers to invest in more hygienic equipment and increased awareness of health and quality issues, but has increased the potential for microbial growth and enzyme activity.

Modern methods of manufacture, plus quality standards demanded by national legislation, EU directives, and the dairy industry in general, have ensured that the raw milk on receipt at the processing factory tastes good, has a low bacterial count, and is free from impurities. This quality is maintained by heat processing followed by cooling to 5°C as soon as possible after receipt of the milk.

Milk used for the manufacture of the dairy products should therefore have low bacterial counts, no off-flavors, and minimal enzymic degradation. Stock rotation and control of temperatures will help to ensure a good quality basic raw material for all short shelf-life dairy products.

7.3.1.2 Cream

The quality of the cream will be dependent on the quality of the milk used for separation and the method of handling before and after pasteurization. Being high in fat the cream is more susceptible to enzyme degradation from lipase which splits the triglycerides into fatty acids, mono and diglycerides, and glycerol. This will result in extreme cases in a rancid taste being evident in the final product. Lipase is largely inactivated by pasteurization. However, as many microorganisms produce lipase, this can still be a problem if the cream is handled unhygienically after processing.

If milk is to be held for a prolonged period prior to separation, it is often pasteurized immediately on receipt. Microbiological quality is monitored by testing for total viable count (TVC) and coliforms.

7.3.1.3 Concentrates

The two methods generally used for the concentration of milk solids are:

1. removal of water by evaporation
2. removal of water by membrane filtration.

For both methods the quality of the concentrated milk is dependent on the quality of the raw milk used, and the handling techniques before and during processing.

The microbiological flora should be low in heat-resistant bacteria (thermodurics) and low in spore formers. Chemically, milk which is to be evaporated must be heat stable in order to withstand the pre-heat treatment process.

After concentration, the temperature of storage and handling methods are critical to the microbiological quality. Depending on the rate of cooling and the concentration factor, lactose crystalization may occur, which could adversely affect the flavor and textural characteristics of the final product.

Quality is monitored microbiologically by checking for spore formers and coliforms and chemically for fat, lactose, protein, total solids, and pH.

7.3.1.4 Cultures

Culture management and the activity and viability of the organisms are vital to final product quality and to the shelf life of fermented product.

Culture preparation and handling demands the highest standards of hygiene combined with attention to detail in the manufacturing techniques and in the selection of plant. The risk of airborne contamination by yeast, molds, bacteria, and bacteriophage must be eliminated.

The cultures, when used, must contain very high numbers of viable and active bacteria in the correct ratio to achieve the desired pH in the appropriate time.

The quality of the bulk culture is maintained by storage at below 5°C for a maximum of 48 hours. The quality is monitored by testing for yeasts, molds, cell balance, and culture activity.

For direct vat inoculation, frozen cell concentrates are used and it is essential that these are stored at very low temperatures and used in accordance with the manufacturer's instructions.

7.3.1.5 Fruit

After harvesting, fruit will remain acceptable for only a limited period of time. The time varies considerably by fruit type and variety, country of origin, harvesting methods, and the method of storage employed by the grower.

Spoilage is caused mainly by microbiological activity (especially when fruit has been damaged) and chemical reactions catalyzed by enzymes which occur naturally in the fruit.

To ensure good quality fruit for further processing it is necessary to either destroy the factors responsible for spoilage or to inhibit them. This is achieved by either heat sterilization (canning) or deep freezing. The processed fruits (canned or frozen) are all regarded as raw materials to produce conserves for inclusion in dairy desserts.

A major area of concern, not directly associated with shelf life but of paramount importance, is the presence of pesticides and other agrochemicals, which are used during growth and post-harvest storage. All fruits should be screened for these. They must be controlled within the limits specified in the legislation governing agrochemical residues.

7.3.1.6 Other ingredients

As for any of the other raw materials used, the source of the ingredient, approval of supplier, the identification of critical control points (CCPs), and hygiene standards are all important factors in the shelf life of the final product.

7.3.2 Processing

An understanding of the synergistic and antagonistic reactions that take place with different processing methods when a variety of functional ingredients are used is necessary when a formulation is being developed. This may have a very important influence on the stability of the final product.

7.3.2.1 Yogurt

Having specified the microbiological, physical, and chemical qualities and the accepted tolerances of the various parameters, the basic yogurt mix is subjected to a relatively severe heat treatment. This achieves four objectives:

1. Destruction of any undesirable organisms including pathogens.
2. Formation of physico-chemical complexes and denaturation of the whey proteins which ultimately affect the texture and water binding capacity of the coagulum.
3. Reduction of dissolved oxygen.
4. Production of amino acids, peptides, and other factors which enhance the growth of the culture organisms.

After addition of the culture, which is traditionally thermophilic, it is essential that the organisms (typically *Lactobacillus delbrueckii* subsp. *bulgaricus* and *Streptococcus salivarius* subsp. *thermophilus)* have ideal conditions in which to grow, i.e., optimum temperature and a medium with all the desired factors, but without any inhibitory substances. The influence of cultures on final product quality and their susceptibility to outside influences must not be underestimated.

After fermentation to the desired pH the yogurt is cooled to below 20°C and often stored for a short period prior to filling.

It is essential that the pumps, heat exchangers, pipelines, and storage tanks are cleaned and sterilized before use. Samples of the yogurt base from each batch should be checked for viscosity, texture, flavor, and the presence of coliforms, yeasts, and molds. Coliforms will not normally survive in the low pH medium, but if initially present will indicate poor or inadequate cleaning of the plant. A typical manufacturing scheme for yogurts is given in Figure 7–7.

7.3.2.2 Fromage frais

Many of the manufacturing principles that apply to yogurt also apply to the manufacture of fromage frais. As with yogurt production, the solids level of the milk is increased and the mix is heat treated and held before cooling to incubation temperatures.

However, the lactic acid–producing cultures used are different. These, being mesophilic organisms, require a lower incubation temperature and give fromage frais its distinct flavor and texture. As the incubation temperature is much lower than that for yogurt, and could also be suitable for the growth of undesirable organisms, great care must be taken in ensuring that the vats and associated equipment are all clean and sterile.

After incubation the coagulum is cooled to inactivate the cultures and prevent a further decrease in pH. At this stage, if the desired texture and viscosity have not been reached, a further concentration of total solids is necessary. This is achieved by first re-heating the mix followed by either mechanical separation or membrane filtration.

Having achieved the correct viscosity and total solids, cream can be blended into the base (i.e., the plain curd, also known as "fromage blanc") to give the required fat level. Fruit is normally added prior to filling.

At each critical stage of the process samples are taken and checked for the presence of coliform, yeast, and molds. The final product is also checked for viscosity, texture, and flavor. The manufacturing process for fromage frais follows the general scheme shown in Figure 7–8.

7.3.2.3 Desserts

Desserts range from the relatively simple jellies, mousses, and custards to layered products such as trifles, cheesecakes, and complex continental desserts such as tiramisu. This combination gives a variety of textures, tastes, colors, and appearances. For shelf-life evaluation it is important that these attributes are maintained through life, e.g., crunchy texture, stability of aeration, absence of syneresis, and leaching of color.

Microbiologically the desserts are more vulnerable to microbiological contamination which may include pathogens due to the more neutral pH of the product and increased handling.

Processing methods for individual components are relatively simple; a simplified trifle-making process is represented in Figure 7–9.

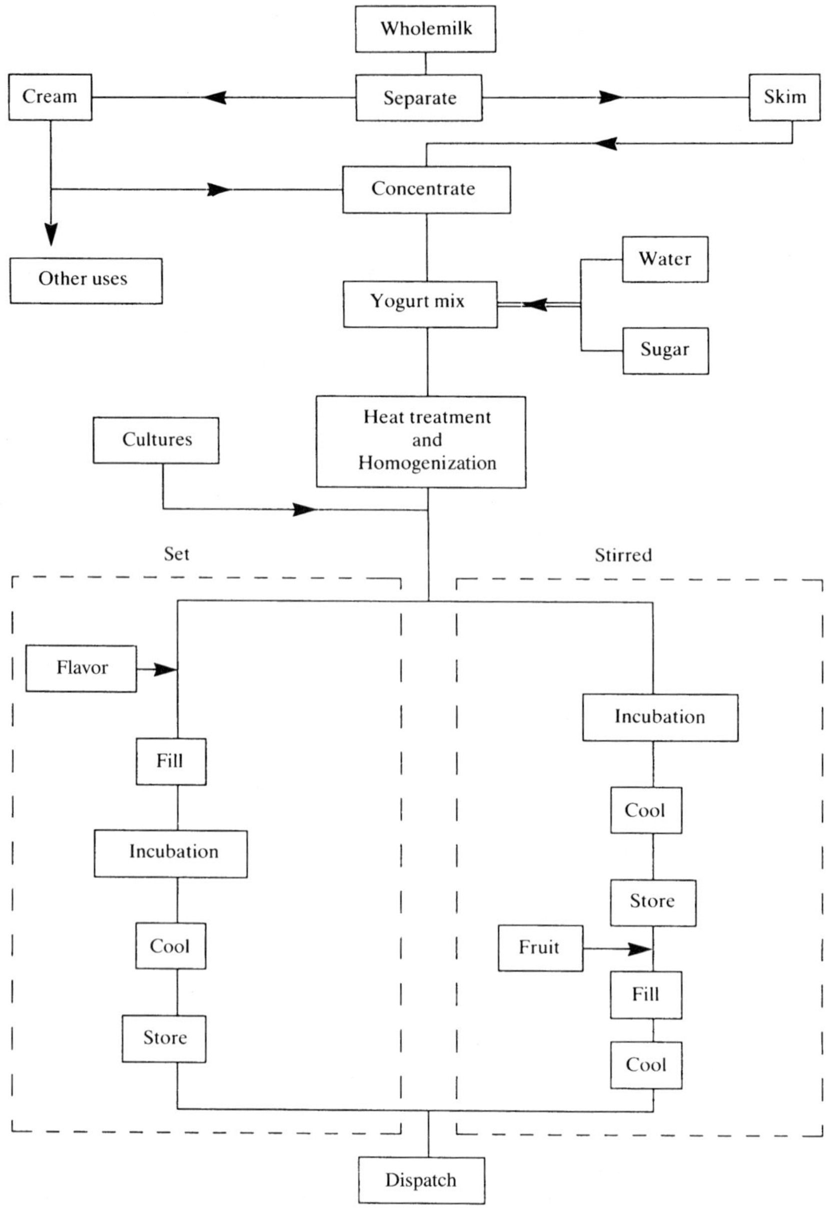

Figure 7–7 A typical manufacturing scheme for yogurts.

(a) Mousse. Traditionally a mousse is high in fat with a high overrun, and this foam must be stable for the duration of its shelf life. Modern trends, however, are for low-fat mousses.

The structure of a mousse is normally maintained by the use of stabilizers or whipping agents. With advancements in functional and novel ingredients different structures can be achieved ranging from a soft open texture with large air cells to a strong almost rubbery texture with very small air cells.

(b) Custards. The selective use of stabilizers and starches will ensure that the desired texture

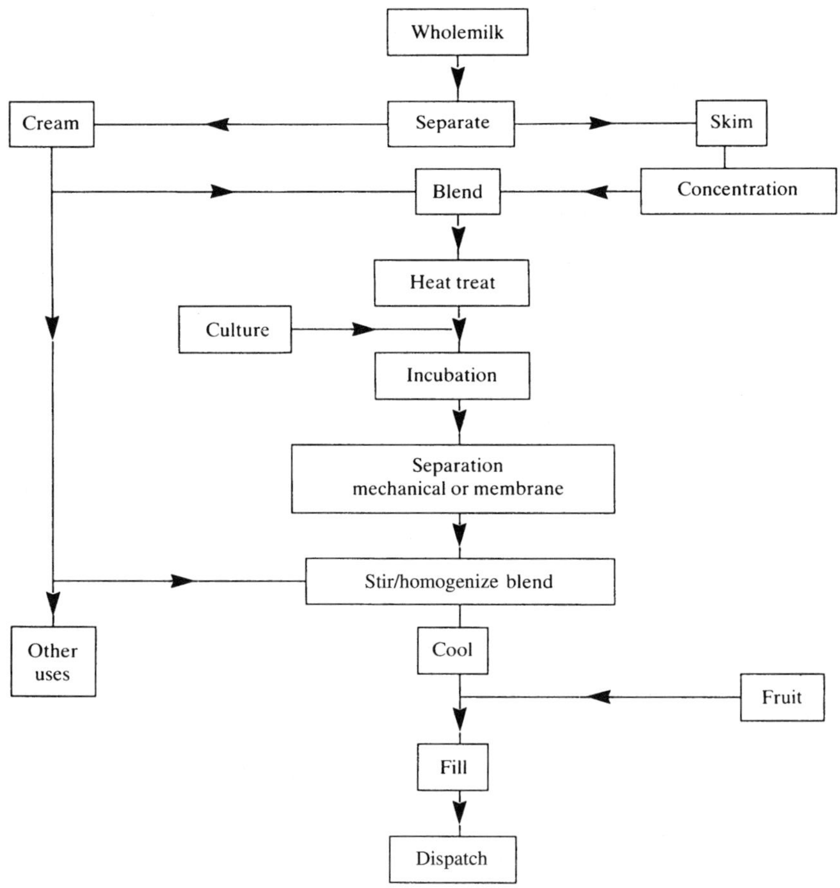

Figure 7–8 The manufacturing process for fromage frais.

is achieved—ranging from creamy to light and short to gelatinous.

The starch and other hydrocolloids must be well dispersed and fully hydrated before being heat treated to the desired temperature. The time and temperature used must ensure that the starch is cooked and the custard is pasteurized. The product can either be filled hot, or if used in combination with other ingredients, cooled to a temperature slightly above the gelation point. This prevents damaging the structure of the other ingredients in a multilayered product.

(c) Jellies. Gelatine can contain a relatively high level of spore formers and the processing conditions must ensure that these are destroyed. The critical area is the storage temperature of the mix prior to filling, or in the case of complex desserts, before addition to the other components of the product. This temperature must be sufficiently high to prevent the growth of microorganisms, but must not adversely affect the color and flavor of the jelly and other components.

The most important aspects of desserts manufacture are the implementation of good manufacturing practices, personal and plant hygiene, and training. Implementation of aseptic techniques during processing, handling, and filling is vital to ensure that the full shelf-life potential of the product is realized. Critical control points should be identified and appropriate testing regimes be in place to ensure that the key standards are met.

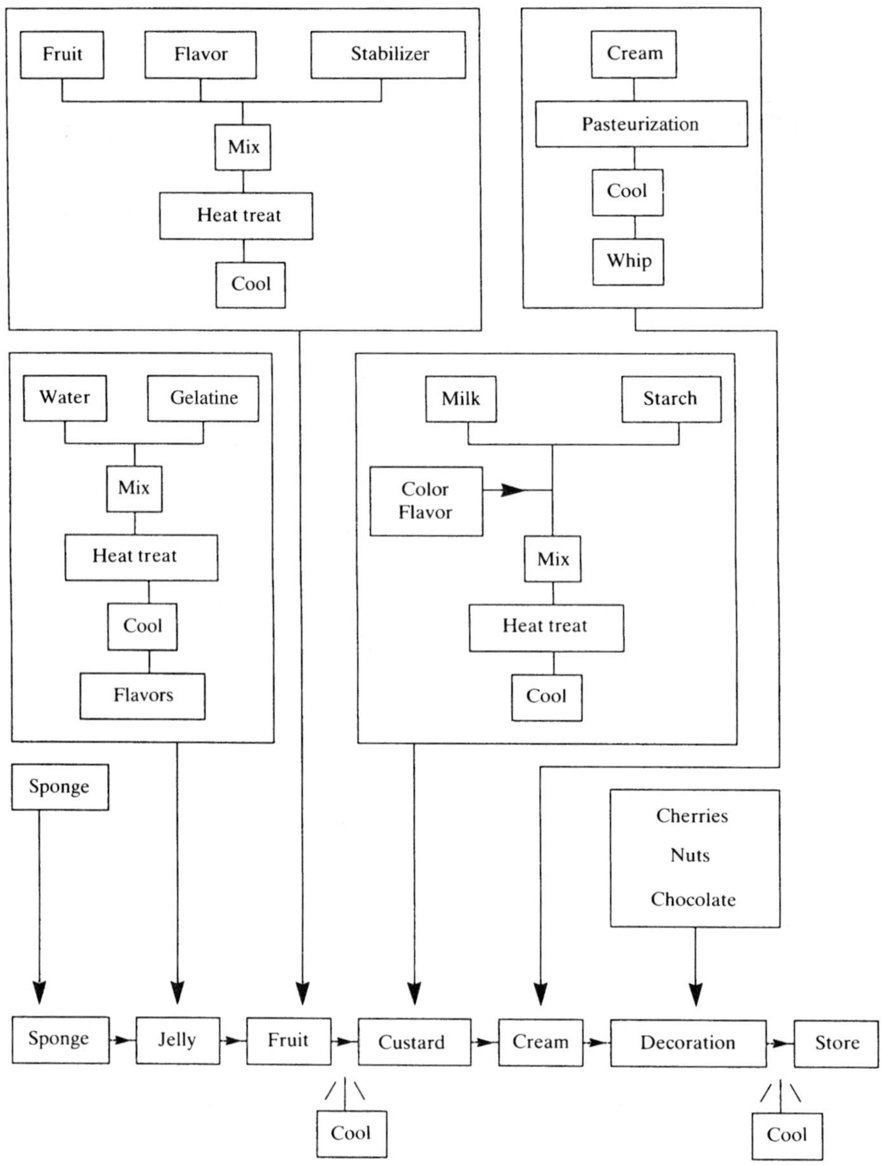

Figure 7–9 A simplified trifle-making process.

7.3.2.4 Fruit conserves

These will be specific for the product in terms of flavor, viscosity, pH, and color. The conserve chosen will depend upon the pH level, sweetness, and the addition rate to the final product.

To produce a fruit conserve which is stable for use in dairy products and free from yeasts, molds, and other undesirable organisms, the mix should be heated to a minimum of 85°C for several minutes.

The fruit conserve will contain stabilizers (starch, pectins, gums) to give the desired viscosity, enhance mouthfeel, and ensure that an even distribution of the fruit throughout the mix is maintained. Sugar is added to improve flavor and in some cases reduce the water activity.

The heating process may damage the color of the mix and the flavor profile of the fruit. Therefore, color and flavoring are usually added aseptically at the end of the process.

It is important to note that the shelf life of the fruit conserve as an ingredient will have a bearing on the shelf life of the final product. The normal shelf life of a conserve is 6–8 weeks; however, its microbiological stability is often significantly greater. The limiting factors are the color and flavor changes. Therefore, if a yogurt is to have a shelf life of 2 weeks, then the fruit conserve must be used at least 2 weeks before the end of its own shelf life.

For large-scale manufacture the fruit conserve is normally delivered in 800–1000 kg stainless steel vessels, for smaller processing operations in 25–40 kg poly bags. The methods adopted to fill these containers, the heat treatment given during processing, and the positive release systems in place by the fruit supplier should ensure that the fruit conserves arrive at the processing factory free from any yeast and mold contamination.

The same standards of hygiene and handling of the container must be implemented at the processing factory when the fruit vessels are connected to and disconnected from the blending station. Similarly the poly bags must be emptied into presterilized vessels where plant hygiene is of paramount importance.

The fruit vessels may be used in production several times within the shelf life of the conserve. Each time the vessel is used a risk of contamination from yeast and mold occurs. To manage the risk factor, good manufacturing practices should be adopted and operators instructed on handling methods, food safety issues and hygiene.

A further safeguard is achieved by measuring the carbon dioxide level within the headspace of the conserve vessel before initial use and at each subsequent visit to the filling area. An increase in carbon dioxide level may indicate activity of yeasts and the vessel should be rejected.

7.3.3 Filling and packaging

The filling operation and choice of packaging materials are critical to the ability of almost all food products to achieve their potential shelf life. In the case of short shelf-life products such as yogurt and dairy desserts, this part of the process is critical due to the relatively delicate physical nature and microbiological sensitivity of the products. The filling and packaging operation is the last direct contact that the manufacturer has with the product.

The incorrect choice of filling mechanism, poor hygiene practices, and the inability of packaging materials to protect the product can result in premature shelf-life failure and customer dissatisfaction. Therefore, all the effort put into the selection of raw materials and developing the correct processing conditions can be wasted by injudicious actions in the final stages of production.

7.3.3.1 Filling

The most important aspect of the filling operation with regard to shelf life is hygiene—the filler must be capable of being thoroughly cleaned and sterilized prior to the commencement of the filling operation. Failure to do this will inevitably result in, at best, a reduced shelf life of a product and, at worst, a food poisoning incident.

Most modern, high-throughput fillers are capable of being fully cleaned in place. However, many older, small-scale fillers need to be stripped down and manually cleaned. With this manual operation it is essential that the filler is reassembled with great care, and that a sterilizing rinse is given prior to the commencement of the filling operation.

Most microbiological shelf-life failure is caused by post-processing contamination at the filling stage either due to poor cleaning, poor design, or environmental factors.

In the case of products such as yogurt and fromage frais which are of relatively low pH (<4.6) the most common spoilage organisms are yeasts. The yeasts involved in the spoilage of dairy products are commonly the lactose-fermenting genera. However, due to the levels of sucrose incorporated into many products, other genera may also be found. Common spoilage yeasts are *Saccharomyces* and *Candida* spp. Molds can also cause the premature spoilage of yogurts and fromage frais. However, as yeast and mold contaminations often occur together, it tends to be the production of gas by yeasts which is first noticed as a "blowing" pack, and therefore, the yeast is identified as the spoilage organism.

The more neutral desserts (pH >4.6), while being subject to premature spoilage by yeasts and molds, are also an ideal medium for bacterial growth. The bacteria most frequently found to contaminate desserts at the filling stage are members of the Enterobacteriaceae group, *Pseudomonas* spp., and *Bacillus* spp. (in particular *B. cereus)* which is commonly associated with cream products.

Bacterial spoilage may take two forms:

1. The souring of products due to the production of acid and off-flavors.
2. The enzymic breakdown of the structure of products due to the activities of exo-enzymes.

7.3.3.2 Packaging

Packaging, both in design and material terms, can have a major influence on the shelf life of the product. Packaging fulfills a number of functions which may include:

- protection of the product from physical damage
- protection of the product from the environment
- protection of the product from changes in the environment
- provision of attractive marketing designs to convey information including the date code to the consumer
- presentation of the product in an easy to handle and store format for distribution to the retailer and consumer.

The packaging of product can therefore influence the shelf life in a number of ways:

(a) Microbiologically. Packaging can carry microorganisms, and therefore must be manufactured and stored hygienically to ensure a minimum microbial loading. This is particularly important in the case of preformed pots which must be stored in a clean, dry, pest-proof environment. Ideally the outer cases should only be removed prior to the containers being taken into the filling room.

For "form-fill-seal" packaging which is wound on a spool, again the outer packaging should only be removed immediately prior to entering the filling room. Additionally a substantial amount of the reel should be discarded before the material is fed into the pot-forming chamber of the filler. Lidding materials should be treated similarly to minimize any microbiological contamination.

For heat-sealed containers where the lid is bonded to the pot the most common sealing problem is caused by product contaminating the lip of the pot and interfering with the bonding process.

"Snap-on" lids must be produced to tight tolerances to ensure an effective seal is obtained. Often a plastic tamper-evident ring or other secondary packaging will assist in maintaining the integrity of the pack.

The seal of the pack must be secure in order to exclude environmental contamination, particularly during the cooling phase when a partial vacuum may be formed.

(b) Organoleptically. The most common "off-flavor" associated with yogurts and dairy desserts which develops with time is caused by the oxidation of the milk fat. Flavor degradation also occurs with time and very often is an oxidative change.

Utilizing the correct type of packaging material which reduces the level of oxygen permeability and light penetration can help to minimize changes induced by these environmental factors.

In neutral or near-neutral pH desserts, the shelf life is significantly shorter due to microbiological considerations, and therefore, oxidation is unlikely to be critical.

In some high-fat desserts, with fat levels in excess of 5%, the flavor changes due to oxidation may be the determining factor.

(c) Visually. As with flavors, color fade or change may occur over a prolonged period. This is particularly noticeable with natural red colors such as beetroot red (betanin). Many factors can influence color change, including oxygen, pH, and the presence of sulfur dioxide, which may be carried over in small quantities from any fruit that is added to the yogurt or desserts.

7.3.4 Storage and distribution

All foods, and dairy desserts are no exception, may spoil during storage and distribution. In general this may be due to:

- growth of microorganisms
- biochemical and chemical changes such as oxidation, rancidity development, and moisture migration
- attacks by vermin, birds, or other pests.

Physical damage of products is prevented by enforcing good standards of storage, handling, and pest control within the distribution network. However, this is outside the scope of this chapter.

The microbiological and biochemical changes are greatly reduced by storing products at low temperatures. Effective temperature control is, therefore, an important measure in preventing the development of spoilage organisms and ensuring that dairy desserts achieve their potential shelf life, as well as minimizing any food poisoning risk.

Within the EU, the Food Safety (Temperature Control) Regulations 1995 formalized this requirement for the industry, such that dairy desserts must be stored and distributed at temperatures of less than 8°C. In practice, the maximum recommended temperature for storage and distribution of all dairy desserts is below 5°C.

In order to achieve the desired temperature standards within the distribution chain the following is recommended.

1. Products should be stacked in such a way that the cold air can circulate freely.
2. Products should be pre-cooled before entering the cold store.
3. Opening of the store or vehicle doors should be minimized.
4. Efficient temperature monitoring and alarm systems should be installed and maintained.
5. A high air flow/velocity within the store should be employed.
6. Products should be loaded onto vehicles via an air lock system wherever possible and under no circumstances should be left out on a loading bank.
7. Vehicles should be pre-cooled prior to loading the products.

The other major factor that can influence shelf life during the storage and distribution operation is light. This is most likely to be a critical factor in the display cabinet. However, it can be suitably controlled by the choice of packaging as discussed in the previous section.

7.3.5 Consumer storage

When determining the shelf life of a product, which in the main depends upon the integrity of the "chill chain" to maintain its desirable properties, full account must be given for the potential abuse of product by the retailer and consumer. During recent years surveys have been commissioned by industry, research bodies, and government agencies to provide accurate data on the way the general public handle and store chilled products after purchase. One of these is *Consumer Handling of Chilled Foods,* published in the UK by the Ministry of Agriculture, Fisheries and Food (MAFF, 1991), which investigated many factors associated with shopping, consumer handling, and attitudes to chilled products.

The temperature regime used for establishing product shelf life must therefore be designed to take into account:

1. The time that the product may be out of refrigeration, between being removed from the retailer's chill cabinet and being placed in the consumer's refrigerator.
2. The number of days it will be stored in the consumer's refrigerator.
3. The likely position in which it will be stored within the consumer's refrigerator (dairy products are often placed near the

top or in the door, both of which are relatively warm positions).
4. The temperature range within which domestic refrigerators operate.

Education of the consumer about the product is important. Clear instructions of storage conditions should be given on the pack, and advertising information can be used to emphasize the importance of temperature control.

At the end of the day, however, it must be accepted that the last link in the "chill chain" for any product is probably the weakest and all the efforts of a manufacturer can be confounded, if the product is handled badly by the consumer.

7.4 EVALUATION METHODS

There are no absolute tests to evaluate shelf life. The methodology used and the process for evaluation will depend on the product, the quality criteria, and the expertise within the company.

Before attempting to evaluate the shelf life it is important to have an understanding of the product composition, and the changes that may be induced by various processing conditions.

Different testing regimes will be required for:

1. low pH products
2. neutral pH products
3. those with high-risk ingredients
4. low water activity products.

In terms of microbiological changes it is important to establish which organisms are likely to be present in the final product, which are likely to affect safety or cause spoilage during shelf life, and the possible sources of contamination.

Microbiological, chemical, and physical changes which affect final product quality can in many instances be detected by sensory analysis. These changes should be either evaluated or considered during the development process.

Sensory analysis can play a major role in evaluating shelf life (caution must be exercised to ensure that no microbiological safety risk is present if the technique is employed).

It is essential that sufficient shelf-life data are collected during the pilot and scale-up trials to identify the CCPs and the factors that will affect the stability of the product. It is recommended that a minimum of three full evaluation tests are carried out on products from factory scale pre-production trials to confirm the shelf life prior to launch of the product. Figure 7–10 shows the main sequence of operations in the development of a new product.

Many complicated cycling regimes to imitate the temperature patterns to which products may be subjected have been suggested and indeed are used for shelf-life determinations. However, for simplicity it is often better to use a two-step system ensuring that the temperatures chosen allow sufficient room for abuse and are not idealistic.

A typical regime for dairy desserts would be storage for 2 days at 4°C followed by storage at 8°C for a period exceeding the target shelf life.

7.4.1 Shelf-life determination—yogurt

The critical factors to consider are:

1. target life
2. temperature of storage and distribution
3. yeast and mold contamination
4. syneresis and texture
5. degradation of flavor and color.

The methods described can be used for shelf-life evaluation of existing products as well as new products. In the case of new products, however, it is recommended that a hazard analysis based on the Hazard Analysis Critical Control Point (HACCP) concept is carried out at the development and pre-production stages.

Sufficient samples must be taken to cover all the planned tests. Samples are stored in temperature-controlled rooms operating at 7°C, 21°C, and 30°C. These are checked at regular intervals through to the target shelf life and beyond as illustrated in Table 7–2.

Microbiologically all samples are checked for coliforms, yeasts, and molds. Although coliforms do not survive for much longer than 24 hours due to the low pH, their initial presence in the product gives an indication of poor plant hygiene and the possibility of other contamination. The testing for coliforms using Violet Red

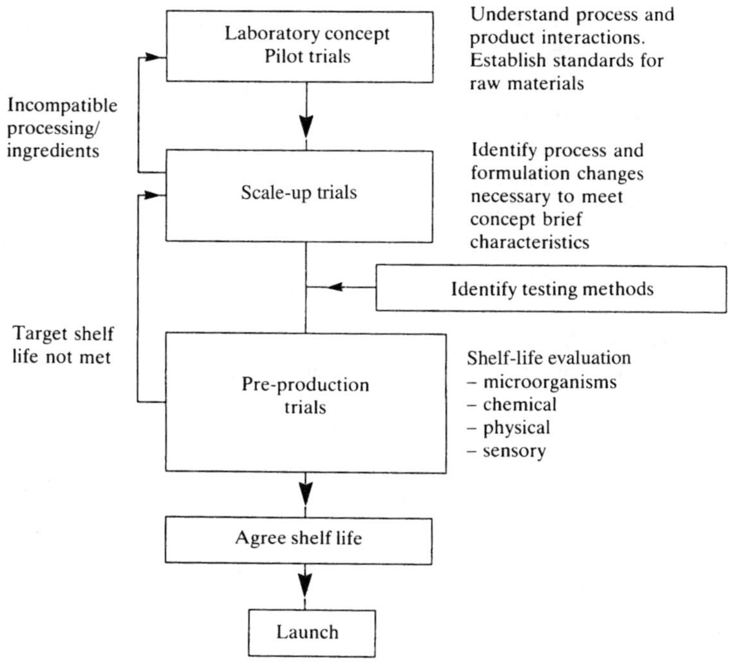

Figure 7–10 The main sequence of operations in the development of a new product.

Bile Agar is regarded as a useful check in establishing the shelf life of the product.

The number of yeasts and molds present in the product can be detected by plating on Rose-Bengal Chloramphenicol Agar where individual colonies can be counted.

Due to the small quantity of product needed to carry out these tests (1 ml or 0.1 ml), the results tend to be unreliable, especially when only low levels of yeasts are present. The tests are, however, a useful step in the identification of yeasts and molds, and hence in establishing the possible cause of contamination.

Storage of yogurt at 21°C and 30°C for shelf-life evaluation evolved as a result of the difficulties experienced with the plate count method.

Table 7–2 Testing schedule—shelf-life determination

No. of days		*0*	*1*	*2*	*3*	*4*	*7*	*14*	*21*	*28*
Coliforms		*								
Y & M colony count		*					*	*		
Incubated samples	7°C				*	*	*	*	*	*
checked for gas	21°C		*	*	*	*	*	*	*	*
production	30°C		*	*	*	*	*	*	*	*
Texture and viscosity			*				*	*	*	
pH/acidity		*	*				*	*	*	*
Flavor profile			*				*	*	*	*

Storage temperature: 7°C unless indicated.
Target shelf life: 21 days.
Day of production: day 0.
* = Sampling occasion.

The presence of yeasts can be detected by gas formation when the product is stored at these temperatures. Gas formation in a pot of yogurt is evident by doming of the lid, or if the product is heavily contaminated or incubated for a longer period, by gas forcing the lid clear of the rim of the pot. Storage at 21°C and 30°C also enables the factory to react quickly to any potential problems.

The 21°C and 30°C tests were developed following trials where samples of "clean" uncontaminated product were inoculated with yeasts from several pots of yogurt known to be contaminated.

From these trials it was concluded that the time taken for the yeasts to produce sufficient gas to dome the lid was an indication of the severity of the contamination. The shorter the time to "gas" the more contaminated the product. The results are summarized in Table 7–3. It was also established that different species of yeasts had different rates of gas production as indicated by the hours before doming in the three trials. These differences should be taken into account in the shelf-life evaluation process.

The differences in the time taken to blow (gas formation between samples incubated at 21°C and 30°C) have also been investigated. Some discrepancies between the results were found; nevertheless, storage of samples at 21°C and 30°C forms an essential part of the shelf-life evaluation of yogurt and indeed of factory control procedures.

The investigation was carried out at a yogurt production site covering a period of three months and involving 5,760 samples (samples were taken every 30 minutes from each filling machine during the production cycle). The results of the investigation are shown in Figures 7–11 and 7–12. The charts show, based on the total number of samples that were found to blow in 2 days at 30°C (Figure 7–11) and in 3 days at 30°C (Figure 7–12), the percentage breakdown of samples kept at 21°C that were found to blow during storage. For example (Figure 7–11), 233 samples were found to blow after 2 days at 30°C. Among the duplicate samples stored at 21°C, 46% were found to blow after 3 days, 19% after 4 days and so on.

A comparison of "blowing" times at 21°C and 7°C demonstrates a clear correlation between the two. The shelf life of the refrigerated product can therefore be predicted by the time taken for samples stored at the elevated temperature to blow. Results are shown in Figure 7–13.

Table 7–3 Yogurt inoculated with spoilage yeasts and samples incubated at 30°C

	No. of yeasts/pot	*Hours before doming*
Trial 1	60 000	40
	6000	48
	600	63
	6	76
	Control (Nil)	>168
Trial 2	360 000	15
	36000	24
	360	46
	4	96
	Control (Nil)	>144
Trial 3	260 000	24
	26000	46
	260	96
	26	120
	Control (Nil)	>144

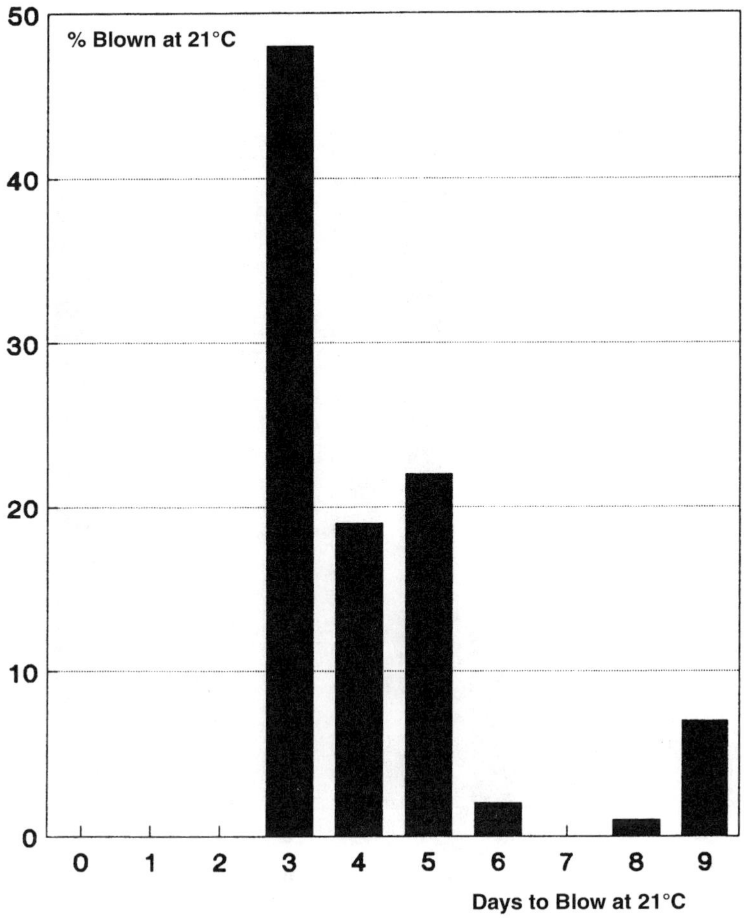

Figure 7–11 Storage of yogurt samples—comparison of gas formation at 30°C and 21°C (223 samples blown in 2 days at 30°C).

The majority of short shelf-life yogurts are "live," i.e., the culture organisms are still viable. Although their metabolic rate at 7°C is relatively low, there is nonetheless some activity. This can be determined during shelf life by pH measurement (see Figure 7–14), by titratable acidity determination, and by taste.

The increase in acidity can alter the flavor profile, and if combined with degradation of fruit flavors, it can often become manifested as off and stale taints. To establish the extent of flavor degradation, trained assessors are used.

The deterioration can be noted and the assessors are able to rank the product according to age fairly consistently. The chemical and biochemical changes that occur over time show that the product does decline with age. From a sensory point of view the shelf life is set at least 1 day prior to the product changes becoming so evident that they are judged to be unacceptable to the consumer. Spider plots (Figure 7–15) can be used to highlight differences between fresh and stored samples.

7.5 CURRENT DEVELOPMENTS AND THE FUTURE

During the last two decades many developments have been made by both industry and re-

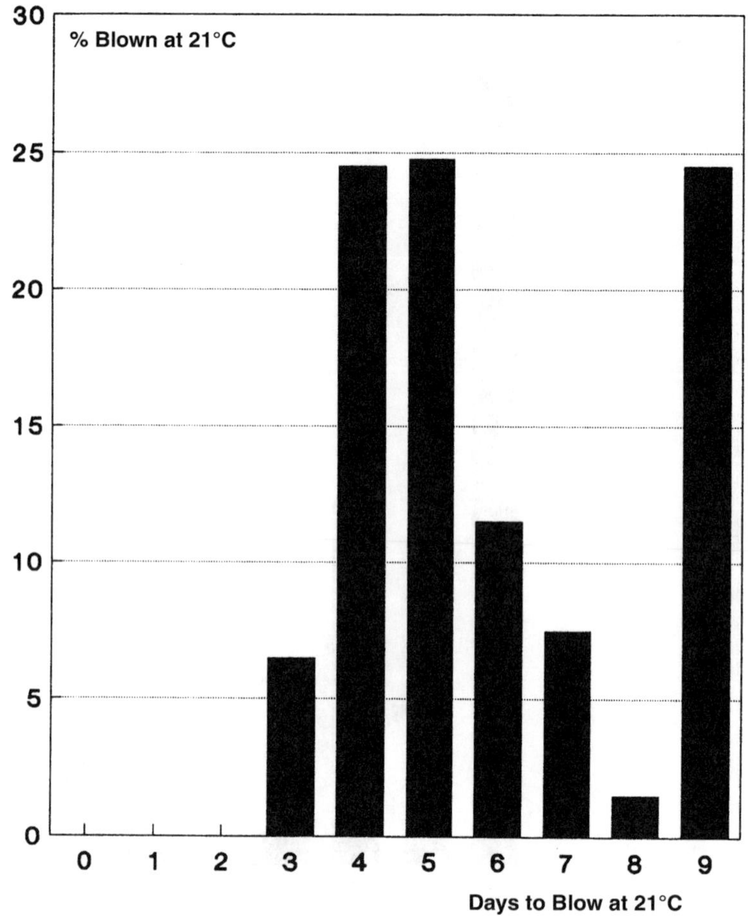

Figure 7–12 Storage of yogurt samples—comparison of gas formation at 30°C and 21°C (341 samples blown in 3 days at 30°C).

search organizations to extend the shelf life of many types of foods. These have included techniques and aids such as:

- irradiation
- modified and controlled atmosphere packaging
- oxygen absorbers
- aseptic processing and packaging developments.

However, few of these developments have been of use in the extension of shelf life of fresh dairy desserts due to cost, adverse effects on the food, or consumer attitudes.

Most progress in this product sector has been achieved by better practices and attention to detail. The key areas have been:

- improved selection and quality of raw materials
- greater knowledge and utilization of functional ingredients
- improved process plant design and hygiene
- the use of techniques such as HACCP and challenge testing to establish safe and quality-related shelf life
- greater use of rapid method techniques for analysis and hygiene monitoring

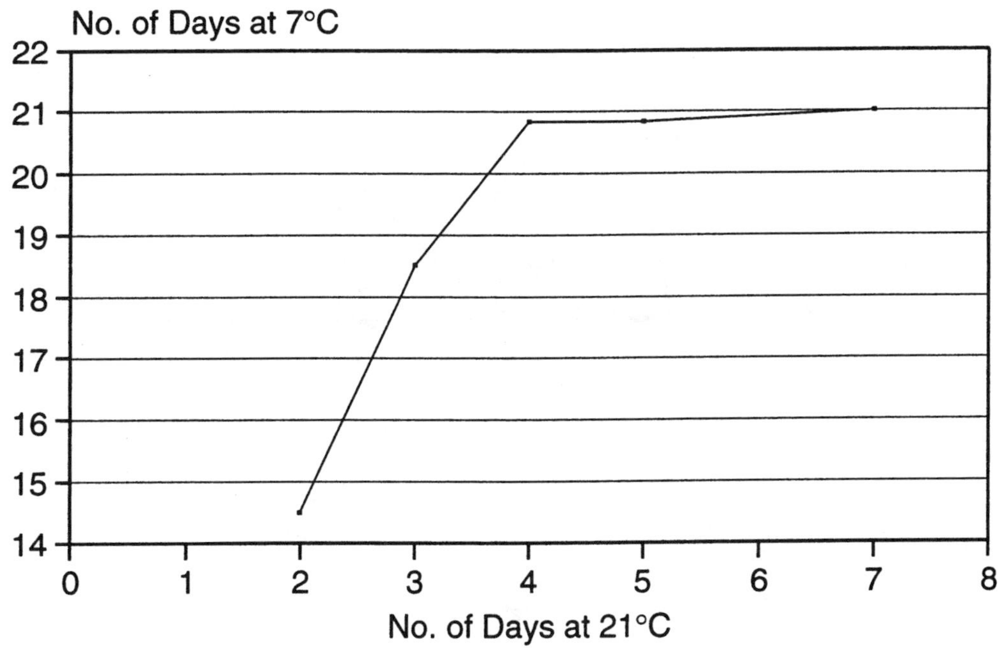

Figure 7–13 Storage of yogurt samples—comparison of blowing times at 21°C and 7°C.

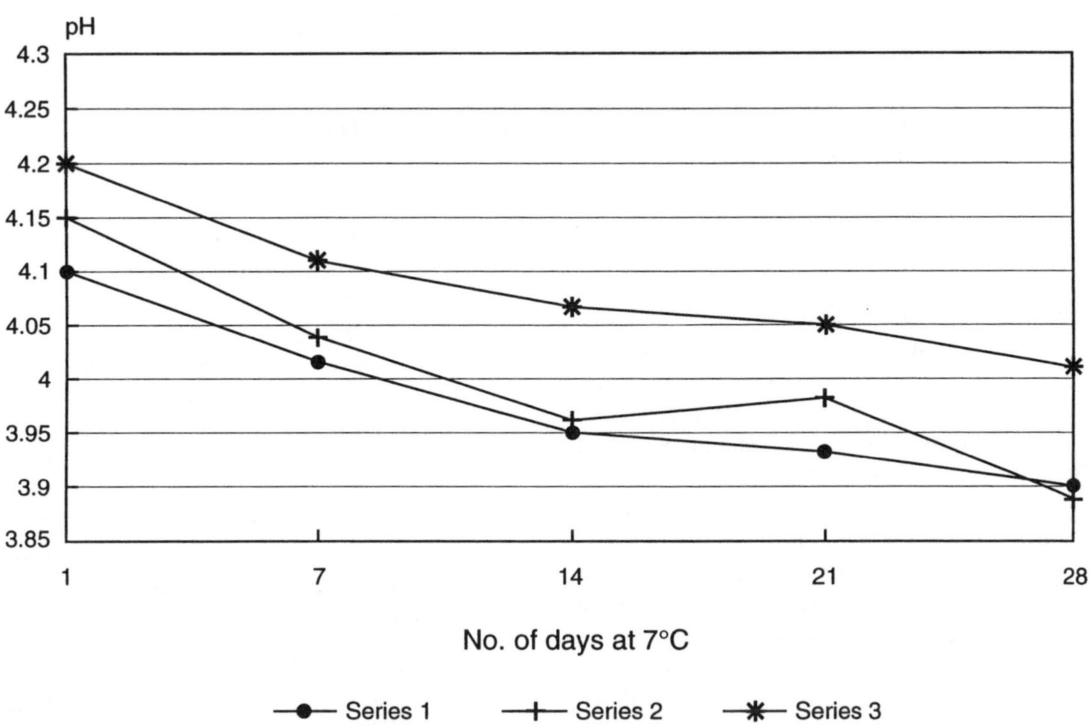

Figure 7–14 Storage of yogurt samples—pH change with time. Series 1, •; series 2, +; series 3, *.

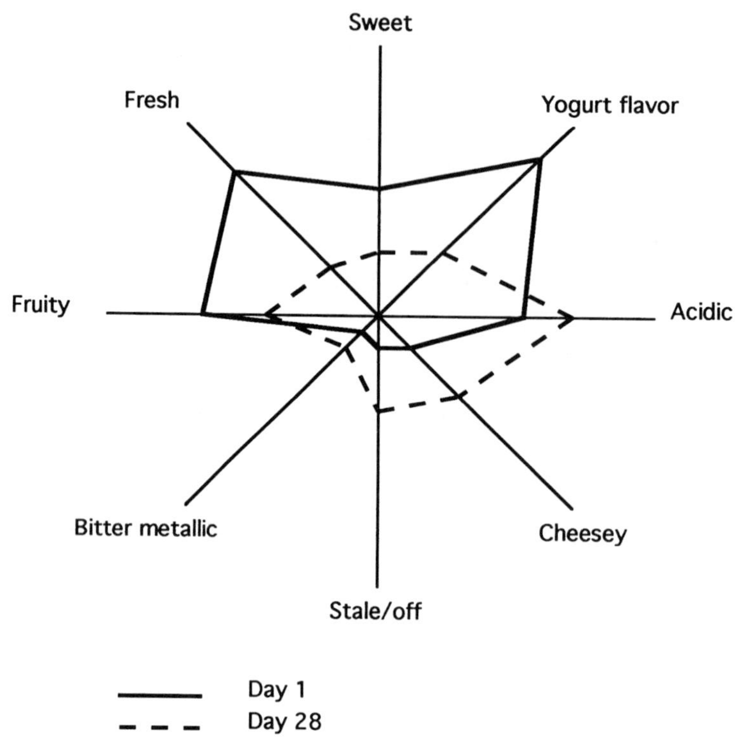

Figure 7–15 Spider plots—differences between fresh and stored samples. —, day 1; – –, day 28.

- improved cooling and refrigeration
- the selection of technologically advanced packaging materials.

One area which has been the subject of much recent work and debate is the use of time–temperature indicators (TTIs). These are markers which may be applied to the outer packaging, which will indicate to the reader the "temperature history" of the product. Clearly they have the potential to indicate the residual safe shelf life of desserts, either to the retailer or the consumer. However, current costs are relatively high and some reliability problems exist.

The changes within the desserts market have been so great within the last 20 years that no one in the 1960s could have foreseen or predicted the size and scope of this market in the 1990s. Equally it is difficult to envisage what further changes will occur during the next 20 years. Certainly for the short term, growth both in terms of volume and variety and the drive for improved quality and safety will inevitably continue. However, it can be predicted confidently that consumers will continue to demand:

1. Less wastage in packaging. Tertiary and possibly even secondary packaging will be gradually reduced unless they are biodegradable or can be recycled.
2. Improved safety and freshness. These are likely to become more important than a long shelf life.
3. Better value for money than in the affluent 1980s.
4. Greater attention to nutritional and health aspects: low cholesterol, low fat, high in unsaturates, low sugars, high fiber, added vitamins, health-promoting minerals, and functional dairy products.

All of these requirements will place further demand on the food technologists as they attempt to meet industry's needs and the consumer's wants.

BIBLIOGRAPHY

Birch, G.G. and Parker, K.J. (eds) (1984) *Control of Food Quality and Food Analysis,* Elsevier Applied Science, Amsterdam, The Netherlands.

Campden and Chorleywood Food Research Association (1997) *Evaluation of Shelf Life for Chilled Foods,* Technical Manual No. 28.

Campden Food and Drink Research Association (1991) *Shelf Life Problems, Technology and Solutions,* Symposium Proceedings.

Dennis, C. and Stringer, M.F. (eds) (1992) *Chilled Foods—A Comprehensive Guide,* Ellis Horwood, Chichester, England.

Marth, E.H. (1978) *Standard Methods for the Examination of Dairy Products,* 14th Ed., American Public Health Association, Washington, DC.

Ministry of Agriculture, Fisheries and Food (1991) *Consumer Handling of Chilled Foods: A Survey of Time and Temperature Conditions.*

Shapton, D. and Shapton, N. (eds) (1991) *Principles and Practices for the Safe Processing of Foods,* Butterworth and Heinemann, Boston, MA.

Sutherland, J.P., Varnam, A.H. and Evans, M.G. (1986) *A Colour Atlas of Food Quality Control,* Wolfe Publishing Limited.

Walker, S.J. (1988) Major spoilage micro-organisms in milk and dairy products. *Journal of Society of Dairy Technology,* **41**, 91–92.

CHAPTER 8

Fresh and Lightly Preserved Seafood

P. Dalgaard

8.1 INTRODUCTION

The annual world production of fish and shellfish, of 95 million tonnes in 1996, has increased very little since 1990 as compared to previously from 1950 to 1990. In developed countries, catches of marine fish have decreased markedly since 1988 and approximately 70% of world catches now originate from developing countries (Anon., 1994, 1996). Globally, the limited availability and the increased transportation of fish raw material and seafood make it most important to minimize seafood losses at all levels from catch, through processing to storage and distribution. The US National Research Council indicated 25% of the world's food supply to be lost due to microbial activity (Anon., 1985). Improved management and food preservation technology are needed and this is an important challenge because the interest in marketing of more natural and less heavily preserved products has increased (Gould, 1996). In this situation, determination and prediction of the shelf life of fresh and lightly preserved seafood have become particularly important to prevent losses due to spoilage. Furthermore, shelf life of seafood can be short and variable (Table 8–3) due to the fact that (1) fish flesh has a high pH, high free amino acid content, and the presence of trimethylamine oxide (TMAO) and also it may harbor a natural psychrotolerant spoilage microflora with the ability to grow rapidly in chilled products; and (2) most fish are caught wild, i.e., hunted and therefore fish raw material often has a lower uniformity than raw material from agriculture (Huss et al., 1997).

Microbial activity is the major general cause of seafood deterioration with chemical spoilage (oxidation and rancidity) and insect problems being of importance in more specific situations. However, structured and scientifically based systems for assessment and management of microbial seafood spoilage have not been developed in the same way as for safety risks related to microbial activity in seafood (Council Directive 91/493/EEC, 1991; Huss, 1994; US FDA, 1995; WTO SPS Agreement; Martin et al., 1997).

This chapter describes methods for the determination, prediction, and extension of shelf life of fresh and lightly preserved seafood and discusses their potential use within quality assurance systems. Microbial seafood spoilage, particularly the concept of specific spoilage organism (SSO), is presented in detail. Sensory and instrumental methods for shelf-life determination and methods for prediction of shelf life as a function of storage condition and product characteristics are reviewed. Finally, shelf-life extension based on targeted inhibition of specific spoilage organisms is mentioned. For reviews

---

Acknowledgments: This chapter has been prepared partly on the basis of a presentation, Ecology of Spoilage and the Specific Spoilage Organism Concept, given by the author in January 1999 at the winter meeting of the Society for Applied Microbiology in London. Thanks are expressed to colleagues at DIFRES, Prof. Allan Bremner, and Prof. Hans Henrik Huss for their valuable comments on the manuscript.

and proceedings concerning spoilage evaluation and technology for shelf-life extension of seafood see Reay and Shewan, 1949; Tarr, 1954; Farber, 1965; Kreuzer, 1971; Connell, 1980; Kramer and Liston, 1987; Huss et al. 1992; Botta, 1995; Huss, 1995; Luten et al., 1997; Hall, 1997; Shahidi et al., 1997; and Ólafsdóttir et al., 1998a.

8.2 MICROBIOLOGY AND THE SPECIFIC SPOILAGE ORGANISM (SSO) CONCEPT

Microbial activity is responsible for spoilage of most fresh and many lightly preserved seafoods. Therefore, it is not surprising and it has long been known that remaining product shelf life depends upon the degree of bacterial contamination and particularly of the growth conditions for spoilage bacteria during storage (Dyer and Dyer, 1950; Tarr, 1954). However, only some microorganisms in seafood are important for spoilage as clearly stated: "Bacterial counts are valueless as a measure of the degree of spoilage of fresh fillets," and "There is a very close correlation between the numbers of psychrophlilic gram-negative organisms on fillets and their keeping time in cold storage" (Castell et al., 1948). In fact, microorganisms in seafood have been divided into "non-spoilers" and "spoilers," the latter being able to grow in products at usual storage temperature and to produce objectionable changes in flavor, texture or color (Castell and Anderson, 1948; Lerke et al., 1965). Later, a general pattern of microbial growth and activity during storage of seafood was named the specific spoilage organism (SSO) concept (Figure 8–1). On newly processed, fresh or lightly preserved seafood SSOs are usually present in very low concentrations and constitute only a minor part of the total microflora. During storage at particular conditions of temperature, atmospheres, % NaCl, a_w, preservatives, etc., SSOs often grow faster than the

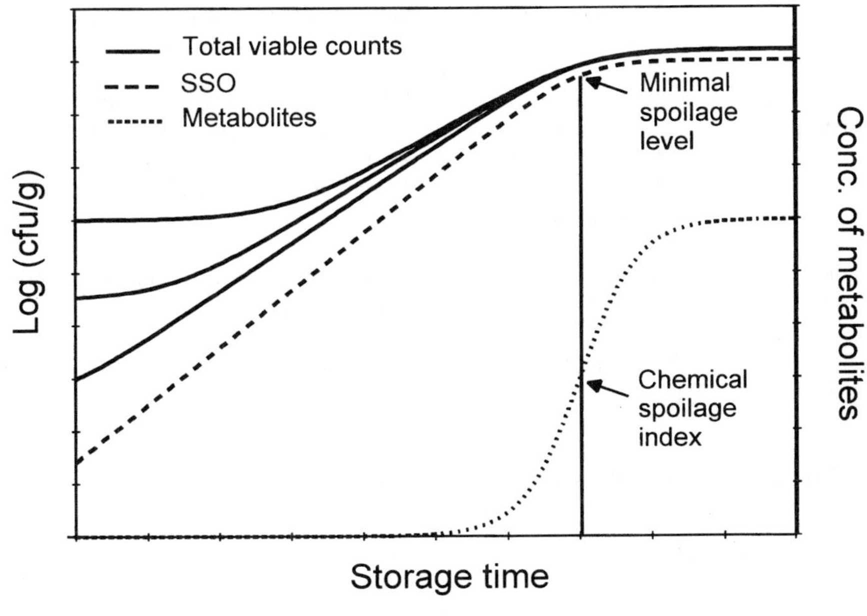

Figure 8–1 Specific spoilage organisms (SSO) concept. Changes in total viable counts (TVC), specific spoilage organisms (SSO), and metabolites produced by SSO. Changes in TVC depending on three different initial levels are shown. Concentration of SSO and of their metabolites at time of sensory rejection can be used as microbiological and chemical indices of spoilage, respectively. Reprinted from Dalgaard, 1998, page 138.

remaining microflora, produce metabolites responsible for off-flavors, and finally result in sensory rejection. This pattern has been observed in different foods and seafoods (Chai et al., 1968; Herbert et al., 1971; Jørgensen et al., 1988; Dainty and Mackey, 1992; Dalgaard, 1998; Koutsoumanis et al., 1998). The spoilage microflora or spoilage association, i.e., the microflora found at the time of sensory product rejection, is often a mixture of microbial species of which some can be completely without importance for spoilage and safety. SSOs, on the other hand, are the group of microorganisms, often of a single species, responsible for chemical changes resulting in sensory product rejection (Huss et al., 1997; Table 8–1).

The SSO concept has important implications with respect to principles and practice of shelf-life evaluation (Figure 8–2). The number of SSOs and the concentration of their metabolites can be used as objective indices of spoilage in shelf-life determination. With SSOs responsible for spoilage a close relation between log-numbers of SSOs and remaining shelf life must be expected. The corresponding correlation for total viable counts (TVC) should be reasonable with low initial TVC values but less close for higher initial TVC levels (Figure 8–1). This figure also shows the possibility of predicting shelf life from initial numbers and growth of SSOs. Identification and isolation of SSOs have allowed development of mathematical growth models and thereby prediction of shelf life under

Table 8–1 Specific spoilage organisms (SSO) and typical spoilage metabolites in different types of seafood

Product	*Typical SSO*	*Typical metabolites*
Fresh chilled fish stored in air	*Shewanella putrefaciens*[1]	TMA[4], H_2S and other sulfur compounds, acetic acid[5]
	Pseudomonas spp.[2]	Biogenic amines, ketones, aldehydes, esters, sulfur compounds but no H_2S[5]
Fresh chilled fish stored in vacuum or MAP	*Photobacterium phosphoreum*[1]	TMA, hypoxanthine, acetic acid, biogenic amines, aldehydes, alcohols[5]
	Lactic acid bacteria[2]	Ammonia, biogenic amines[5]
	Brochothrix thermosphacta[2]	Ammonia, organic acids[5]
Fresh fish stored at > 10°C in air	*Vibrionaceae*, *Enterobacteriaceae*	TMA, hypoxanthine, volatile sulfur compounds, acetic acid[5]
Cooked and brined shrimps in MAP at 15–25°C	*Enterococcus faecalis*	Ammonia, acetic acid, biogenic amines[5]
Sugar-salted herring	halophilic anaerobes,[3] osmotolerant yeasts	Indole, H_2S, acids[5]

(1) Typical of marine, temperate water fish.
(2) Typical of freshwater fish and fish from warmer waters.
(3) Not identified.
(4) Trimethylamine.
(5) Information about metabolites produced by different groups of SSO in seafood is limited and that several specific compounds, i.e., potential chemical indicators of spoilage may exist.

Source: Reprinted from Luten et al., *Seafood from Producer to Consumer: Integrated Approach to Quality,* Copyright (1997), p. 421, with permission from Elsevier Science.

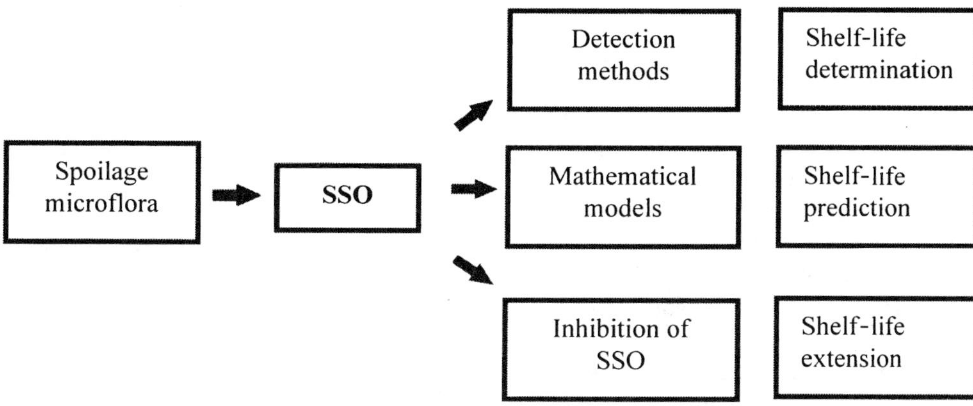

Figure 8–2 Application of knowledge about specific spoilage organisms (SSO) for determination, prediction and extension of shelf life.

both constant and fluctuating storage conditions. Finally the SSO concept can be important in shelf-life extension. When a single group of microorganisms is responsible for product spoilage, characteristics of the SSO, as determined in pure culture studies, can lead to new preservation methods targeted at inhibition of the SSO. In this way, new preservation techniques can be particularly mild as they will inhibit SSOs without necessarily influencing the total microflora in a product.

Application of the SSO concept for determination, prediction and extension of shelf life requires knowledge about an SSO and its spoilage domain, i.e., the range of conditions like temperature, atmosphere, a_w, and pH within which an SSO is responsible for product spoilage (Dalgaard, 1995a). Spoilage domains are most important for the practical application of the SSO concept because microbial seafood spoilage is dynamic and spoilage microflora, SSO, and spoilage reactions may change as a function of storage conditions and product characteristics. On the basis of intrinsic factors of fish and shellfish, it is, in some cases, possible to explain microbiological developments at particular storage conditions. For example, in fresh seafood, habitats of the live fish or shellfish (temperate or tropical; marine or fresh water) influence spoilage microflora and SSO (Table 8–1).

Due to the poikilotherm nature of fish and shellfish, microorganisms on outer surfaces of fish reflect the microbiology of the water where they live. However, some fishes with developed intestinal systems have high levels of specific gut microorganisms. *P. phosphoreum*, for example, can be the dominant gut microorganism in cod and from there it can contaminate fish fillets during processing and become the SSO of chilled packed products (Dalgaard, 1995b; Dalgaard et al., 1997b). Most fish and shellfish from marine waters and some species from fresh water contain TMAO and have a high positive redox potential. Many seafood spoilage bacteria including *S. putrefaciens*, *P. phosphoreum*, most *Enterobacteriaceae* and *Vibrionaceae* utilize TMAO in anaerobic respiration (Barret and Kwan, 1985), and this facilitates their growth in fishery products, particularly in vacuum or modified atmosphere packed (MAP) products (Table 8–1). Nevertheless, *Enterococcus faecalis* spoiled cooked and brined MAP shrimps (*Pandalus borealis*) stored at 15–25°C without reducing the content of ~25 mg-N TMAO/100 g (Dalgaard and Jørgensen, in press). In sugar-salted herring, a high redox potential reduced growth of a strict anaerobic spoilage bacterium and thereby contributed to a long shelf life (Knøchel and Huss, 1984a, 1984b; Table 8–3). Nevertheless, the general effect of redox potential on shelf life of seafood remains unclear. A *post rigor* pH of fish flesh, between 5.9 and 6.7, is high compared to meats and reflects a generally lower level of carbohydrates in fish muscles. At pH ~6 growth of *Shewenalla putrefaciens* is reduced substantially. This probably influences shelf life and the relative importance of this organism and of *Pseudomonas* spp. in the spoilage of aerobically stored fresh fish with low pH. However, the general importance of pH and related carbohydrate levels in spoilage of fresh seafood needs further study (Tarr, 1954; Gram and Huss, 1996; Koutsoumanis and Nychas, 1999).

Unfortunately, information about effects of product characteristics and storage conditions on spoilage microbiology and shelf life is incomplete, particularly for lightly preserved seafood. It is not yet possible, in general, to predict the spoilage microflora and spoilage reactions that become limiting for shelf life of new products or for products stored under new conditions. Storage trials and detailed analyses are needed to determine spoilage reactions and only then can microbiological or other methods be used appropriately for determination, prediction, or extension of shelf life. Packaging has a pronounced effect on SSO in seafood (Table 8–1) and this effect clearly depends on the freshness of the raw fish. Data in Table 8–1 correspond to good manufacturing practice; if raw fish is partly spoiled under aerobic condition before packaging, SSO obviously will reflect this situation.

Identification of SSO in storage trials relies on comparison of characteristics of naturally contaminated seafood and isolated microorganisms. ***Spoilage potential***, i.e., the ability of an isolate to produce off-flavors in a relevant medium and under realistic storage conditions has been used for decades to identify SSO (Castell and Anderson, 1948). More recently, ***spoilage activity***, i.e., cell concentration required for off-flavor production, and yield factors, e.g., amount of TMA produced per cell, was used for SSO identification (Dalgaard, 1995b). Comparison of profiles of microbial metabolites is another promising technique for SSO identification but so far it has not been used much in seafood. In fact, information about microbial metabolites in seafood spoilage is limited as indicated in Table 8–1.

8.3 DETERMINATION OF SHELF LIFE

Methods for realistic determination of product shelf life are essential for (1) the setting up of processing and distribution conditions, that prevent product rejection, (2) assuring compliance with guidelines and specification in purchase agreements, and (3) most important, preventing disappointed consumers.

8.3.1 Sensory methods

Sensory properties of seafood change during storage and a series of phases can be identified in most products. Shelf life is the time in storage at which the sensory properties of a product correspond to those indicative of spoilage. Sensory methods used in quality control must be objective and in this section a few objective sensory methods applied for determination and prediction of seafood shelf life are described. For reviews of other objective and subjective sensory methods used in product development and consumer studies see Connell and Shewan, 1980; York and Sereda, 1993; and Nielsen, 1998.

Regulations in the European Union require freshness grading of most fish to be marketed within the community (Council Directive 103/76, 1976; see Howgate et al., 1992, and Council Regulation No. 2406/96 for more recent amendments). Grading, using EU schemes, is usually carried out at auctions by trained personnel. Whole and gutted fish are assessed for appearance, odor of skin, outer slime, eyes, gills, and peritoneum. Fish are placed in the grades Extra (**E**), **A**, **B**, or Unfit (**C**) on the basis of schemes for different groups of species. Each characteristic is marked on a **0 to 3** basis, with **3** being completely fresh and **0** being spoiled. Finally, fish are graded from average scores (AS). Grade E (for Extra) has AS > 2.7. Grade A has 2.7 < AS < 2.0, Grade B has 2.0 < AS < 1.0. Fish with AS below 1.0 are graded unfit for human consumption. For determination and prediction of shelf life for use in research and industry, EU schemes have disadvantages as information on remaining shelf life cannot be obtained directly from the final freshness grades. It has also been suggested that the EU schemes are too complicated and therefore are not actually followed in practice (Connell and Howgate, 1986; Howgate, 1987).

A detailed objective sensory scoring system for use in laboratory studies was developed in 1948 at Torry Research Station in Scotland. The Torry scheme was developed to evaluate the effects of different biological and technological factors of spoilage and "quality factors" (today usually named quality attributes) in wet whitefish stored in ice (Shewan et al., 1953; Ehrenberg and Shewan, 1953). This scheme relies on assessment of raw appearance, raw odor, raw texture, cooked odor, cooked texture, and cooked flavor, typically by 4 to 9 trained assessors. Each quality factor is scaled by 5 or 10 descriptive terms and corresponding numerical scores. Scores decrease with storage time, and end of shelf life has been determined as corresponding to different values between 4.0 and 6.0, for factors scored as 0–10 (Spencer and Baines, 1964; Cann et al., 1983; Gibson, 1985). At constant storage conditions scores tend to follow a straight line, which facilitates determination of shelf life and rates of spoilage. However, S-shaped curves have been reported (Ehrenberg and Shewan, 1955; Shewan, 1962; Spencer and Baines, 1964).

Torry scheme scores for cooked flavor alone have been used in shelf-life determination and it seems logical, therefore, to consider a less detailed sensory scheme. In fact, numerous systems for assessment of one or several quality attributes divided into from 2 to 10 descriptive terms and numerical scores have been suggested. Table 8–2 shows a scheme used at DIFRES in Lyngby, Denmark. This scheme includes classification as accept or reject, grades, and scoring by descriptive terms and 10 points. Average scores, usually obtained by 5 to 10 assessors, decrease with storage time, and end of shelf life has been defined as an average score of 4 (Figure 8–3). S-shaped curves for quality scores make it difficult to calculate rates of spoilage. The S-shape is most pronounced when shelf life is extended by preservatives, modified atmosphere packaging, or extremely hygienic processing (Huss et al., 1974; Bremner, 1985). It is worth noting that assessors evaluate changes in quality attributes and not in personal preference, i.e., the system is objective despite using the term "acceptable" in Table 8–2.

Table 8–2 Classification and scoring system for sensory assessment of odor and flavor of raw and cooked seafood

		Grade		*Score*
Acceptable	No off-odor/flavor	I	Odor/flavor characteristic of species, very fresh, seaweedy	10, 9, 8
			Loss of odor/flavor	7
			Neutral	6
	Slight off-odors/flavor	II	Slight off-odors/flavors such as mousy, garlic, bready, sour, fruity, rancid	5, 4
		Limit of acceptability		
Reject	Severe off-odor/flavor	III	Strong off-odors/flavors such as stale cabbage, NH_3, H_2S or sulfides	3, 2, 1

Source: Adapted from Nielsen, 1995 with permission from the publisher, Food and Agriculture Organization of the United Nations, *Quality and Quality Changes in Fresh Fish. Technical Paper 348.*

The three-grade system shown in Table 8–2 has been used frequently for sensory evaluation of various seafoods. Appearance, odor, taste, and texture of raw or cooked products are usually evaluated and expressed as a single overall value (I, II, or III). Shelf life can be determined as the time when the average percentage of samples in Grade III reaches a defined level, e.g., 50% (Figure 8–9). It is also possible to determine shelf life and rates of spoilage from the means of grade scores, which form straight lines when plotted against storage time. As an example, data from the two experiments shown in Figure 8–9 had correlation coefficients > 0.94 when evaluated in this way. The three-grade system can be used for various types of seafood, i.e., it is not limited to specific species of whole fish. Furthermore, little training of assessors is required and useful data for shelf-life determination can be obtained (Matches, 1982; Sado, 1993; Dalgaard et al., 1998; Koutsoumanis and Nychas, 1999).

The demerit point system was developed at CSIRO in Hobart, Australia, in the 1980s (Bremner, 1985; Bremner et al., 1987). The system was developed for simple and rapid, species independent, evaluation of whole or gutted fish by sight, smell, and touch. A relatively large number of attributes are evaluated in sequence and each is scored 0 to 3 by either novice or experienced assessors. Demerit points of a sample, i.e., the sum of all attribute scores, increase linearly with storage time, and rates of spoilage at constant storage condition can easily be calculated (Figure 8–3). More recently, the system has been used in Europe and renamed the quality index method (QIM). QIM has been used successfully for several fresh and thawed whole fish species but development of species-specific QIM schemes is needed. For fish fillets, application of QIM is more difficult. QIM schemes have been developed and used for anchovy, brill, cod (fresh and frozen), haddock, flounder, herring, hoki, plaice, redfish, saithe, salmon, sar-

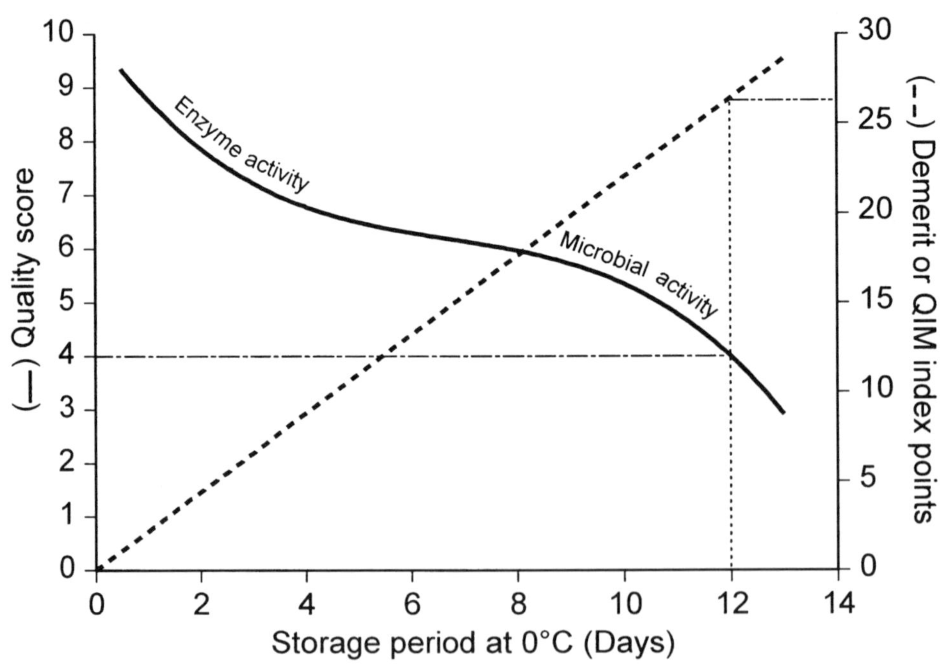

Figure 8–3 Typical evolution of sensory scores as determined by the scheme shown in Table 8–2 and by the Quality Index Method (QIM). *Source:* Reproduced and modified from Nielsen, © 1995 with permission from the publisher, Food and Agriculture Organization of the United Nations, *Quality and Quality Changes in Fresh Fish. Technical Paper 348.*

dines, sole, spotted trevalla, turbot, whiting, and four species of warm water fishes from Australia (Bremner, 1985; Branch and Vail, 1985; Larsen et al., 1992; Jónsdóttir, 1992; Ryder et al., 1993; Luten and Martinsdóttir, 1998; Hyldig and Nielsen, 1998).

Different QIM schemes use differing total numbers of index points and a general value corresponding to end of shelf life has not been defined. This problem could possibly be circumvented by using a percentage of the maximum number of index points as a measure for end of shelf life. With a few fish species, linear increases in demerit points have been observed at different temperatures. Consequently, shelf life can be predicted at different and fluctuating temperatures (Branch and Vail, 1985). However, additional studies are required to determine the range of applicability for many of the species-specific QIM schemes. It is particularly necessary to show whether slopes of QIM index point lines at different temperatures correspond to available temperature spoilage models and also if QIM values, corresponding to sensory rejection, are independent of storage temperatures. QIM is a promising sensory method but if the EU schemes are too complicated for practical use, as indicated by Connell and Howgate (1986), this may also be true for the detailed species-specific QIM schemes. It is hoped that future developments in this area will result in QIM schemes applicable for groups of seafood.

Objective sensory methods are essential for inspection/grading, quality control, and shelf-life determination of seafood. Sensory methods have the disadvantages of being (1) expensive and time consuming, particularly when numerous trained assessors are needed, and (2) difficult to calibrate. Consequently, instrumental methods are needed to supplement sensory assessments, e.g., for specifications as part of purchase agreements. It is also worth noting that results of objective sensory methods relying on trained assessors correlate poorly with consumer acceptance of the products. Connell and Howgate (1971) found no clear difference in consumer acceptance of cod and haddock stored for 2 and 8 days in ice. More recently, Juhl and Poulsen (1999) evaluated modified atmosphere packed cod and plaice fillets and found no difference in consumer acceptance between products stored 2, 6, 9, and 12 days at 2°C. These consumer studies clearly indicate shelf-life prediction, i.e., prediction of time to overt spoilage, as more important than detailed evaluation of sensory attributes during processing and distribution.

8.3.2 Instrumental methods

In quality control and seafood inspection, instrumental methods are needed to supplement sensory analyses. Such instrumental methods should be causally related to sensory changes or at least correlated with sensory analyses over some critical range. Furthermore, instrumental methods should be convenient to use, rapid, and inexpensive. For application as part of quality assurance programmes, e.g., for system set-up and verification, similar properties of instrumental methods are needed but in this case speed and ease of use are less important. It has been stated that no single method is generally reliable for assessment of freshness and spoilage of seafood (Tarr, 1954; Oehlenschläger and Sørensen, 1998). Nevertheless, numerous methods seem appropriate as long as their limitations in terms of products, preserving parameters, and storage conditions are understood and respected. Such microbiological, chemical/biochemical, and other instrumental methods will be discussed below.

8.3.2.1 Microbiological methods

Early studies found total viable counts (TVC), also called aerobic plate counts, to correlate poorly with the degree of fish spoilage (Castell et al., 1948; Huss et al., 1974). Nevertheless, TVC are used in microbiological standards and regulations of some European countries, Japan, and the USA (Huss, 1994). Measurements of TVC are also used extensively in purchase agreements between buyers and vendors of seafood.

Recently, counts of four different SSOs, i.e., *S. putrefaciens, P. phosphoreum, B. thermosphacta,* and lactic acid bacteria, were shown to correlate closely ($0.90 <$ corr. coef. < 0.99) with

Table 8–3 Shelf life of packed and unpacked seafoods at different temperatures

Product	*Storage temperature*	*Packaging*	*Shelf life (days)*	*References*
Fresh seafood				
White-fleshed, average size, and caught in temperate and cold waters	0°C	air	12–18	Connell and Howgate (1986)
Cod	0°C	VP[a]	13–14	Dalgaard et al. (1993)
Cod	0°C	MAP[b]	16–20	Dalgaard et al. (1993)
Cod	–2°C	air	26	Merritt (1969)
Cod	–2.5°C	MAP	36	Einarsson (1992)
Caught in warm waters	0°C	air	18–35	Connell and Howgate (1986)
Tilapia	4°C	air	9	Reddy et al. (1994)
Tilapia	4°C	MAP	> 25	Reddy et al. (1994)
Large halibut, tuna, and similar fish	0°C	air	21–22	Connell and Howgate (1986)
Salmon	0°C	VP	19	Donald and Gibson (1992)
Salmon	15°C	VP	3	Donald and Gibson (1992)
Dark-fleshed, small fish				
Low fat	0°C	air	6–9	Connell and Howgate (1986)
High fat with large amounts of feed	0°C	air	4–6	Connell and Howgate (1986)
Shellfish, temperate and cold waters	0°C	air	6–10	Connell and Howgate (1986)
Pink shrimps	0°C	air	3–4	Barnett et al. (1978)
Pink shrimps	–1°C	air	4–5	Barnett et al. (1978)
Shellfish in warm waters	0°C	air	8–12	Connell and Howgate (1986)
Shrimps	0°C	air	8	Fatima et al. (1988)
Shrimps	–3°C	air	16	Fatima et al. (1988)
Lightly preserved				
Cold-smoked salmon	2°C	VP	50–60	Cortesi et al. (1997)
	5°C	VP	20–56	Truelstrup Hansen (1995)
	10°C	VP	~ 20	Cortesi et al. (1997)
Hot smoked cod (lightly salted)	0.5°C	air	24–26	Tinker et al. (1985)
	12°C	air	17–18	Tinker et al. (1985)
Hot smoked mackerel (lightly salted)	0°C	MAP	8	Cann et al. (1983)
	10°C	MAP	5–6	Cann et al. (1983)

continues

Table 8–3 continued

Product	Storage temperature	Packaging	Shelf life (days)	References
Brined roe (lump fish)	5°C	VP	75–90	Basby et al. (1998)
Cooked and brined shrimps	0°C	MAP	230–>308	Dalgaard and Jørgensen (in press)
	25°C	MAP	4–6	Dalgaard and Jørgensen (in press)
Sous vide cooked				
Cod	0°C	VP	28	Gibson (1994)
Cod fillets	3°C	VP	15–21	Ben Embarak (1994)
Fermented				
Som fak, low-salt, from Thailand	4°C	air	18	Østergaard et al. (1998)
Som fak, low-salt, from Thailand	29°C	air	7	Østergaard et al. (1998)
Semipreserved and preserved				
Cod (salted)	0°C	air	500	Gibson (1994)
Sugar salted herring	5°C	brine	>360	Knøchel and Huss (1984a)
Frozen				
Lean fish, whole or fillet blocks	–18°C	glazed	120–240	Connell and Howgate (1986)
	–30°C	glazed	240–720	Connell and Howgate (1986)
Fatty fish	–18°C		90–120	Connell and Howgate (1986)
	–30°C		180–360	Connell and Howgate (1986)

(a) Vacuum packed
(b) Modified atmosphere packed

remaining product shelf life and they always correlated closer than log numbers of TVC (Capell et al., 1998; Dalgaard, 1998; Koutsoumanis et al., 1998). It was concluded that microbiological measurements could be used to evaluate remaining shelf life and when such measurements are needed, counts of SSO, whenever known, should be used to supplement or replace classical TVC measurements. As an example, one SSO per gram (log cfu/g = 0) corresponded to 17 days remaining shelf life in whole cod stored aerobically and to 18–19 days in MAP cod fillets. Similar TVC corresponded to 4 and > 7 weeks remaining shelf life, respectively (Figure 8–4). These latter values are unrealistic due to lack of a causative relation between TVC and remaining shelf life. For microbiological evaluation of fresh, frozen, and cold-smoked fish ICMSF (1986) recommended TVC limits of 5×10^5 cfu/g for marginally acceptable samples and 10^7 cfu/g for unacceptable samples. However, these TVC limits seem indeed inappropriate for several seafoods. For example, MAP cod fillets with TVC of 10^6 cfu/g can be of excellent sensory quality and have remaining shelf life of 1–2 weeks at 0°C (Figure 8–4). Also, for lightly preserved products such as cold-smoked salmon TVC can remain at ~10^7 cfu/g for 1–2 weeks, or more, in products of acceptable sensory quality (Truelstrup Hansen, 1995).

For microbiological levels to be compared with critical limits in shelf-life determination, the sampling method needs to be specified, as sampling from fish surfaces, i.e., skin, or from fish flesh substantially influences microbiological counts (Table 8–4). See ICMSF (1986) for recommended sampling procedures.

Many marine microorganisms, including some which can be abundant in seafood, are psychrotolerant and heat labile; they require NaCl for growth. These properties should be respected when TVC is enumerated in seafood. Clearly, pour plating with agar of ~ 45°C, use of standard plate count agar without NaCl, and incubation of plates at 30–37°C must be avoided. ICMSF (1986) recommended incubation of plates at 25°C for TVC enumeration. At DIFRES even this temperature was found to be too high for TVC determination in MAP seafood, where *P. phosphoreum* dominated the spoilage microflora. Spreadplating on prechilled plates of Long and Hammer's agar with 1% NaCl and incubated at 15°C for 5–7 days is a most useful general procedure for TVC determi-

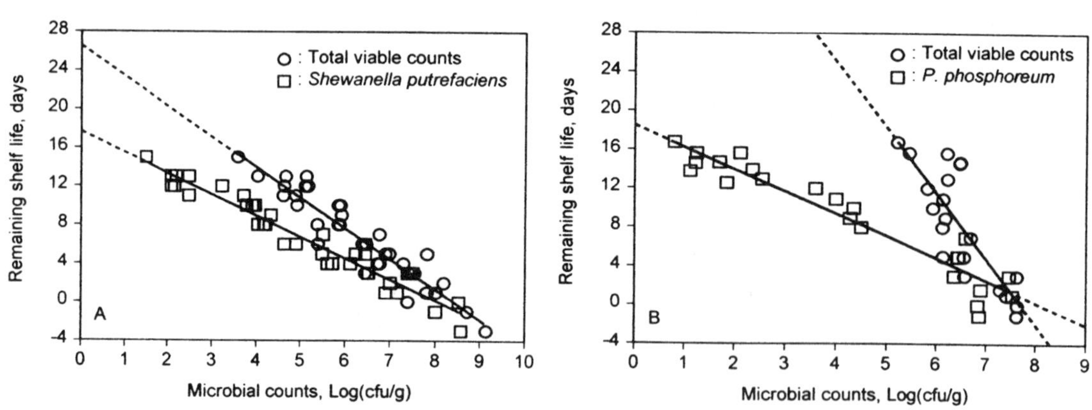

Figure 8–4 Relation between log-numbers of TVC and the SSO, *Shewanella putrefaciens* and *Photobacterium phosphoreum* in cod stored aerobically (A) and in modified atmosphere packaging (B), respectively. *Source:* Data from Jørgensen et al., Microbiological Quality and Shelf Life Prediction of Chilled Fish, *International Journal of Food Microbiology,* Vol. 6, pp. 295–307, © 1988 and Dalgaard et al., Importance of Photobacterium Phosphoreum in Relation to Spoilage of Modified Atmosphere-Packed Fish Products, *Letters of Applied Microbiology,* Vol. 24, pp. 373–378, © 1997b.

Table 8–4 Microbiological counts on surfaces and in flesh at end of shelf life of different fishes

	TVC		*H_2S-producers*		
Fish species	*Surface log(cfu/cm²)*	*Flesh log(cfu/g)*	*Surface log(cfu/cm²)*	*Flesh log(cfu/g)*	*References*
Orange roughy	7–7.5	4.5–5	4.5–5	1.5	Scott et al. (1986)
Smooth oreo dory	7.5–8	4.5	6.5–7	3.5–4	Scott et al. (1992)
Hoki	~8	4.5	6.5	~3	Ryder et al. (1993)

nation in fresh and lightly preserved seafood (Castell et al., 1948; ICMSF, 1986; van Spreekens, 1974; Dalgaard et al., 1993, 1996, 1997b). Lerke and Farber (1969) used a very rapid direct microscopic counting method for TVC and shelf-life determination of sole stored at 5.0°C. This technique, however, is only useful for TVC levels greater than 10^5 cfu/g.

Media are available for enumeration of several of the seafood SSOs. *B. thermosphacta* can be determined by spreadplating on streptomycin sulphate thallous acetate actidione (STAA) agar (25°C, 2–3d) (Gardner, 1966). *Enterococcus faecalis* is enumerated by Slanetz and Bartley agar (35°–44°C) (Slanetz and Bartley, 1957). Non-aciduric lactic acid bacteria, which are common in seafood, can be determined using sodium nitrite-actidion-polymyxin (NAP) agar with pH 6.7 (25°C, 3d) (Davidson and Cronin, 1973). For enumeration of *P. phosphoreum* in fresh fish a conductance method relying primarily on CO_2 resistance and TMAO reduction is available (Dalgaard et al., 1996). Levels of 10^7 cfu/g can be detected in about 12 hours but up to 50 hours are required for enumeration of < 10 cfu/g, which is usually found in freshly processed fish. *Pseudomonas* spp. can be enumerated on cetrimide-fusidin-cephaloridine (CFC) agar (25°C, 3d) (Mead and Adams, 1977). More rapid enumeration of *Pseudomonas* spp. is possible by using a modified CFC medium and impedance measurements (Salvat et al., 1997). *S. putrefaciens* has been enumerated by pour plating in iron agar (3 d, 20–25°C) (Levin, 1968; Gram et al., 1987). Iron agar is not selective but at least in chilled fresh fish the H_2S-producing bacteria, which form black colonies, are almost exclusively *S. putrefaciens*.

Results from several differential microbiological methods have been found to correlate with numbers of TVC, SSO, or remaining product shelf life. These differential methods typically relied on TMAO reduction with time-to-changes-in-conductance or time-to-dye-reduction being measured (Gibson, 1985; Gibson and Ogden, 1987; Jørgensen et al., 1988; Gram, 1992).

The classical microbiological methods described above are slow, with results being available after 12 hours to 5 days. Modern microbiological methods such as polymerase chain reaction (PCR), oligonucleotide probes, or antibody techniques can provide responses within one working day. However, due to their lack of sensitivity when applied in food, qualitative or semi-quantitative responses, and high cost, these methods are unlikely to be used for practical seafood shelf-life evaluation in the near future.

In some seafood, application of microbiological methods for shelf-life evaluation may also be limited because off-flavors develop at approximately the same time in sterile and in naturally contaminated blocks of fish flesh. This has been observed in bastard trumpeter, yellow-eyed mullet, and snapper, whereas sterile muscle of cod can be chill stored for many weeks without development of off-odors (Herbert et al., 1971; Herbert and Shewan, 1975; Fletcher and Statham, 1988a, 1988b; Fletcher and Hodgson, 1988).

In summary, microbiological methods can be used for determination of remaining shelf life in many fresh seafoods, but further studies are required for identification of SSOs and their spoilage domain, particularly in lightly preserved seafoods. Slow microbiological methods are inappropriate for on-line or at-line measurements in seafood processing. However, specific and sensitive, but not necessarily rapid, microbiological method can be used to determine typical levels of contamination and rates of growth of SSOs as a function of, e.g., storage conditions. Once that type of information has been generated, the primary factor to measure on- or at-line in quality control will be the product temperature.

8.3.2.2 Chemical and biochemical methods

Fields et al. (1968) discussed microbiological metabolic by-products and suggested the following criteria for a chemical indicator of food quality: (1) the compound must be present at low levels or absent in sound foods, (2) with increased spoilage there must be an increase in the amount of the indicator, (3) the compound should make it possible to differentiate low-quality raw material from poor processing conditions, and (4) the indicator should be produced by the dominant spoilage flora. These authors also mentioned that a spoilage index must be as reliable as sensory criteria and should indicate stages of spoilage that cannot be established definitely by sensory assessments. Although suggested for microbiological metabolites only, these criteria are relevant to consider when discussing instrumental methods for shelf-life determination.

Adenine nucleotides. During storage of fresh and lightly preserved seafood, adenosine triphosphate (ATP) is degraded into adenosine diphosphate (ADP), adenosine monophosphate (AMP), inosine monophosphate (IMP), inosine (Ino), and hypoxanthine (Hx). IMP is formed by autolytic enzymes whereas spoilage bacteria contribute to Ino and Hx formation. Hx has a bitter taste, which may be part of off-flavors in stale fish. Saito et al. (1959) suggested the K-value = ([Ino] + [Hx]) × 100/([ATP] + [ADP] + [AMP] + [IMP] + [Ino] + [Hx]) as a measure for fish freshness. In most fish, K-values increase linearly during the first days of chilled storage and therefore can be used as an index of freshness. In Japan a K-value of 20% is used as critical limit for fish to be consumed raw. In some fish species, K-values reach maximal levels before sensory rejection and therefore cannot be used in general as index of spoilage. However, in species like trout and tuna K-values increase linearly until spoilage is noted (Price et al., 1991; Gill, 1995a; Henehan et al., 1998). Both rates of Hx production and Hx concentrations at sensory rejection vary substantially between fish species. Nevertheless, Hx concentration and the Hx-index (= log([Hx] + 5)) have been useful for shelf-life determination in specific seafoods (see, e.g., Burt et al., 1976; Price et al., 1991; Ryder et al., 1993).

As an index of quality, ratios of catabolites like the K-value have been suggested to be less prone to fish-to-fish or species-to-species variability than single compounds but more extensive analyses are required. The K_i-value = ([Ino] + [Hx]) × 100/([IMP] + [Ino] + [Hx]), which is quantitatively similar to the K-value, reduces this problem. For determination of adenine nucleotides in seafood, numerous chromatographic and enzyme-based methods are available. Development of new biosensors for simple, rapid, and cost-efficient determinations has been an active field of research and improved methods for practical use in process control and seafood inspection will most likely become commercially available (Gill, 1992, 1995a; Henehan et al., 1998; Ghosh et al., 1998; Carsol and Mascini, 1998; Park and Kim, 1999; Niu and Lee, 1999).

Volatile amines. Microbial activity in seafood typically results in production of trimethylamine (TMA), if TMAO is present, and in ammonia produced from amino acid metabolism. Activity of autolytic enzymes produces dimethylamine (DMA) from TMAO and ammonia by degradation of nucleotides. Total volatile nitrogen (TVN) and TMA are most useful indices of spoilage in fresh and lightly preserved seafood (see, e.g., Ehrenberg and Shewan, 1955; Hebard et al., 1982; Oehlenschläger, 1992, 1997). TVN includes primarily ammonia, TMA, and DMA. In several studies, fish without off-flavors contained < 2 mg-N TMA/100 g and < 12 mg-N TVN/100 g, fish with some off-flavors but without being spoiled organoleptically contained 2–10 mg-N TMA/100 g and 12–20 mg-N TVN/100 g, and spoiled fish contained > 10 mg-N TMA/100 g and > 20 mg-N TVN/100 g (Banks et al., 1980; Hebard et al., 1982). Codex Alimentarius proposed TMA as a chemical spoilage index of marine fishes and the European Commission (Council Regulation No. 95/149/EEC) specified TVN to be used if sensory evaluation indicates doubt about fish freshness. Critical limits from 25 to 35 mg-N TVN/100 g were proposed for different fish species. Nevertheless, it is uncertain if volatile amines alone cause off-flavors (Davis and Gill, 1936; Stansby, 1962). In fact, sugar-salted herring may contain approximately 75 mg-N TVN/100 g and about 35 mg-N TMA/100 g without appearing spoiled organoleptically (Knøchel and Huss, 1984a, 1984b). This, however, can be due to a product pH of less than 6, as the amine and ammonia-like odor of TMA increases substantially at pH values above 6.7 (Castell and Triggs, 1955). It is also worth noting that the TMA concentration at the point of sensory rejection of cod depends on storage conditions. Under MAP storage a higher TMA concentration is found at spoilage because the SSOs under these conditions, *P. phosphoreum*, do not produce sulfur compound in the same way as *S. putrefaciens,* which is the SSO responsible for spoilage of aerobically stored cod (Dalgaard, 1995b).

In fish containing TMAO, production of TMA will be proportional to growth of the SSO as indicated by the dotted line in Figure 8–1. Little TMA is produced until SSOs reach high numbers and this compound can be used only as an index of spoilage and not as an index of freshness. A TMA index (= log(1 + mg-N TMA/100 g)) and a similar TVN index were found to provide a straight line relation between amine levels and seafood storage time, but this does not improve the value of amines as a freshness indicator (Ehrenberg and Shewan, 1955; Burt et al., 1976). Very little ammonia is produced during storage of chilled whitefish, but due to higher free amino acid content much more ammonia can be formed in chilled cephalopods like squid, in crustaceans, and in some dark-fleshed fish like herring (Gill, 1995b). Clearly, the relative value of TMA and TVN as indices of spoilage is species dependent. Formation of DMA from TMAO is due to the indigenous enzyme TMAO dimethylase. It is a slow process primarily of importance in frozen fish with very long shelf life (Table 8–3). For reviews of the numerous methods available for determination of volatile amines in seafood see Gill, 1992 and Oehlenschläger, 1997.

Biogenic amines. Many seafood spoilage bacteria produce one or more of the biogenic amines, e.g., agmatine, cadaverine, histamine, putrescine, spermidine, spermine, and tyramine. Production of biogenic amines in seafood is species dependent, in part due to large variations in concentrations of the corresponding free amino acid substrates. Mietz and Kamas (1977) suggested a quality index (= ppm histamine + ppm putrescine + ppm cadaverine)/(1 + ppm spermidine + ppm spermine) for canned tuna. End of shelf life corresponded to an index value of 10. This index also allowed a high percentage of rockfish fillets, lobster tails, salmon steaks, and shrimps to be graded in agreement with sensory evaluation (Mietz and Karmas, 1978). More recently, Veciana-Nogués et al. (1997) suggested a different biogenic amine index (BAI) (= ppm histamine + ppm cadaverine + ppm tyrosine + ppm putrescine) which they found more valuable for grading tuna. End of shelf life was defined as a BAI value of 50 ppm. Biogenic amines are heat stable and therefore, together with organic acids, are appropriate for evaluation of raw material used in canned seafood (Hillig, 1956; Gill, 1995b). Another quality index relying on biogenic amines and pH has been suggested for cold-smoked salmon (Jørgensen and Dalgaard, 1999). Values of this multiple compound quality index, developed by partial least squares regression, corresponded to the percentage of samples rejected by a sensory panel. Various chromatographic methods are available for determination of biogenic amines and different biosensors responding to individual compounds or groups of biogenic amines in seafood have been developed (Chemnitius et al., 1992; Gill, 1995b; Male et al., 1996; Draisci et al., 1998).

Volatile sulfur compounds. Volatile sulfur compounds have very unpleasant odors and they are obvious chemical indices of spoilage. Some early studies found hydrogen sulfide in spoiled seafood (Farber, 1965) and Herbert et al. (1971) detected around 150 ppb hydrogen sulfide in cod at sensory rejection. However, quantification of volatile sulfides has been laborious and most studies reported qualitative data only. With recent developments in gas sensors and electronic noses (see Strachan and Nickolson, 1992 and the discussion below) volatile sulfides show promise as indicators of spoilage in various seafoods. However, formation of volatile sulfur compounds like other microbial metabolites depends on processing and storage conditions even for a particular fish species. As an example, *S. putrefaciens* produced volatile sulfides in aerobically stored cod. But hydrogen sulfide, methanthiol, dimethylsulfide, dimethyldisulfide, and organic acids, most likely responsible for the extremely unpleasant off-odors determined in sous vide cooked and chill stored cod fillets, were produced by *Clostridium sardiniense.* Nevertheless, volatile sulfides were not produced in fresh modified atmosphere packaging (MAP) cod fillets spoiled by *P. phosphoreum* (Dalgaard, 1995b; Ben Embarak and Dalgaard, unpublished).

Other compounds. Organic acids, e.g., acetic, butyric, formic, pivaleric, propionic, and succinic acid, can develop during storage of different fresh and lightly preserved seafoods. Organic acids were suggested together with histamine for the evaluation in spoilage of tuna. However, reliable critical limits for individual or for combinations of organic acids corresponding to end of shelf life remain to be determined and validated for different seafoods (Hillig, 1956; Farber, 1965; Truelstrup Hansen, 1995; Dalgaard et al., 1998; Koutsoumanis and Nychas, 1999).

On the basis of almost 40 volatile compounds, determined by headspace gas chromatography, multivariate statistical methods allowed grading of Pacific pink salmon and packed Pacific ocean perch into three categories determined by sensory analyses (Girard and Nakai, 1994; Shamaila et al., 1995). These experiments are most interesting but further studies are needed for more exact shelf-life determination on the basis of such multiple compound analyses.

Electronic noses. Gas sensor array systems also known as "electronic noses" respond to groups of volatile compounds and may be used for evaluation of food spoilage (Ólafsdóttir et al., 1997a, 1997b; Blixt and Borch, 1999; Hurst, 1999). These measurements require little sample preparation and time of analysis is relatively short, approximately 10–90 minutes. Different gas sensors, including electrochemical, metal oxide, and organic polymers, are available. In seafood at constant storage temperatures, responses of individual electrochemical sensors correlated closely with sensory QIM scores and TVN values (Ólafsdóttir et al., 1997a, 1997b, 1998b). With salmon fillets, different shrimp species, and tuna, multivariate statistical analysis showed close correlation between responses from 12 conducting polymer sensors and sensory grades, TVN values, and levels of microorganisms (Luzuriaga and Balaban, 1999a, 1999b; Newman et al., 1999). However, responses from six metal oxide sensors when analyzed by an artificial neural network provided only 65% correct allocation of whiting samples into five grades as determined by sensory evaluations (Smelt et al., 1998). Winquist et al. (1998) analyzed volatile compounds produced in cooked shrimps and cod roe by using 10 metal oxide semiconductor sensors. Samples were stored at 4°C with or without pre-heating at room temperature for 3 hours. Sensor responses changed with storage time but surprisingly no clear difference between the two treatments was observed. Stability of correlation between sensory data and electronic nose responses represents a problem for practical application of gas sensors in seafood shelf-life evaluation. Further studies of sensors, calibration schemes, and statistical data evaluation procedures are required for optimal application of electronic noses in seafood shelf-life evaluation and to determine whether shelf-life prediction, for example at fluctuating temperatures, will be possible.

Visual spoilage indicators. Retail packaging of seafood under vacuum or in modified atmosphere has increased in Europe during the 1990s. The opportunities for consumers to carry out classical marketplace evaluation by smell and touch are limited in such products and visual indicators of spoilage compounds within packs could be useful. FreshTag™ (Cox Recorders, USA) is an example of an adhesive color indicator for detection of volatile amines in packed seafood. However, as discussed above, various other compounds can indicate seafood spoilage. Therefore, it seems likely that technology of gas phase biosensors (Bárzana, 1995; Hong and Park, 1999) will be used to develop other visual indicators for detection of volatile compounds, e.g., sulfur compounds, acids, alchohols, aldehydes, ketones, etc. Clearly, successful application of visual indicators of this type is limited in the same way as other indices of spoilage and it must be confirmed that the visual indicator can determine spoilage for normal variation in product composition and under abuse storage conditions.

8.3.2.3 Other instrumental methods

Measurements of dielectric properties have been used for almost 40 years for quality grading and remaining shelf-life determination of various fishes (Oehlenschläger and Nesvadba, 1998). The Intellectron Fishtester, the Torrymeter, and the RT-Freshtester represent instruments with increasing degrees of sophistication. Readings of all instruments reflect dielectrical properties of fish, and they decrease with storage time, almost following a straight line. Based on these rapid and non-destructive measurements the RT-Freshtester allows automatic grading of 60–70 fish samples per minute. Electrical properties of fish are not directly responsible for sensory spoilage. It is therefore to be expected that numerous factors influence the relation between such measurements and seafood spoilage. In fact, these instruments need calibration depending on season and fish handling, and they are not useful for grading of frozen/thawed fish, partially frozen, i.e., superchilled fish, fish chilled in refrigerated seawater, and fish fillet and they are less useful for fatty fish. These problems and a high cost have very much limited the practical use of these instruments.

Near infrared (NIR) diffuse reflectance is another technology where measurements are rapid and thereby has potential for on-line quality grading. Sigernes et al. (1998) found a correlation coefficient of 0.95 between storage time of cod fillets in ice and near infrared diffuse reflectance (400–1100 nm) data. However, in a more extensive study a similar approach provided a very low correlation coefficient of only 0.62 between NIR data (1000–2222 nm) and sensory (QIM) scores of thawed cod (Jørgensen and Jensen, 1997). It remains uncertain if the range of applicability of this technology with respect to fish species, processing, and storage conditions will allow it to be used in practice.

Very rapid instrumental methods as discussed above may be used for freshness grading as part of on-line process control. However, it should be kept in mind that fish freshness can also be determined by using a traceability system with recording of product temperature from the moment of catch. This alternative will in many cases be easier to manage, more cost-efficient, and certainly more reliable for fish species where the normal shelf life is known, for example, for storage in ice.

8.4 SHELF-LIFE PREDICTION

Transportation of seafood and raw fish increases globally and this results in the need for methods to evaluate and predict the effects of storage and distribution conditions. Quality attributes and remaining shelf life can be determined, for example, after processing. However, unless changes in the measured attributes can be predicted as a function of storage conditions, the assessment will be of little value for products as experienced by the end users.

The short and variable shelf life of many types of seafood makes it particularly important to predict effects of product characteristics and storage conditions. Shelf life of seafood has been predicted by (1) kinetic models for growth of SSOs at different environmental conditions, (2) empirical models for relative rates of spoilage, and (3) simple models for linear relations between indices of spoilage and remaining product shelf life at different temperatures as discussed below. The effect of storage times and temperatures has been determined to be cumulative and this has allowed various predictive models to be used for time–temperature integration (Charm et al., 1972; McMeekin et al., 1993; Dalgaard et al., 1997a; Rönnow et al., 1999).

8.4.1 Kinetic models for growth of SSO

Mathematical models for the effect of temperature, atmosphere, and water activity on growth of several SSOs are available. These kinetic models allow growth of the SSOs and the remaining product shelf life to be predicted from the point of catch or harvest through processing, storage, and distribution. However, shelf-life prediction by kinetic models requires some information: (1) the SSOs and their spoilage domain in the specific product, (2) the initial numbers of the SSOs, (3) a relevant spoilage criterion such as the level of SSOs at sensory rejection, and (4) recording of relevant conditions, e.g., temperature of storage (Dalgaard, 1997).

Growth of SSOs in seafood is usually exponential and only limited when high cell concentrations are reached (Figure 8–1). This growth pattern allowed a simple square root model (equation 8.1) to predict the influence of temperature on growth of *S. putrefaciens* and on shelf life of chilled and aerobically stored fresh fish (Ratkowsky et al., 1982; Dalgaard, 1993, 1999).

$$\sqrt{\mu_{max}} = b(T - T_{min}) \qquad (8.1)$$

In equation (8.1) μ_{max} is the maximum specific growth rate, T is temperature in °C, and T_{min} the theoretical minimum growth temperature. A more extensive model including the effect of temperature and carbon dioxide concentration is available for *P. phosphoreum,* which is the SSO of several fresh MAP seafoods (Dalgaard et al., 1997a, 1997b). Product validation studies carried out during summer and winter showed this model to accurately predict both growth of the SSO and shelf life of MAP cod fillets (Figure 8–5).

Neumeyer et al. (1997a) developed a model for the effect of temperature and water activity on growth of psychrotrophic *Pseudomonas* spp. of dairy origin. Validation studies showed their model to predict accurately the growth of naturally occurring pseudomonas in fish mince (Neumeyer et al., 1997b) and in several species of fresh fish (Dalgaard, 1999). Recently, another model was suggested for growth of *Pseudomonas* spp. and for shelf-life prediction of Mediterranean sea bream at 0–15°C (Koutsoumanis et al., 1999). Of the other SSOs indicated in Table 8–1 growth models have been developed for *B. thermosphacta* (McClure et al., 1993) and lactic acid bacteria (Muermans et al., 1993; Devlieghere et al., 1999) but these models have not yet been validated in seafood.

Kinetic models that predict both growth of SSO and shelf life accurately, i.e., within ± 25%, can be developed on the basis of laboratory experiments using liquid media. This, however, requires careful selection of SSO, media, growth systems, and experimental protocols (McMeekin et al., 1993; Dalgaard, 1997; Dalgaard et al., 1997a). Nevertheless, product validation studies in seafood are always needed to confirm the performance of predictive models. Kinetic models can be evaluated by using the bias factor, which determines the average deviation between observed and predicted responses (Ross, 1996). For users of predictive models this index of performance is convenient. Before using a model they just need to verify that the bias factor values as determined in relevant validation studies are within 0.75–1.25.

8.4.2 Empirical models for relative rates of spoilage

Temperature is the single most important factor influencing shelf life of fresh and lightly preserved seafood. Spoilage reactions in seafood may change at abuse temperatures, e.g., >10–15°C. Furthermore, shelf life of different types of fresh and lightly preserved seafood varies

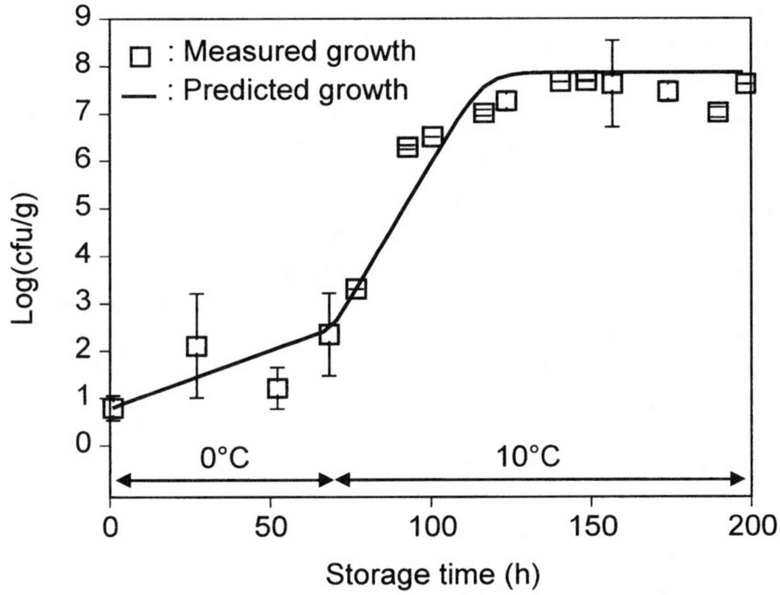

Figure 8–5 Observed and predicted growth of *P. phosphoreum* in modified atmosphere packed cod fillets stored at varying temperatures. *Source:* Reprinted from *International Journal of Food Microbiology,* Vol. 38, P. Dalgaard et al., Application of an Iterative Approach for Development of a Microbial Model Predicting the Shelf Life of Packed Fish, p. 176, Copyright (1997), with permission from Elsevier Science.

substantially (Table 8–3). Despite the complex and dynamic nature of seafood spoilage, it has been possible to develop simple and entirely empirical relative rate of spoilage models to predict shelf life at different temperatures. Storage of fresh fish in ice is common practice and Nixon (1971) defined relative rates of spoilage (RRS) as the keeping time at 0°C divided by the keeping time at T(°C). The effect of temperature on RRS was later shown to be similar for different types of seafood, as discussed below. Mathematical RRS models are convenient, as they allow shelf life to be predicted at different storage temperatures if shelf life has been determined only at a single known temperature. RRS models are easy to use, and require no knowledge about spoilage reactions. Spencer and Baines (1964) suggested the first RRS temperature model, RRS = 1 + c × T, for shelf life of cod stored between −1°C and +25°C. In this straight line model T is temperature (°C) and c a constant. Olley and Ratkowsky (1973) searched a universal temperature model for RRS of food and developed the OR-curve, which differed both from a straight line and from the classical Arrhenius model for the effect of temperature. On the basis of a microbiological growth model equation (8.1) and a T_{min}-value of −10°C, Ratkowsky et al. (1982) developed a square root spoilage model for the effect of temperature on RRS, equation (8.2). RRS-values predicted by this model correspond closely to the OR-curve between 0°C and 15°C.

$$\sqrt{RSS} = \frac{b(T - T_{min})}{b(0 - T_{min})} = 1 + 0.1 \times T \quad (8.2)$$

In equation (8.2) T is storage temperature (°C). Average RRS values of different fresh seafood from cold and temperate waters, including superchilled and modified atmosphere packed products, fit the square root spoilage model between −3°C and +15°C (Dalgaard and Huss, 1997). Superchilling can extend shelf life of seafood substantially (Table 8–3), and equation

(8.2) can be used to predict this effect between 0°C and –3°C. As an example, shelf life of a product that keeps 14 days at 0°C is predicted to be 22 days at –2°C.

Fresh tropical fish keep longer at 0°C than fresh fish from cold and temperate waters, and the effect of temperature on shelf life is also different for the two types of products (Table 8–3). Dalgaard and Huss (1997) evaluated this situation with fish from fresh, brackish, and marine waters and suggested an exponential tropical spoilage model equation (8.3). This exponential model differs from the square root spoilage models (Figure 8–6) and corresponds to an apparent activation energy of 80 kJ mol^{-1}. RRS for temperate and tropical fish differ primarily at storage temperatures > 20°C. To evaluate effects of delayed or no icing on tropical fish typically caught at ~25°C, the difference between equations (8.2) and (8.3) can be of considerable practical importance.

$$\mathrm{Ln(RSS)} = 0.12 \times \mathrm{T} \qquad (8.3)$$

With lightly preserved seafood, only few data are available for the effect of temperature on RRS of different types of products. Dalgaard and Jørgensen (in press) studied cooked and brined MAP shrimps and found the effect of temperature on RRS more pronounced than previously observed with other seafoods in equation (8.4) (Figure 8–6). The slope parameter of 0.15 in the exponential RRS model in equation (8.4) corresponds to an apparent activation energy of 103 kJ mol^{-1}. The reference temperature of 0°C applied to calculation of RRS in fresh fish may be inappropriate for lightly preserved products and a different reference temperature, e.g., 5°C can be used in equation (8.4).

$$\frac{\text{shelf life at } T_{ref}}{\text{shelf life at } T} = \exp[0.15 \times (\mathrm{T} - \mathrm{T}_{ref})] \qquad (8.4)$$

With other lightly preserved seafood, such as hot smoked cod and hot smoked mackerel, the effect of temperature on shelf life is very modest and corresponds to an apparent activation energy of only ~20 kJ mol^{-1} (Figure 8–6; Cann et al., 1983; Tinker et al., 1985). Clearly, temperature does not have a universal effect on shelf life

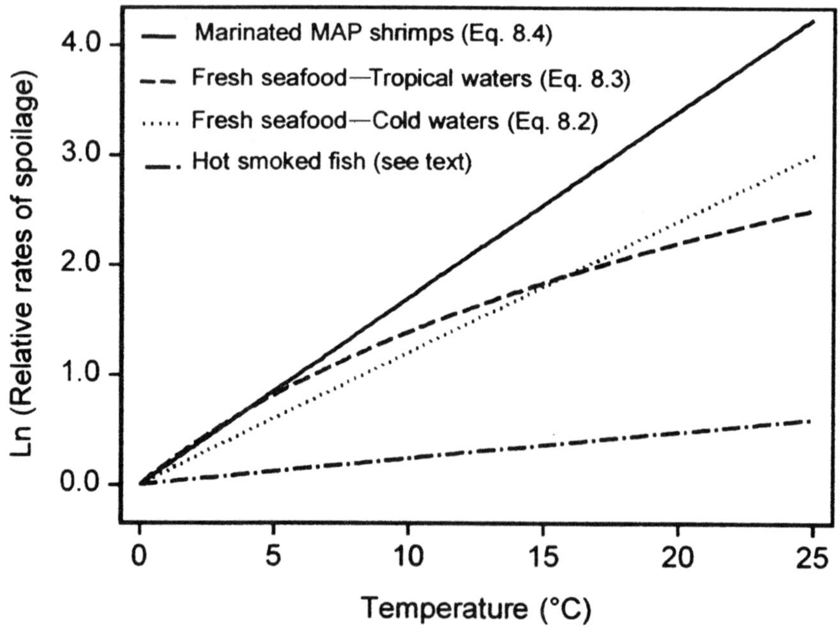

Figure 8–6 Predicted relative rates of spoilage (RRS) for different seafoods stored between 0°C and 25°C.

of different seafoods (Figure 8–6), and this must be taken into account when RRS models are used for shelf-life prediction.

8.4.3 Linear models for indices of spoilage

Spencer and Baines (1964) predicted remaining shelf life of cod on the basis of different indices of quality used in combination with an RRS model. During storage of cod, Torry scheme scores for raw odor and cooked flavor, TMA index values, and TVC values were determined and used for calculation of linear rates of spoilage. Shelf lives at different temperatures were then predicted from differences between initial and final values of the indices of quality and from RRS. Later, Branch and Vail (1985) used demerit/QIM points as index of quality and predicted shelf life at different temperatures on the basis of rates of spoilage determined by the square root spoilage model equation (8.2). The combination of the sensory QIM method and RRS models seemed promising, but as already mentioned in section 8.3.1, further evaluation of QIM responses at different temperatures is needed before this approach can be used with different seafoods. Gibson (1985) used the Spencer and Baines approach to predict shelf life of vacuum packed or modified atmosphere packed cod fillets stored at 0, 5, and 10°C. Quality index values were determined by a differential microbiological conductance method. This method determines the time required for microorganisms in homogenized fish samples to reach high numbers and change conductance in a TMAO containing medium. The rates of change in these values were used together with the square root model equation (8.2) for shelf-life prediction. Recently, the linear relation between log(cfu/g) of the SSO, *P. phosphoreum*, and remaining shelf life of MAP cod fillets was combined with an RRS model (Figure 8–7; equation (8.5); Dalgaard, 1998). Again, the Spencer and Baines approach allowed the development of a very simple model for prediction

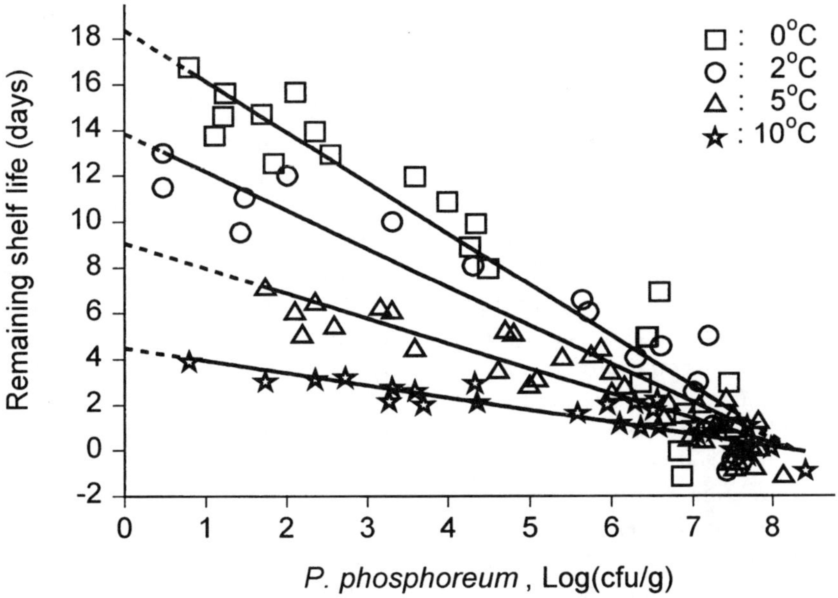

Figure 8–7 Remaining shelf life of MAP (40–60% CO_2) cod fillets at different temperatures. Reprinted from Dalgaard, 1998, page 170. *Source:* P. Dalgaard, Phobacterium Phosphoreum—a Microbial Parameter for Prediction of Remaining Shelf Life in MAP Cod Fillets. In *Methods to Determine the Freshness of Fish in Research Industry,* Olafsdottir et al (eds.). Copyright © 1998 IIF/IIR. All rights reserved.

of shelf life on the basis of both initial product "quality" and storage temperature.

$$\text{remaining shelf life (days)} = \frac{2.22 \times [8.27 - \log(P.\ phosphoreum/g)]}{\exp(0.14 \times T)} \quad (8.5)$$

In equation (8.5) T is storage temperature (°C). The models discussed above for shelf-life prediction of various fresh and lightly preserved seafoods can be used, e.g., to evaluate of the effect of temperature abuse. For this and other applications, user-friendly softwares substantially increase the practical usefulness of the models as discussed in the following section.

8.4.4 Application software and time–temperature integration

Mathematical models for shelf-life prediction can be included in computer software, and this facilitates their practical use in several ways. Application software allows users without detailed mathematical knowledge to apply models and thereby obtain information from the usually large amount of data collected during model development. More sophisticated application softwares allow models to be used for evaluation of fluctuating product and storage conditions. This is particularly important where the software including relevant shelf-life models allows the effect of actual product temperature profiles, recorded by data loggers, to be evaluated by electronic time–temperature integration (TTI).

Nixon suggested and patented a temperature function integrator (Nixon, 1971, 1977). Temperatures of fish in holds on board vessels were recorded and evaluated by a sigmoidal RRS curve corresponding to the OR-curve. The integrated effect of time and temperature on shelf life was then transferred by wireless data communication, e.g., to a receiver on land. Later, different time–temperature integration devices were developed on the basis of the square root spoilage model equation (8.2). McMeekin et al. (1993) presented an excellent review of these devices and on time–temperature integration in general. More recently, a computer software with models for growth of SSOs from seafood has been developed (Dalgaard, 1999). The Seafood Spoilage Predictor (SSP) software was developed specifically to predict shelf life of fish and fishery products at constant and fluctuating temperatures. SSP version 1 included two RRS models (see equations (8.2) and (8.3)) and kinetic models for growth of *P. phosphoreum* and *S. putrefaciens*. This software can evaluate simple temperature profiles entered manually into the software as well as product temperature profiles recorded by different temperature loggers. Remaining shelf life at 0, 5, and 10°C and microbial growth are predicted by the software (Figure 8–8). SSP was developed at the Danish Institute for Fisheries Research (DIFRES) in Lyngby, and it can be downloaded free of charge from the homepage of the microbiology group at DIFRES (http://www.dfu.min.dk/micro/ssp/).

The Food Spoilage Predictor (FSP) software developed at University of Tasmania includes a model for psychrotrophic pseudomonas and allows shelf life of some seafoods to be predicted (Table 8–1). FSP was developed to evaluate product temperature profiles recorded by Gemini Data Loggers like Tinytag®. FSP is commercially available. Food MicroModel (Anon., 1997) includes growth models for *B. thermosphacta* and *Lactobacillus plantarum* but the ability of these models to predict shelf life of seafood remains to be determined.

Time–temperature integration (TTI) tags are another type of device for evaluating the influence of temperature on shelf life. TTI tags are typically simple, inexpensive devices, showing an easily measurable change that reflects their temperature history (Taoukis et al., 1991). TTI tags can be developed as adhesive labels to be placed in contact with individually or bulk packed food. Storage temperature influences shelf life of different fresh and lightly preserved seafoods very differently (Figure 8–6). Selection of TTI tags with appropriate temperature kinetics is therefore of utmost importance for successful application of this technology. It has been suggested that TTI tags be selected on the basis of values for temperature characteristics in spoilage models, i.e., activation energies (E_a) and T_{min}-values. However, no single temperature model is appropriate for different seafoods and

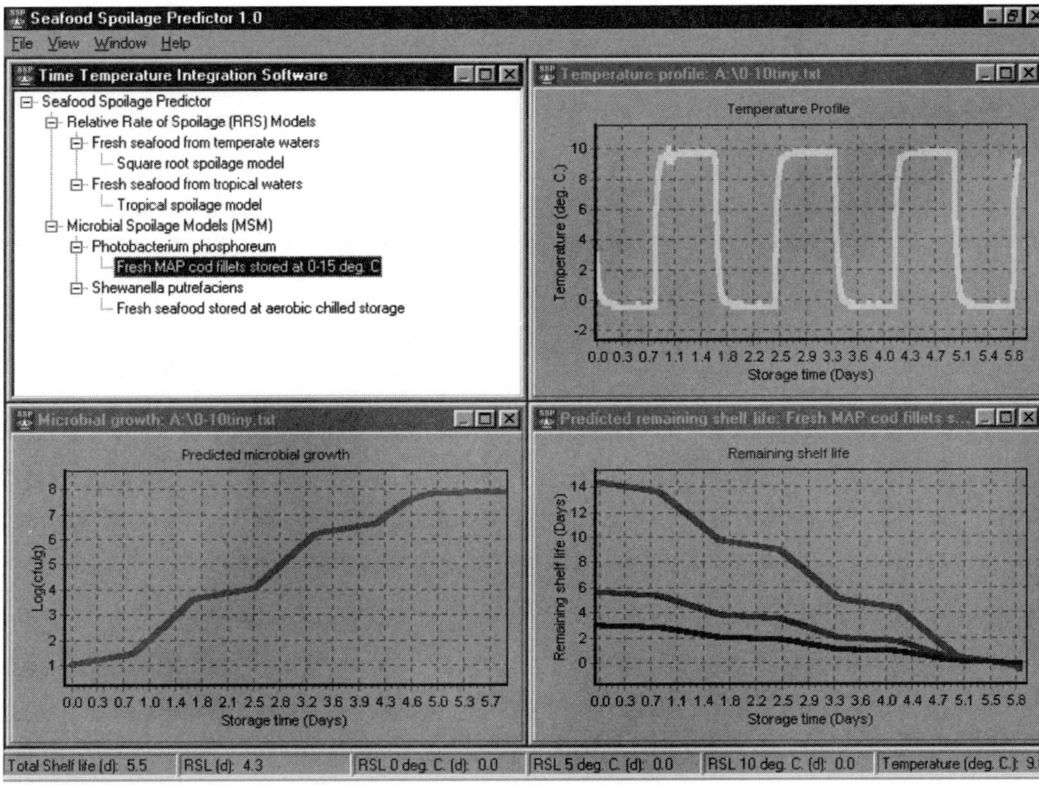

Figure 8–8 Output from the Seafood Spoilage Predictor (SSP) software showing the effect of a temperature profile, on growth of *Photobacterium phosphoreum,* and on remaining shelf life of modified atmosphere packed cod fillets at 0°C (top curve), 5°C (middle curve) and 10°C (bottom curve). The SSP software is available free of charge at http://www.dfu.min.dk/micro/ssp/.

TTI tags. Comparison of curves for relative rates of tag responses and RRS, therefore, is required in order to select TTI tags with the correct characteristics (Taoukis and Labuza, 1989; McMeekin et al., 1993; Dalgaard, 1996).

Temperature characteristics of TTI tags to some extent reflect the reactions used for construction of the tags. MonitorMark™ from 3M™ with E_a of 33–50 kJ mol^{-1} relies on diffusion reactions. These tags may be useful for some lightly preserved seafood, but a high price limits their application in this field. Fresh-Scan® and Fresh-Check® from Life Lines™ are based on polymerization reactions. Their activation energy of 75–113 kJ mol^{-1} is appropriate for many types of seafood, and the linear decrease in reflectance of these tags was shown to correlate well with sensory properties of both fresh and irradiated cod fillets (Tinker et al., 1985, 1987). VITSAB® (Visual Indicator Tag Systems AB) produce tags based on enzymatic hydrolysis of a lipid substrate. VITSAB tags are available with a wide range of activation energies. They may be useful for many fresh and lightly preserved seafoods except products with very low activation energies, e.g., some hot smoked seafoods and for products with very long shelf life, e.g., cooked and brined MAP shrimps (Table 8–3). VITSAB tags have been developed with one or three "windows" to indicate end of shelf life or different degrees of freshness during storage and distribution (Rönnow et al., 1999).

Application of mathematical shelf-life models and TTI tags is increasing in the fishery sector although these techniques are still not frequently used. However, there are several reasons to believe that their application will increase substantially in near future. The price of loggers for collection of temperature data is dropping and at the same time loggers have become more powerful and user friendly, and wireless technology is becoming affordable for use in seafood processing. Furthermore, focus on traceability in food manufacturing is increasing (Moe, 1998). This is primarily due to a series of issues related to food safety, but establishment of systems for traceability of products and data in the seafood supply chain will very much increase the value of mathematical spoilage models. With product temperature recorded from the moment of catch and throughout the chill chain, shelf-life models can be used to determine remaining shelf life for a particular lot at any time. Combined use of traceability systems and mathematical shelf-life models will most likely reduce the need for product evaluation by sensory and instrumental methods in the future.

8.5 EXTENSION OF SHELF LIFE

This section briefly discusses shelf-life extension of fresh and lightly preserved seafood on the basis of the specific spoilage organism concept (Figures 8–1 and 8–2). For reviews on preservation methods and shelf-life extension of seafood see Pedrosa-Menabrito and Regenstein (1989), Russell and Gould (1991), Haard (1992) and Ashie et al. (1996).

The SSO concept indicates a single group of microorganisms is responsible for product spoilage. Consequently, by reducing the growth of this particular group of microorganisms shelf life should increase. In fact, by reducing growth of the SSO in MAP cod fillets shelf life was extended proportionally (Figure 8–9). The low level of EDTA used in that study had no effect

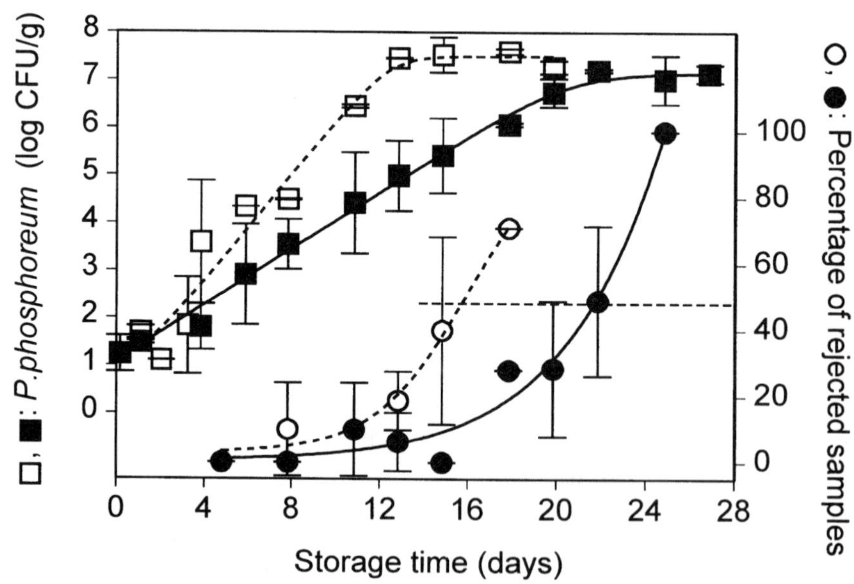

Figure 8–9 Inhibition of *Photobacterium phosphoreum* and corresponding extension of shelf life for modified atmosphere packed cod fillets at 0°C. Open and solid symbols indicate addition of 0 ppm and 500 ppm $Na_2CaEDTA$, respectively. *Source:* Reproduced from Dalgaard et al., © 1998 with permission from *Journal of Food Protection.* Copyright held by the International Association of Milk Food and Environmental Sanitarians, Inc., Des Moines, Iowa.

on the remaining microflora and it was concluded that any procedure inhibiting the SSO would extend shelf life of that product. Consequently, it should be possible to develop mild preservation procedures on the basis of particular characteristics of different SSOs (Table 8–1).

P. phosphoreum is sensitive to freezing and as another example of preservation targeted on an SSO, Guldager et al. (1998) inactivated this spoilage microorganism by simple frozen storage of cod. Compared to fresh MAP fillets, shelf life of thawed MAP cod fillets stored at 2°C was extended from 11–12 days to more than 20 days and no growth of *P. phosphoreum* was found in the previously frozen products. With aerobically stored fish, reduction of total viable counts, e.g., by strict hygienic production, mild treatments of fish with antibiotics, chlorine, or heat had very little effect on shelf life (Castell et al., 1948; Huss et al., 1974; Vaz-Pires et al., 1994; Huss, 1995). This is in contrast to the considerable effect obtained by inhibiting SSO in MAP cod but the difference is a direct consequence of the SSO concept. In fact, treatments that reduce the total microflora should not necessarily be expected to extend shelf life (Figure 8–1). Targeted inhibition of SSO is a particularly interesting approach to achieve mild preservation. Targeted inhibition of spoilage organisms has been very little studied (Gram, 1991), possibly due to lack of knowledge about SSOs and their spoilage domains in seafood. However, with increased interest in mild and natural preservation methods it is logical to use information about SSOs and their characteristics in this way.

8.6 CONCLUDING REMARKS

The specific spoilage organism (SSO) concept was shown to have considerable potential for practical applications with respect to determination, prediction, and extension of seafood shelf life. Most examples have concerned modified atmosphere packed cod; further studies on identification of SSOs and their spoilage domain in different seafoods are indeed required. At present, SSOs and their spoilage domain as well as changes in quality attributes in general are determined in storage trials with existing products. In the future, it should also be attempted to predict the type of microorganisms or spoilage reaction that eventually will limit shelf life of new products or products stored under new conditions. Based on that type of information it will be possible to use microbiological, chemical/biochemical, and sensory methods much more efficiently in quality assurance and product development.

REFERENCES

Anon. (1985) *An Evaluation of the Role of Microbiological Criteria for Foods and Food Ingredients. Food and Nutrition Board, National Research Council*, National Academic Press, Washington, DC.

Anon. (1994) Review of the state of world marine fishery resources, *Food and Agricultural Organization (FAO) Fish.* Technical Paper No. 335. FAO, Rome, Italy.

Anon. (1996) *FAO Yearbook, Fishery Statistics Capture Production,* Vol 82. FAO, Rome, Italy.

Anon. (1997) *Food MicroModel—User Manual v. 2.5*, Food MicroModel Ltd, Surrey, England.

Ashie, I.N.A., Smith, J.P. and Simpson, B.K. (1996) Spoilage and shelf life extension of fresh fish and shellfish. *Crit. Rev. Food Sci. Nut.*, **36**, 87–121.

Banks, H., Nickelson II, R. and Finne, G. (1980) Shelf life studies on carbon dioxide packed finfish from the Gulf of Mexico. *J. Food Sci.*, **45**, 157–162.

Barnett, H.J., Nelson, R.W., Hunter, P.J. and Groninger, H. (1978) Use of carbon dioxide dissolved in refrigerated brine for the preservation of pink shrimp. *Mar. Fish. Rev.*, **40**, 25–28.

Barret, E.L. and Kwan, H.S. (1985) Bacterial reduction of trimethylamine oxide. *Annu. Rev. Microbiol.*, **39**, 131–149.

Bárzana, E. (1995) Gas phase biosensors. *Adv. Biochem. Eng./Biotechnol.*, **53**, 1–15.

Basby, M., Jeppesen, V.F. and Huss, H.H. (1998) Spoilage of lightly salted lumpfish (*Cyclopterus lumpus*) roe at 5°C. *J. Aquat. Food Prod. Technol.*, **7**, 23–34.

Ben Embarak, P.K. (1994) *Microbial Safety and Spoilage of Sous Vide Fish Products* (*PhD Thesis*). Technological Laboratory, Danish Ministry of Agriculture and Fisheries, Technical University, Lyngby, Denmark.

Blixt, Y. and Borch, E. (1999) Using an electronic nose for determining the spoilage of vacuum-packaged beef. *Int. J. Food Microbiol.*, **46**, 123–134.

Botta, J.R. (1995) *Evaluation of Seafood Freshness Quality*, VCH Publishing Inc., New York, NY.

Branch, A.C. and Vail, A.M.A. (1985) Bringing fish inspection into the computer age. *Food Technol. Aust.*, **37**, 352–355.

Bremner, H.A. (1985). A convenient easy-to-use system for estimating the quality of chilled seafood, in *Fish Processing Bulletin No. 7*, Department of Scientific and Industrial Research, Wellington, New Zealand, pp. 59–70.

Bremner, A.H., Olley, J. and Vail, A.M.A. (1987) Estimating time-temperature effects by a rapid systematic sensory method, in *Seafood Quality Determination* (eds D.E. Kramer and J. Liston). Elsevier, Amsterdam, The Netherlands, pp. 413–435.

Burt, J.R., Gibson, D.M., Jason, A.C. and Sanders, H.R. (1976) Comparison of methods of freshness assessment of wet fish. II. Instrumental and chemical assessments of boxed experimental fish. *J. Fd. Technol.*, **11**, 73–89.

Cann, D.C., Smith, G.L. and Houston, N.C. (1983) *Further Studies on Marine Fish Storage Under Modified Atmosphere Packaging*, Torry Research Station, MAFF, Aberdeen, United Kingdom.

Capell, C., Vaz-Pires, P. and Kirby, R. (1998) Use of counts of hydrogen sulphide producing bacteria to estimate remaining shelf life of fresh fish, in, *Methods to Determine the Freshness of Fish in Research and Industry* (eds Ólafsdóttir et al.). Int. Inst. Refrig. Paris, France, pp. 175–182.

Carsol, M.A. and Mascini, M. (1998) Development of a system with enzyme reactors for the determination of fish freshness. *Talanta,* **47**, 335–342.

Castell, C.H. and Anderson, G.W. (1948) Bacteria associated with spoilage of cod fillets. *J. Fish. Res. Bd. Canada,* **7**, 370–377.

Castell, C.H. and Triggs, R.E. (1955) Spoilage of haddock in the trawlers at sea: The measurement of spoilage and standards of quality. *J. Fish. Res. Bd. Canada,* **12**, 329–341.

Castell, C.H., Anderson, G.W. and Pivnick, H. (1948) Relation of bacterial counts to quality of cod fillets. *J. Fish. Res. Bd. Canada,* **7**, 378–388.

Chai, T., Chen, C., Rosen, A. and Levin, R.E. (1968) Detection and incidence of specific species of spoilage bacteria on fish. *Appl. Microbiol.*, **16**, 1738–1741.

Charm, S.E., Learson, R.J., Ronsivalli, L.J. and Schwartz, M. (1972) Organoleptic technique predicts refrigeration shelf life of fish. *Food Technol.*, **26**, 65–68.

Chemnitius, G., Suzuki, M., Isobe, K., Kimura, J., Karube, I. and Schmid, R.D. (1992) Thin-film polyamine biosensor: substrate specificity and application to fish freshness determination. *Analyt. Chim. Acta,* **263**, 93–100.

Connell, J.J. (1980) *Advances in Fish Science and Technology,* Fishing News Books Ltd, Farnham, England.

Connell, J.J. and Howgate, P. (1971) Consumer evaluation of fresh and frozen fish. Fish inspection and quality control, in *Fish Inspection and Quality Control* (ed R. Kreuzer). FAO, Fishing News (Books) Ltd., London, England, pp. 155–159.

Connell, J.J. and Howgate, P.F. (1986) Fish and fish poducts, in *Quality Control in the Food Industry* (ed S.M. Herschdoerfer). Academic Press, London, England, pp. 347–405.

Connell, J.J. and Shewan, J.M. (1980) Sensory and non-sensory assessment of fish, in *Advances in Fish Science and Technology* (ed J.J. Connell). Fishing News Books Ltd., Farnham, England, pp. 56–65.

Cortesi, M.L., Sarli, T., Santoro, A., Murru, N. and Pepe, T. (1997) Distribution and behaviour of *Listeria monocytogenes* in three lots of naturally-contaminated vacuum-packed smoked salmon stored at 2 and 10°C. *Int. J. Food Microbiol.*, **37**, 209–214.

Council Directive 103/76/EEC of 19 January 1976 laying down common marketing standards for certain fresh or chilled fish. *Official Journal of the European Communities* No. L 20, 28.01.1976, 29–34.

Council Directive 91/493/EEC of 22 July laying down the health conditions for the production and the placing on the market of fishery products. *Official Journal of the European Communities* No. L 268, 24.09.1991, 15.

Dainty, R.H. and Mackey, B.M. (1992) The relationship between the phenotypic properties of bacteria from chill-stored meat and spoilage processes. *J. Appl. Bact.*, **73**, 103S–114S.

Dalgaard, P. (1993) *Evaluation and Prediction of Microbial Fish Spoilage (PhD Thesis)*, Technological Laboratory, Danish Ministry of Fisheries, Lyngby, Denmark.

Dalgaard, P. (1995a) Modelling of microbial activity and prediction of shelf life for packed fresh fish. *Int. J. Food Microbiol.*, **26**, 305–317.

Dalgaard, P. (1995b) Qualitative and quantitative characterization of spoilage bacteria from packed fish. *Int. J. Food Microbiol.*, **26**, 319–333.

Dalgaard, P. (1996) Predictive modelling and time-temperature integration. *Refrig. Sci. Technol Proc.*, **2**, 409–419.

Dalgaard, P. (1997) Predictive microbiological modelling and seafood quality, in *Seafood from Producer to Consumer, Integrated Approach to Quality* (eds J.B. Luten, T. Børresen and J. Oehlenschläger). Elsevier, Amsterdam, The Netherlands, pp. 431–443.

Dalgaard, P. (1998) *Photobacterium phosphoreum*—a microbial parameter for prediction of remaining shelf life in MAP cod fillets, in *Methods to Determine the Freshness of Fish in Research and Industry* (eds Ólafsdóttir et al.). Int. Inst. Refrig., Paris, France, pp. 166–174.

Dalgaard, P. (1999) Modelling of seafood spoilage. *Refrigeration Science and Technology Proceedings,* **EUR 18816**, 143–151.

Dalgaard, P. and Huss, H.H. (1997) Mathematical modelling used for evaluation and prediction of microbial fish spoilage, in *Seafood Safety, Processing and Biotechnology* (eds F. Shahidi, Y. Jones, and D.D. Kitts). Technomic Publishing Company Inc., Lancaster, PA, pp. 73–89.

Dalgaard, P. and Jørgensen, L.V. (in press) Modified atmosphere packed cooked and brined shrimps have shelf life over 7 months at 0°C, but spoil in 4–6 days at 25°C. *Int. J. Food Sci. Technol.*

Dalgaard, P., Gram, L. and Huss, H.H. (1993) Spoilage and shelf life of cod fillets packed in vacuum or modified atmospheres. *Int. J. Food Microbiol.*, **19**, 283–294.

Dalgaard, P., Mejlholm, O. and Huss, H.H. (1996) Conductance method for quantitative determination of *Photobacterium phosphoreum* in fish products. *J. Appl. Bact.*, **81**, 57–64.

Dalgaard, P., Mejlholm, O. and Huss, H.H. (1997a) Application of an iterative approach for development of a microbial model predicting the shelf life of packed fish. *Int. J. Food Microbiol.*, **38**, 169–179.

Dalgaard, P., Mejlholm, O., Huss, H.H. and Christiansen, T.J. (1997b) Importance of *Photobacterium phosphoreum* in relation to spoilage of modified atmosphere-packed fish products. *Lett. Appl. Microbiol.*, **24**, 373–378.

Dalgaard, P., Munoz, L.G. and Mejlholm, O. (1998) Specific inhibition of *Photobacterium phosphoreum* extends the shelf life of modified-atmosphere-packed cod fillets. *J. Food Prot.*, **61**, 1191–1194.

Davidson, C.M. and Cronin, F. (1973) Medium for the selective enumeration of lactic acid bacteria from foods. *Appl. Environ. Microbiol.*, **26**, 439–440.

Davis, W.L. and Gill, E. (1936) Investigations on fishy flavor. *J. Soc. Chem. Ind.*, **55**, 141T–146T.

Devlieghere, F., Van Belle, B. and Debevere, J. (1999) Shelf life of modified atmosphere packed cooked meat products: A predictive model. *Int. J. Food Microbiol.*, **46**, 57–70.

Donald, B. and Gibson, D.M. (1992) Preliminary observations on the flora of fresh vacuum packed salmon steaks, in *Quality Assurance in the Fish Industry* (eds H.H. Huss, M. Jacobsen and J. Liston). Elsevier, Amsterdam, The Netherlands, pp. 107–114.

Draisci, R., Volpe, G., Lucentini, L., Cecillia, A., Federico, R. and Palleschi, G. (1998) Determination of biogenic amines with an electrochemical biosensor and its application to salted anchovies. *Food Chem.*, **62**, 225–232.

Dyer, W.J. and Dyer, F.E. (1950) Amines in fish muscle. VI. Spoilage in freshly cut cod fillets. *J. Fish. Res. Bd. Canada*, **7**, 580–584.

Ehrenberg, A.S.C. and Shewan, J.M. (1953) The objective approach to sensory tests of food. *J. Sci. Food Agric.*, **4**, 482–490.

Ehrenberg, A.S.C. and Shewan, J.M. (1955) Volatile bases and sensory quality-factors in iced whitefish. *J. Sci. Food Agric.*, **6**, 207–217.

Einarsson, H. (1992) Predicting the shelf life of cod (*Gadus morhua*) fillets stored in air and modified atmosphere at temperatures between –4°C and +16°C, in *Quality Assurance in the Fish Industry* (eds H.H. Huss, M. Jacobsen and J. Liston). Elsevier, Amsterdam, The Netherlands, pp. 479–488.

Farber, L. (1965) Freshness tests, in *Fish as Food* (ed G. Borgstrom). Academic Press, New York, NY, pp. 65–126.

Fatima, R., Khan, M.A. and Qadri, R.B. (1988) Shelf life of shrimp (*Penaeus merguiensis*) stored in ice (0°C) and partially frozen (–3°C). *J. Sci. Food Agric.*, **42**, 235–247.

Fields, M.L., Richmond, B.S. and Baldwin, R.E. (1968) Food quality as determined by metabolic by-products of microorganisms. *Adv. Food Res.*, **16**, 161–229.

Fletcher, G.C. and Hodgson, J.A. (1988) Shelf life of sterile snapper (*Chrysophrys auratus*). *J. Food Sci.*, **53**, 1327–1332.

Fletcher, G.C. and Statham, J.A. (1988a) Deterioration of sterile chill-stored and frozen trumpeter fish (*Latridopsis forsteri*). *J. Food Sci.*, **53**, 1336–1339.

Fletcher, G.C. and Statham, J.A. (1988b) Shelf life of yellow-eyed mullet (*Aldrichetta forsteri*) at 4°C. *J. Food Sci.*, **53**, 1030–1035.

Gardner, G.A. (1966) A selective medium for the enumeration of *Microbacterium thermosphactum* in meat and meat products. *J. Appl. Bact.*, **29**, 455–460.

Ghosh, S., Saker, D. and Misra, T.N. (1998) Development of an amperometric enzyme electrode biosensor for fish freshness detection. *Sensors and Actuators B-Chemical*, **53**, 58–62.

Gibson, D.M. (1985) Predicting the shelf life of packed fish from conductance measurements. *J. Appl. Bact.*, **58**, 465–470.

Gibson, D.M. (1994) Preservation technology and shelf life of fish and fish products, in *Shelf Life Evaluation of Foods* (eds C.M.D. Man and A.A. Jones). Blackie Academic and Professional, Glasgow, Scotland, pp. 72–84.

Gibson, D.M. and Ogden, I.D. (1987) Estimating the shelf life of packed fish, in *Seafood Quality Determination* (eds D.E. Kramer and J. Liston). Elsevier, Amsterdam, The Netherlands, pp. 437–451.

Gill, T.A. (1992) Biochemical and chemical indices of seafood quality, in, *Quality Assurance in the Fish Industry* (eds H.H. Huss, M. Jacobsen and J. Liston). Elsevier, Amsterdam, The Netherlands, pp. 377–388.

Gill, T. (1995a) Autolytic changes, in *Quality and Quality Changes in Fresh Fish* (ed H.H. Huss). FAO, Rome, Italy, pp. 39–50.

Gill, T. (1995b) Biochemical and chemical methods, in *Quality and Quality Changes in Fresh Fish* (ed H.H. Huss). FAO, Rome, Italy, pp. 139–148.

Girard, B. and Nakai, S. (1994) Grade classification of canned pink salmon with static headspace volatile patterns. *J. Food Sci.*, **59**, 507–512.

Gould, G.W. (1996) Methods for preservation and extension of shelf life. *Int. J. Food Microbiol.*, **33**, 51–64.

Gram, L. (1991) Inhibition of mesophilic spoilage by *Aeromonas* spp. on fish by salt, potassium sorbate, liquid smoke, and chilling. *J. Food Prot.*, **54**, 436–442.

Gram, L. (1992) Evaluation of the bacteriological quality of seafood, in *Quality Assurance in the Fish Industry* (eds H.H. Huss, M. Jacobsen and J. Liston). Elsevier, Amsterdam, The Netherlands, pp. 269–282.

Gram, L. and Huss, H.H. (1996) Microbial spoilage of fish and fish products. *Int. J. Food Microbiol.*, **33**, 121–137.

Gram, L., Trolle, G. and Huss, H.H. (1987) Detection of specific spoilage bacteria from fish stored at low (0°C) and high (20°C) temperatures. *Int. J. Food Microbiol.*, **4**, 65–72.

Guldager, H.S., Bøknæs, N., Østerberg, C., Nielsen, J. and Dalgaard, P. (1998) Thawed cod fillets spoil less rapidly than unfrozen fillets when stored under modified atmosphere at 2°C. *J. Food Prot.*, **61**, 1129–1136.

Haard, N.F. (1992) Technological aspects of extending prime quality of seafood: A review. *J. Aquat. Food Prod. Technol.*, **1**, 9–27.

Hall, G.M. (1997) *Fish Processing Technology*, Blackie Academic and Professional, Glasgow, Scotland.

Hebard, C.E., Flick, G.J. and Martin, R.E. (1982) Occurrence and significance of trimethylamine oxide and its derivatives in fish and shellfish, in *Chemistry and Biochemistry of Marine Food Products* (eds R.E. Martin, G.J. Flick, C.E. Hebard, and D.R. Ward). AVI Publishing Company, Westport, CT, pp. 149–304.

Henehan, G., Proctor, M., Abu-Ghannam, N., Wills, G., and McLoughlin, J.V. (1998). Adenine nucleotides and their metabolites as determinants of fish freshness, in, *Methods to Determine the Freshness of Fish in Research and Industry* (eds Ólafsdóttir et al.). Int. Inst. Refrig., Paris, France, pp. 266–272.

Herbert, R.A. and Shewan, J.M. (1975) Precursors of the volatile sulfides in spoiling North Sea cod (*Gadus morhua*). *J. Sci. Food Agric.*, **26**, 1195–1202.

Herbert, R.A., Hendrie, M.S., Gibson, D.M. and Shewan, J.M. (1971) Bacteria active in the spoilage of certain sea foods. *J. Appl. Bact.*, **34**, 41–50.

Hillig, F. (1956) Volatile acids, succinic acid, and histamine as indices of decomposition in tuna. *Ass. Off. Agri. Chem.*, **39**, 773–800.

Hong, S.-I. and Park, W.-S. (1999) Development of color indicators for Kimchi packaging. *J. Food Sci.*, **64**, 255–257.

Howgate, P. (1987) Fish inspection and quality control in Europe, in *Seafood Quality Determination* (eds D.E. Kramer and J. Liston). Elsevier, Amsterdam, The Netherlands, pp. 605–613.

Howgate, P., Johnston, A. and Whittle, J. (1992) *Multilingual Guide to EC Freshness Grades for Fishery Products*. Torry Research Station, Aberdeen, Scotland.

Hurst, W.J. (1999) *Electronic Noses and Sensor Array Based Systems. Design and Application*, Technomic Publishing Co. Inc., Lancaster, PA.

Huss, H.H. (1994) *Assurance of Seafood Quality*, Technical Paper No. 334. FAO Fisheries, Rome, Italy.

Huss, H.H. (1995) *Quality and Quality Changes in Fresh Fish,* Technical Paper No. 348. FAO Fisheries, Rome, Italy.

Huss, H.H., Dalgaard, P. and Gram, L. (1997) Microbiology of fish and fish products, in *Seafood from Producer to Consumer, Integrated Approach to Quality* (eds J.B. Luten, T. Børresen and J. Oehlenschläger). Elsevier, Amsterdam, The Netherlands, pp. 413–430.

Huss, H.H., Dalsgaard, D., Hansen, L., Ladefoged, H., Pedersen, A. and Zittan, L. (1974) The influence of hygiene in catch handling on the storage life of iced cod and plaice. *J. Food Technol.*, **9**, 213–221.

Huss, H.H., Jacobsen, M. and Liston, J. (1992) *Quality Assurance in the Fish Industry*, Elsevier, Amsterdam, The Netherlands.

Hyldig, G. and Nielsen, J. (1998) A rapid sensory method for quality management, in *Methods to Determine the Freshness of Fish in Research and Industry* (eds Ólafsdóttir et al.). Int. Inst. Refrig, Paris, France, pp. 297–305.

ICMSF (1986) Sampling plans for fish and shellfish, in *Microorganisms in Foods*, Blackwell Scientific Publications, Oxford, England, pp. 181–196.

Jónsdóttir, S. (1992) Quality index method and TQM-system, in *Quality Issues in the Fish Industry* (eds Ólafsson and Ingphórsson). The Research Liaison Office, University of Iceland, Reykjavik, Iceland, pp. 81–94.

Jørgensen, B.M. and Jensen, H.S. (1997) Can near-infrared spectrometry be used to measure quality attributes in frozen cod? in *Seafood from Producer to Consumer, Integrated Approach to Quality* (eds J.B. Luten, T. Børresen and J. Oehlenschläger). Elsevier, Amsterdam, The Netherlands, pp. 491–496.

Jørgensen, B.R., Gibson, D.M. and Huss, H.H. (1988) Microbiological quality and shelf life prediction of chilled fish. *Int. J. Food Microbiol.*, **6**, 295–307.

Jørgensen, L.V. and Dalgaard, P. (1999) Quality index for vacuum-packed cold-smoked salmon developed by multivariate regression of biogenic amines and pH, in *Food Microbiology and Food Safety into the Next Millennium* (eds A.C.J. Tuijtelaars, R.A. Samson, F.M. Rombouts and S. Notermans). Foundation Food Micro '99, Wageningen, pp. 52–55.

Juhl, H.J. and Poulsen, C.S. (1999) Fresh fish—quality in retail sale (In Danish). *Fødevareindustrien,* **June**, 36–39.

Knøchel, S. and Huss, H.H. (1984a) Ripening and spoilage of sugar salted herring with and without nitrate. I. Microbiological and related chemical changes. *Int. J. Food Microbiol.*, **19**, 203–213.

Knøchel, S. and Huss, H.H. (1984b) Ripening and spoilage of sugar salted herring with and without nitrate. II. Effect of nitrate. *Int. J. Food Microbiol.,* **19**, 215–224.

Koutsoumanis, K. and Nychas, G.J.E. (1999) Chemical and sensory changes associated with microbial flora of Mediterranean boque (*Boops boops*) stored aerobically at 0, 3, 7 and 10°C. *Appl. Environ. Microbiol.,* **65**, 698–706.

Koutsoumanis, K., Stamatiou, P.A. and Nychas, G.J. (1999) Kinetic modelling of microbial fish spoilage, in *Food Microbiology and Food Safety into the Next Millennium* (eds A.C.J. Tuijtelaars, R.A. Samson, F.M. Rombouts and S. Notermans). Foundation Food Micro '99, Wageningen, pp. 923–927.

Koutsoumanis, K., Taoukis, P., Drosinos, E.S. and Nychas, G.J.E. (1998) Lactic acid bacteria and *Brochothrix thermosphacta*—the dominant spoilage microflora of Mediterranean fresh fish stored under modified atmosphere packaging conditions, in *Methods to Determine the Freshness of Fish in Research and Industry* (eds Ólafsdóttir et al.). Int. Inst. Refrig, Paris, France, pp. 158–165.

Kramer, D.E. and Liston, J. (1987) *Seafood Quality Determination,* Elsevier, Amsterdam, The Netherlands.

Kreuzer, R. (1971) *Fish Inspection and Quality Control,* FAO, Fishing News (Books) Limited, London, England.

Larsen, E., Heldbo, J., Misby, C. and Nielsen, J. (1992) Development of a method for quality assessment of fish for human consumption based on sensory evaluation, in *Quality Assurance in the Fish Industry* (eds H.H. Huss, M. Jacobsen, and J. Liston). Elsevier, Amsterdam, The Netherlands, pp. 351–358.

Lerke, P. and Farber, L. (1969) Direct bacterial counts as a rapid freshness test for fish fillets. *Appl. Microbiol.,* **17**, 197–200.

Lerke, P., Adams, R. and Farber, L. (1965) Bacteriology of spoilage of fish muscle: III. Characterization of spoilers. *Appl. Microbiol.,* **13**, 625–630.

Levin, R.E. (1968) Detection and incidence of specific species of spoilage bacteria on fish. I Methodology. *Appl. Microbiol.,* **16**, 1734–1737.

Luten, J. and Martinsdóttir, E. (1998) QIM: A European tool for fish freshness evaluation in the fishery chain, in *Methods to Determine the Freshness of Fish in Research and Industry* (eds Ólafsdóttir et al.). Int. Inst. Refrig., Paris, France, pp. 287–296.

Luten, J.B., Børresen, T. and Oehlenschläger, J. (1997) *Seafood from Producer to Consumer, Integrated Approach to Quality*. Developments in Food Science 38. Elsevier, Amsterdam, The Netherlands.

Luzuriaga, D.A. and Balaban, M.-O. (1999a) Electronic nose evaluation of salmon fillets stored at different temperatures, in *Electronic Noses and Sensor Array Based Systems. Design and Applications* (ed W.J. Hurst). Technomic Publishing Co., Inc., Lancaster, PA, pp. 162–170.

Luzuriaga, D.A. and Balaban, M.O. (1999b) Evaluation of the odor of decomposition in raw and cooked shrimps: Correlation of electronic nose readings, odor sensory evaluation and ammonia levels, in *Electronic Noses and Sensor Array Based Systems. Design and Applications* (ed W.J. Hurst). Technomic Publishing Co., Inc., Lancaster, PA, pp. 177–184.

Male, K.B., Bouvrette, P., Luong, H.T. and Gibbs, B. (1996) Amperometric biosensor for total histamine, putrecine and cadaverine using diamine oxidase. *J. Food Sci.,* **61**, 1012–1016.

Martin, R.E., Collette, R.L. and Slavin, J.W. (1997) *Fish Inspection, Quality Control, and HACCP—A Global Focus*, Technomic Publishing Co. Inc., Lancaster, PA.

Matches, J.R. (1982) Effects of temperature on the decomposition of pacific coast shrimps (*Pandalus jordani*). *J. Food Sci.,* **47**, 1044–1047, 1069.

McClure, P.J., Baranyi, J., Boogard, E., Kelly, T.M. and Roberts, T.A. (1993) A predictive model for the combined effect of pH, sodium chloride and storage temperature on the growth of *Brochotrix thermosphacta. Int. J. Food Microbiol.,* **19**, 161–178.

McMeekin, T.A., Olley, J., Ross, T. and Ratkowsky, D.A. (1993) *Predictive Microbiology: Theory and Application,* Research Studies Press Ltd., Tauntton.

Mead, G.C. and Adams, B.W. (1977) A selective medium for the rapid isolation of *Pseudomonas* associated with poultry meat spoilage. *Br. Poult. Sci.,* **18**, 661–667.

Merritt, J.M. (1969) *Refrigeration on Fishing Vessels*, Fishing News (Books) Ltd., London, England.

Mietz, J.L. and Karmas, E. (1977) Chemical quality index of canned tuna as determined by high-pressure liquid chromatography. *J. Food Sci.,* **42**, 155–158.

Mietz, J.L. and Karmas, E. (1978) Polyamine and histamine content of rockfish, salmon, lobster, and shrimps as an indication of decomposition. *J. AOAC,* **61**, 139–145.

Moe, T. (1998) Perspectives on traceability in food manufacture. *Trends Food Sci. Technol.,* **9**, 211–214.

Muermans, M.L.T., Stekelenburg, F.K., Zwietering, M.H. and Huis in't Veld, J.H.J. (1993) Modelling of the microbiological quality of meat. *Food Control,* **4**(4), 216–221.

Neumeyer, K., Ross, T. and McMeekin, T.A. (1997a) Development of a predictive model to describe the effect of temperature and water activity on the growth of spoilage pseudomonads. *Int. J. Food Microbiol.,* **38**, 45–54.

Neumeyer, K., Ross, T., Thomson, G. and McMeekin, T.A. (1997b) Validation of a model describing the effect of temperature and water activity on the growth of psychrotrophic pseudomonads. *Int. J. Food Microbiol.,* **38**, 55–63.

Newman, D.J., Luzuriaga, D.A. and Balaban, M.O. (1999) Odor and microbiological evaluation of raw tuna: Correlation of sensory and electronic nose data, in *Electronic Noses and Sensor Array Based Systems. Design and Applications* (ed W.J. Hurst). Technomic Publishing Co., Inc., Lancaster, PA, pp. 170–176.

Nielsen, J. (1995) Sensory methods, in *Quality and Quality Changes in Fresh Fish* (ed H.H. Huss). FAO, Rome, Italy, pp. 130–139.

Nielsen, J. (1998) Sensory analysis of fish, in *Methods to Determine the Freshness of Fish in Research and Industry*. (eds Ólafsdóttir et al). Int. Inst. Refrig., Paris, France, pp. 279–286.

Niu, J.J. and Lee, J.Y. (1999) Renewable-surface graphite-ceramic enzyme sensors for the determination of hypoxanthine in fish meat. *Analytical Communications,* **36**, 81–83.

Nixon, P.A. (1971) Temperature integration as a means of assessing storage conditions, in *Report on Quality in Fish Products, Seminar No. 3,* Fishing Industry Board, Wellington, New Zealand, pp. 34–44.

Nixon, P.A. (1977) Temperature function integrator, U.S. Patent 4,061,033.

Oehlenschläger, J. (1992) Evaluation of some well established and some underrated indices for the determination of freshness and/or spoilage of ice stored wet fish, in *Quality Assurance in the Fish Industry* (eds H.H. Huss, M. Jacobsen, and J. Liston). Elsevier, Amsterdam, The Netherlands, pp. 339–350.

Oehlenschläger, J. (1997) Volatile amines as freshness/spoilage indicators. A literature review, in *Seafood from Producer to Consumer, Integrated Approach to Quality* (eds J.B. Luten, T. Børresen, and J. Oehlenschläger). Elsevier, Amsterdam, The Netherlands, pp. 571–586

Oehlenschläger, J. and Nesvadba, P. (1998) Methods for freshness measurements based on electrical properties of fish tissue, in *Methods to Determine the Freshness of Fish in Research and Industry* (eds Ólafsdóttir et al.). Int. Inst. Refrig., Paris, France, pp. 363–368.

Oehlenschläger, J. and Sørensen, N.K. (1998) Criteria of seafood freshness and quality aspects, in *Methods to Determine the Freshness of Fish in Research and Industry* (eds Ólafsdóttir et al.). Int. Inst. Refrig., Paris, France, pp. 30–35.

Ólafsdóttir, G., Hagnadóttir, A. and Martinsdóttir, E. (1998b) Application of gas sessors to evaluate freshness and spoilage of various seafoods, in *Methods to Determine the Freshness of Fish in Research and Industry* (eds Ólafsdóttir et al.). Int. Inst. Refrig., Paris, France, pp. 100–109.

Ólafsdóttir, G., Martinsdóttir, E. and Jónsson, E.H. (1997a) Gas sensor and GC measurements of volatile compounds in capelin (*Mallotus villosus*), in *Seafood from Producer to Consumer, Integrated Approach to Quality* (eds J.B. Luten, T. Børresen, and J. Oehlenschläger). Elsevier, Amsterdam, The Netherlands, pp. 507–520.

Ólafsdóttir, G., Martinsdóttir, E. and Jónsson, E.H. (1997b) Rapid gas sensor measurements to determine spoilage of capelin (*Mallotus villosus*). *J. Agric. Food Chem.,* **45**, 2654–2659.

Ólafsdóttir, G., Luten, J., Dalgaard, P., Careche, M., Verrez-Bagnis, V., Martinsdóttir, E. and Heia, K. (1998a) *Methods to Determine the Freshness of Fish in Research and Industry*. Int. Inst. Refrig., Paris, France.

Olley, J. and Ratkowsky, D.A. (1973) Temperature function integration and its importance in the storage and distribution of flesh foods above the freezing point. *Food Technol. Aust.,* **25**, 66–73.

Østergaard, A., Embarak, P. K. B., Yamprayoon, J., Wedell-Neergaard, C., Huss, H. H. and Gram. L. (1998) Fermentation and spoilage of *som fak*, a Thai low-salt product. *Trop. Sci.,* **38**, 105–112.

Park, I.S. and Kim, N. (1999) Simultaneous determination of hypoxanthine, inosine and inosine 5'-monophosphate with serially connected three enzyme reactors. *Analy. Chim. Acta,* **394**, 201–210.

Pedrosa-Menabrito, A. and Regenstein, J.M. (1990) Shelf life extention of fresh fish—A review. Part II—Preservation of fish. *J. Food Quality,* **1311**, 129–146.

Price, R.J., Melvin, E.F. and Bell, J.W. (1991) Postmortem changes in chilled round, bled and dressed albacore. *J. Food Sci.,* **56**, 318–321.

Ratkowsky, D.A., Olley, J., McMeekin, T.A. and Ball, A. (1982) Relation between temperature and growth rate of bacterial cultures. *J. Bacteriol.,* **149**, 1–5.

Reay, G.A. and Shewan, J.M. (1949) The spoilage of fish and its preservation by chilling. *Adv. Food Res.,* **2**, 343–398.

Reddy, N.R., Schreiber, C.L., Buzard, K.S., Skinner, G.E. and Armstrong, D.J. (1994) Shelf life of fresh tilapia fillets packed in high barrier film with modified atmosphere. *J. Food Sci.,* **59**, 260–264.

Rönnow, P., Simpson, R. and Otwell, S. (1999) The use of enzymatic TTI to monitor time-temperature exposure in distribution of chilled seafood. *Refrig. Sci. Techn. Proc.,* **EUR 18816**, 308–315.

Ross, T. (1996) Indices for performance evaluation of predictive models in food microbiology. *J. Appl. Bact.,* **81**, 501–508.

Russell, N.J. and Gould, G.W. (1991) *Food preservatives*, Blackie Academic and Professional, Glasgow, Scotland.

Ryder, J.M., Fletcher, G.C., Stec, M.G. and Seelye, R.J. (1993) Sensory, microbiological and chemical changes in hoki stored in ice. *Int. J. Food Sci. Technol.,* **28**, 169–180.

Sado, P.N. (1993) Quality attributes: Raw material & finished products, in *Smoked Seafood Products* (eds G.M. Pigott, J.B. Peters, and P. Peyton). College of Ocean & Fishery Sciences, University of Washington, Seattle, WA, p. 37.

Saito, T., Arai, K. and Matsuyoshi, M. (1959) A new method for estimating the freshness of fish. *Bull. Jap. Soc. Sci. Fish.,* **24**, 749–750.

Salvat, G., Rudelle, S., Humbert, F., Colin, P. and Lahellec, C. (1997) A selective medium for the rapid detection by an impedance technique of *Pseudomonas* spp. associated with poultry meat. *J. Appl. Microbiol.,* **83**, 456–463.

Scott, D.N., Fletcher, G.C., Charles, J.C. and Wong, R.J. (1992) Spoilage changes in the deep water fish, smooth oreo dory during storage in ice. *Int. J. Food Sci. Technol.,* **27**, 577–587.

Scott, D.N., Fletcher, G.C., Hogg, M.G. and Ryder, J.M. (1986) Comparison of whole with headed and gutted orange roughy stored in ice: Sensory, microbiological and chemical assessment. *J. Food Sci.,* **51**, 79–86.

Shahidi, F., Jones, Y. and Kitts, D.D. (1997) *Seafood Safety, Processing, and Biotechnology,* Technomic Publishing Co., Inc., Lancaster, PA.

Shamaila, M., Skura, B.J. and Nakai, S. (1995) Monitoring quality of packed Pacific ocean perch (*Sebastes alutus*) in storage. *J. Aquat. Food Prod. Technol.,* **4**, 35–57.

Shewan, J.M. (1962) The bacteriology of fresh and spoiling fish and some related chemical changes. *Rec. Adv. Food Sci.,* 167–193.

Shewan, J.M., Macintosh, R.C., Tucker, C.G. and Ehrenberg, A.S.C. (1953) The development of a numerical scoring system for the sensory assessment of the spoilage of wet white fish stored in ice. *J. Sci. Food Agric.,* **4**, 283–298.

Sigernes, F., Heia, K., Wold, J.P., Eilertsen, G., Esaiassen, M. and Sørensen, N.K. (1998). Assessment of fish (cod) freshness by VIS/NIR spectroscopy, in *Methods to Determine the Freshness of Fish in Research and Industry* (eds Ólafsdóttir et al.). Int. Inst. Refrig., Paris, France, pp. 369–382.

Slanetz, L.W. and Bartley, C.H. (1957) Numbers of enterococci in water, sewage, and feces determined by the membrane filter technique with an improved medium. *J. Bacteriol.,* **74**, 591–595.

Smelt, A.J., Scheerman, P. and Luten, J.B. (1998) Optimisation of a multi gas-sensor system, training of the artificial neural network and use of the electronic nose system for quality assessment of whiting (*Merlangius merlangus*), in *Methods to Determine the Freshness of Fish in Research and Industry* (eds Ólafsdóttir et al.). Int. Inst. Refrig., Paris, France, pp. 110–118.

Spencer, R. and Baines, C.R. (1964) The effect of temperature on the spoilage of wet white fish. Storage at constant temperatures between –1°C and 25°C. *Food Technol.,* **18**, 769–772.

Stansby, M.E. (1962) Specifications on fishy odours and flavours. *Food Technol.,* **April**, 28–32.

Strachan, N.J.C. and Nickolson, F.J. (1992) Gill air analysis as an indicator of cod freshness and spoilage. *Int. J. Food Sci. Technol.,* **27**, 261–269.

Taoukis, P.S. and Labuza, T.P. (1989) Applicability of time-temperature indicators as shelf life monitors of foods products. *J. Food Sci.,* **54**, 783–788.

Taoukis, P.S., Fu, B. and Labuza, T.P. (1991) Time-temperature integrators. *Food Technol.,* **45**, 70–82.

Tarr, H.L.A. (1954) Microbiological deterioration of fish post mortem, its decomposition and control. *Bact. Rev.,* **18**, 1–15.

Tinker, B.L., Learson, R.J. and Slavin, J.W. (1987) Industrial application of a computerized time-temperature monitoring system for measuring the freshness of chilled fish, in *Seafood Quality Determination* (eds D.E. Kramer and J. Liston). Elsevier, Amsterdam, The Netherlands, pp. 403–411.

Tinker, B.L., Slavin, J.W., Learson, R.J. and Ampola, V.G. (1985) Evaluation of automated time-temperature monitoring system in measuring the freshness of chilled fish. *Refrig. Sci. Techn. Proc.,* **1985–4**, 281–291.

Truelstrup Hansen, L. (1995) *Quality of Chilled Vacuum Packed Cold-Smoked Salmon (PhD Thesis),* Danish Institute for Fisheries Research, Department of Seafood Research.

US FDA (1995) Proceedings for the Safe and Sanitary Processing and Importing of Fish and Fishery Products, 21 CFR, Parts 123 and 124.

van Spreekens, K.J.A. (1974) The suitability of a modification of Long and Hammer's medium for the ennumeration of more fastidious bacteria from fresh fishery products. *Arch. Lebensmittelhyg.,* **25**, 213–219.

Vaz-Pires, P., Capell, C. and Kirby, R. (1994) Low-level heat-treatmenmt to extend shelf-life of fresh fish. *Int. J. Food Sci. Technol.,* **29**, 405–413.

Veciana-Nogués, M.T., Marine-Font, A. and Vidal-Carou, M.C. (1997) Biogenic amines as hygienic quality indicators of tuna. Relationships with microbial counts, ATP-related compounds, volatile amines, and organoleptic changes. *J. Agric. Food Chem.,* **45**, 2036–2041.

Winquist, F., Sundgren, H. and Lundström, I. (1998) Electronic noses for food control, in *Biosensors for Food Analysis* (ed A.O. Scott). The Royal Society of Chemistry, Cambridge, England, pp. 170–179.

WTO SPS Agreement. The Agreements on the Application of Sanitary and Phytosanitary Measures and Technical Barriers to Trade. World Trade Organization.

York, R.K. and Sereda, L.M. (1994) Sensory assessment of quality in fish and seafoods, in *Seafoods: Chemistry, Processing Technology and Quality* (eds F. Shahidi and J.R. Botta). Blackie Academic and Professional, Glasgow, Scotland, pp. 233–262.

Chapter 9

Ambient Packaged Cakes

H.P. Jones

9.1 INTRODUCTION

The market of "ambient packaged cakes" covers the range of cakes sold through the supermarket rather than the baker's shop. The major distinction between the two categories is the difference in shelf-life requirements. The small baker makes most goods daily and more complex confectionery products with a shelf life of several days once or twice a week. The packaged cakes sold in the supermarket today normally have a minimum shelf life of three weeks. The need for longer life is necessary because of the time required to distribute to stores with a subsequent requirement for a reasonable period of display on the shelves. Additionally, supermarket products have to be properly packaged to protect them through distribution and in-store handling.

In the UK, the ambient packaged cake market is served by a handful of manufacturers who can produce large quantities of cakes to consistent standards. These companies must have the technical resources to design products with the required attributes. The term "ambient" is necessary to distinguish this market sector from the "frozen" and "chill" sectors which operate with very different shelf-life requirements. Chilled cakes, in particular fresh cream–filled ones, normally have one or two days of life and are merchandised in refrigerated display cabinets. Frozen cakes usually have a shelf life of many months while in the freezer but only 1 to 2 days at ambient once thawed.

A wide range of ingredients and manufacturing techniques are used in the production of cakes. Besides the many ingredients and different processes employed to make cakes such as sponges, fruit cakes and pastries, flour confectionery overlaps with other product sectors of the food industry, namely, chocolate and sugar confectionery. The latter have very different composition and product characteristics; and as such can complicate shelf-life considerations when combined with flour-based products.

One of the key tasks of the cake/food technologist, therefore, is to design cakes in which the characteristics of individual components are not adversely affected during distribution and subsequent storage.

9.2 PRODUCT CLASSIFICATION

Ambient cakes may be classified in several ways such as by formulation or process type, but for the purpose of shelf-life evaluation, the classes listed in Table 9–1 will be used.

One of the challenges of the ambient cake technologist is to formulate products which will retain their moistness, texture, flavor, and appearance for as long as possible. To do this, an in-depth and up-to-date knowledge of ingredients, packaging materials, production processes, and the technology of moisture control is necessary.

Ambient cakes belong to the category of intermediate moisture food products, i.e., with about 20% moisture and water activity (a_w) within the range 0.65–0.90. With these water activities cakes can be preserved for many weeks in a variety of ways, i.e., canning, vacuum packaging,

Table 9–1 Classification of ambient cakes

Type	*Main features*
Sponge cake	Low fat level
Fruit cake	Medium fat level and fruit
Plain cake, e.g., Madeira	High fat level
Chocolate-coated cake	Moisture barrier present
Fruit pies	High and low moisture components
Pastries	Multi-component products
Cake bar	Cake of any type, individually wrapped

gas flushing, and alcohol injection. Canning is sometimes used and will preserve cakes for many months (e.g., army rations) but there are drawbacks which limit its use. The high product temperature reached in the canning process excludes all but the simplest of oven finished cakes, as it can result in pudding-like eating quality. Vacuum packaging can damage the fragile structure of most cakes and so only the very robust type can be packed in this way. The preservation of cakes therefore relies mainly on formulation and control of the inside-pack atmosphere by the use of suitable packaging and, in certain long-life applications, modified atmosphere packaging. The modern cake factory and the systems employed in it are designed in such a way that only raw materials of the right and specified quality are used and that production processes are carried out correctly and hygienically, so that finished products have the specified moisture content and a_w, ensuring the assigned shelf lives are achieved consistently.

9.3 EQUILIBRIUM RELATIVE HUMIDITY AND MOLD-FREE SHELF LIFE

The growth of microorganisms in foods is, in part, determined by the availability of water that will support microbial growth. This is measured by the water activity (a_w) of the food. The equilibrium relative humidity (ERH) of a cake is the humidity at which rates of evaporation and condensation of water are equal. It is the relative humidity which is eventually reached in the atmosphere surrounding the cake when it is closely enclosed by a perfect water barrier. At this equilibrium, the terms ERH and a_w are interchangeable. In the bakery trade, the term ERH is more appropriate and preferred to a_w since microbial growth tends to develop on the exposed surface of the product. The minimum ERHs (and their approximate corresponding sucrose concentrations) permitting growth of microorganisms are shown in Figure 9–1. As can be seen, most molds are only able to grow at an ERH above 80%. Table 9–2 shows a range of bakery products and their approximate ERHs (Cauvain and Seiler, 1992).

Using experimental data, the Flour Milling and Baking Research Association (FMBRA) (now part of Campden and Chorleywood Food Research Association [CCFRA]) in the UK was able to demonstrate the relationship between ERH and the mold-free shelf life (MFSL) of cakes at various storage temperatures (Figure 9–2). It is apparent that there is a linear relationship between ERH and the logarithm of the MFSL within the range 74–90% ERH. This relationship is given by the following expressions for the storage temperatures of 27°C and 21°C, respectively (Seiler, 1976).

$$\text{Log}_{10}\ \text{MFSL (days at 27°C)} = 6.42 - (0.065 \times \text{ERH\%})$$

$$\text{Log}_{10}\ \text{MFSL (days at 21°C)} = 7.91 - (0.081 \times \text{ERH\%})$$

From this work it has been possible to estimate the MFSL of cakes based on their formulations, using the above relationship and since

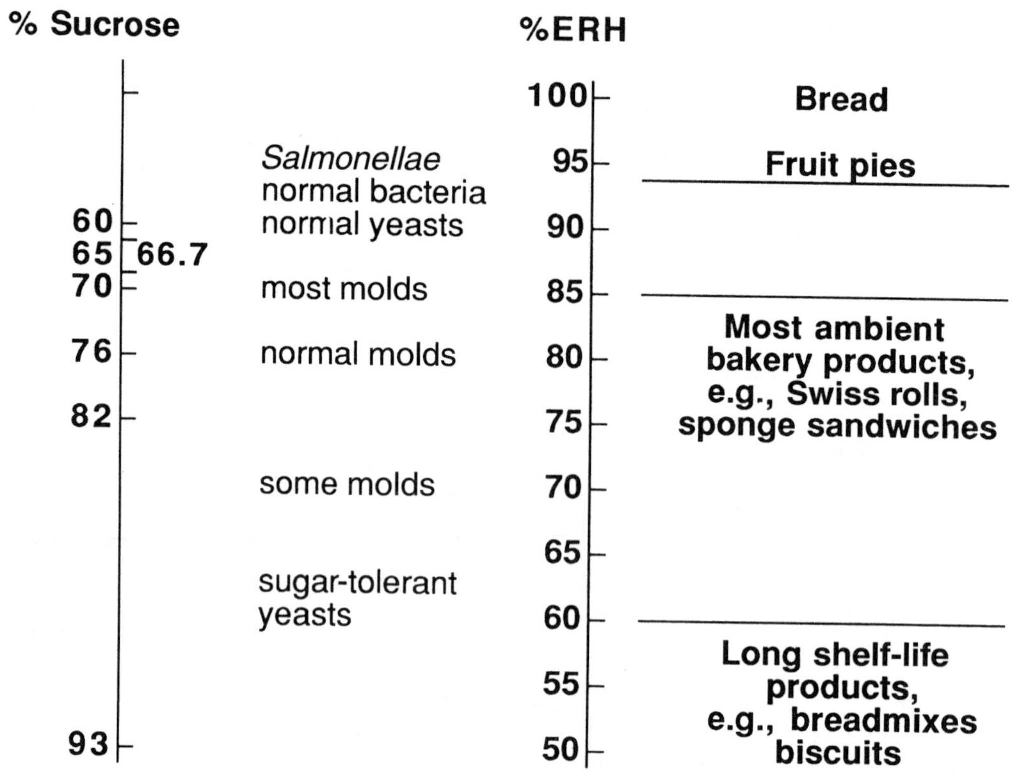

Figure 9–1 Minimum ERHs (and their approximate corresponding sucrose concentrations) permitting growth of microorganisms.

1990, with the help of FMBRA's award-winning ERH CALC™ computer-based program (see section 9.6).

9.4 PRODUCT CHARACTERISTICS

9.4.1 Sponge cakes

This product group is made up of lines such as Swiss rolls, sponge sandwiches, and the foundations for a variety of gateaux. The formulations are usually very low in fat, often containing no added fat. This is necessary to enable the sponge to be rolled, cut, and assembled into a variety of shapes without cracking or breaking. Although the fat content of the sponge is low, the final product may be enriched with buttercream or synthetic cream, with fat content of around 45%. The characteristics of a sponge are light texture, delicate flavor, and good cutting properties.

As well as the total product ERH that needs to be controlled, it is important to ensure that the ERHs of individual components in a multi-component product are also balanced as far as possible to minimize moisture migration from one component to another, unless this is desired. Sponge products lose moisture very rapidly when exposed to the atmosphere because of the large surface area to weight ratio and low fat content. To prevent moisture loss, they are usually wrapped in films that can provide a good moisture vapor barrier, as soon as they are sufficiently cooled. This ensures that the correct moisture level is retained in the product. A shelf life of 3 weeks can usually be obtained by optimizing the formulation and choosing the correct wrapping material.

Table 9–2 A range of bakery products and their approximate ERHs

Product	*ERH%*
Biscuits	5–15
Bread crumb	96–98
Bread crust	30–50
Plain cakes, e.g., Madeira	80–90
Fruited cakes	73–75
Christmas puddings	70–77
Jam	70–80
Marzipan	70–80
Dried fruit	65–70
Glacé cherries	75

Source: Cauvain and Seiler, 1992.

9.4.2 Fruit cakes

The rich fruit cake eaten at Christmas and the more humble slab cake are British specialties noted for their moistness and good keeping qualities. Fruit contents can range from 30% to 100% of the batter weight and consequently a strong cake structure is required to support the fruit and prevent it from sinking during baking. High protein flour, egg, and the judicious use of fat provide the crumb strength to maintain an even distribution of fruit and to hold the cake together when cut. Both defatted soya flour and egg albumen can be used to supplement the protein level.

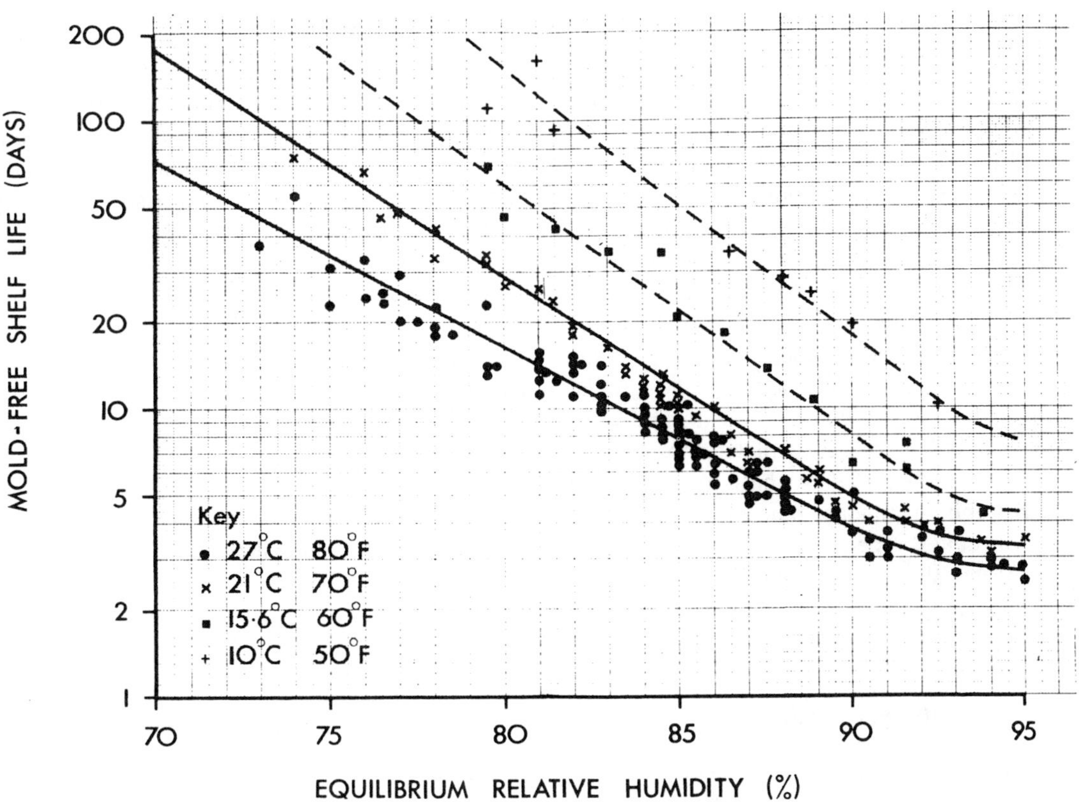

Figure 9–2 MFSL, in days, at various storage temperatures in relation to the ERH of the product. (Source: Cauvain and Seiler, 1992).

The addition of dried fruit helps to increase the shelf life by reducing the ERH of the cake. Dried fruit contains only 16% moisture but in combination with the reducing sugars in it produces very moist eating quality, characteristic of a cake with a much higher moisture content. A shelf life of many weeks can be obtained because of this and it can be significantly extended by adding alcohol to the product before or after baking. Alcohol has been shown to act as a mold inhibitor and retard crumb staling (Guy et al., 1983; Ooraikul, 1991).

Fruit cakes lose moisture slowly even when unwrapped, and it is normal practice to leave slab cakes to stand overnight to allow moisture to equilibrate and fats to set before cutting and wrapping. Heavily fruited cakes soaked in alcohol will remain in good condition for 1 year or more if stored in an airtight container.

9.4.3 Plain cakes

This group comprises many typically British cakes such as Madeira, Battenburg, and Victoria sandwich. There are also many American formulations which are popular today such as Death by Chocolate and Fudge Cake.

The inclusion of chocolate in a formulation increases the fat content further. High fat levels result in lower volume cake than sponge cake but produce a richer, moister eating cake which can be eaten without the addition of buttercream or other fillings. It is common for cake foundations of this type to contain added ingredients such as nuts, cherries, chocolate chips, and so on. They are often used as a base for converting into Battenburg, fondant fancies, and iced celebration cakes. Special account needs to be taken of the ERH of post-bake finishes as moisture migration can spoil the appearance and eating characteristics of decorations.

9.4.4 Chocolate coated cakes

Chocolate coated cakes have become a significant subsector within the ambient cake market and they do have particular technical properties in respect of shelf life. Any of the cake formulations so far discussed can be used as a foundation over which chocolate can be enrobed. Chocolate not only adds a unique flavor and texture but also encloses the product in a fatty barrier which reduces the rate of moisture loss from the cake and thereby assists in shelf-life extension. There are a number of special considerations to be taken account of when developing cakes for chocolate coating. First, there is the need to avoid "crumbing" as it will contaminate chocolate circulating in the enrober. Second, the foundation must not shrink excessively during life as this will leave a fragile chocolate shell which is likely to disintegrate in the pack or when the product is cut or eaten. It is necessary to use a chocolate covering which will cut cleanly without shattering and yet not be too soft to handle in warm weather.

9.4.5 Fruit pies

Fruit pies form one of the largest product sectors within the ambient cake market and are mainly produced in the mini-pie format. The product make-up is approximately 50% pastry and 50% fruit filling and products are formulated to keep the pastry dry and crisp eating, and the filling moist.

Fruit pies are normally oven finished and packed directly after cooling. They are not film wrapped but packed into cartons thereby allowing a gradual loss of moisture to the atmosphere throughout their storage life.

9.4.6 Pastries

This product group encompasses a wide variety of products, which include the following:

- jam tarts
- puff pastries
- Danish pastries
- Viennese tarts
- cherry Bakewell tarts.

Pastries come in many shapes and sizes and their make-up is often complex, as they can have up to six different components. The following fillings and decorations are commonly applied:

- synthetic creams
- jams and jellies
- cake fillings, e.g., Bakewell
- icings
- chocolate
- nuts
- glacé cherries.

The complex nature of their composition poses particular problems in preserving the texture and appearance of components throughout the life of the product. Most of the products mentioned in this group are usually wrapped in impermeable films.

9.4.7 Product characteristics—a summary

From the previous outlines of product characteristics and past experience, a few generalizations can be made concerning product factors that are conducive to longer shelf life. While these are probably not new, it is useful to put them together so that they may help future product design.

1. Complete enrobing in chocolate or chocolate covering helps to retain quality inside (e.g., chocolate-covered Swiss rolls vs. ordinary Swiss rolls).
2. High fruit content: The low ERH and the presence of hydroxymethyl furfural (HMF) in dark fruit cakes contribute to a longer shelf life.
3. Low moisture content (e.g., Jaffa cakes).
4. Jam and cream fillings are better than jam alone or cream alone (e.g., in sponge sandwiches).
5. High sugar content (e.g., Jamaica ginger cake).
6. The presence of preservatives (e.g., sorbic acid) or antimicrobial ingredients (e.g., alcohol).

9.5 FACTORS AFFECTING SHELF LIFE

The most common causes of quality deterioration in ambient cakes are

- microbiological spoilage (visual growth or fermentation)
- crumb staling (dryness and the development of off-flavors)
- rancidity (off-flavor development)
- crystalization of sugars (e.g., grittiness in cream fillings)
- syneresis of jams and jellies
- off-flavors and odors other than rancidity
- chocolate bloom (or loss of gloss in bakers' coatings)
- structural weakness
- color fade, change, and transfer
- moisture migration and loss.

9.5.1 Microbiological spoilage

The microorganisms most commonly found in ambient packaged cakes are molds and yeasts. The ability of these microorganisms to grow and spoil the food is determined by a number of factors such as the availability of water, the environmental temperature, and the pH of the system. In general, the risk of microbial spoilage can be significantly reduced by (1) developing formulations to an appropriate ERH, (2) using a suitable mold inhibitor, and (3) optimizing the pH where possible (Cauvain and Young, 1990; Anon., 1997).

There are few ingredients used in the manufacture of ambient cakes which are generally recognized as actual or potential sources of harmful microorganisms; the most common potentially "sensitive" ingredients are eggs, cocoa powder, and desiccated coconut. However, these are usually pasteurized, dehydrated, and subsequently baked in the products, processes which render them microbiologically safe. Also, they are carefully controlled by the suppliers and cake manufacturers who ensure they remain uncontaminated. When coconut products and cocoa powder are used in "cold" processes to make coatings and fillings, for instance, the manufacturer is well advised to have them checked for pathogens by a competent laboratory.

9.5.2 Crumb staling

Staling is a phenomenon related to changes in the state of cooked starch and recognized in baked products as a dry crumbly texture, often accompanied by an unpleasant flavor and odor (Guy et al., 1983).

9.5.3 Rancidity

The oxidation of fats can lead to off-flavors and odors but this has not been a major problem for the ambient cake manufacturer whose products have a shelf life of around 3 weeks, provided the fats used are fresh and have been stored correctly. However, increase in the use of butter/butter oil over the years has resulted in a higher incidence of rancidity development (Bennion, 1972; Frampton, 1994). Furthermore, some cakes (also from continental Europe) are now sold with shelf lives of 3 to 6 months, which tends to increase the potential for oxidative rancidity.

9.5.4 Crystalization of sugars

This fault may occur in jams, icings, caramel, various cake decorations, and fillings. The causes are frequently related to incorrect formulation and poor processing. Basically, if moisture is free to migrate from a layer of caramel, for example, into an adjoining component, it can result in graining in the caramel (Lees, 1973). Likewise, the loss of moisture from jams and icings can lead to coarse crystal growth in them (Paul, 1972). Indeed, even in the absence of any free moisture transfer, water vapor will migrate as long as there is a sufficient gradient. Of course, moisture migration can generally be minimized by judicious formulation, reducing as far as practicable the difference between the ERHs of the various components.

9.5.5 Syneresis of jams and jellies

This phenomenon can often be seen in jam-filled cakes and manifests itself as a dark, moist syrupy area above or below the jam layer. It is caused by the liquid phase in the jam separating from the gel structure and migrating into the cake crumb. The appearance is unsightly and the eating is cloying and unpleasant. The main causes are normally due to the wrong choice of gelling agents or incorrect processing during jam boiling (Anon., 1984). Excessive pumping of bulk jam after it has set, which can happen immediately prior to filling and depositing, damages the gel structure and encourages syneresis in the finished product.

9.5.6 Color fade, change, and transfer

The widespread rejection of artificial colors in foods in the late 1980s has increased the use of natural colors in the baking industry. Many of the food colors now used are derived from plants, e.g., anthocyanins, which are unstable under certain processing and/or environmental conditions. Ultraviolet barrier films or cartons can be used, for example, to overcome instability to light. Some natural colors are unstable in acid environments and the original colors can be significantly changed as shelf life progresses (Newstubb and Henry, 1988).

9.5.7 Off-flavors and odors other than rancidity

Off-flavors and odors in cakes can be caused by a variety of reasons, some of the most common ones being:

1. The use of old or badly stored ingredients (e.g., dried fruit) can introduce stale/unpleasant flavors which will accentuate as the product ages.
2. Packaging odors can be transmitted to products from printing inks, recycled board, and so on (Anon., 1960).
3. Jam fermentation due to microbial contamination produces alcohol notes which are unacceptable.
4. Taints can come from cleaning chemicals if hygiene procedures are incorrectly carried out.

9.5.8 Chocolate bloom

Chocolate bloom is a common fault in chocolate confectionery products and may also be found in chocolate-covered cakes. It is recognized by the dull grayish appearance of the chocolate coating. The four most frequent causes are poor chocolate tempering, incorrect cooling causing condensation and subsequent sugar crystalization, incompatibility between fats in the chocolate and the product enrobed, and temperature abuse during storage of the finished product (Minifie, 1988).

9.5.9 Structural weakness

Many cake formulations are designed to produce light, tender cakes, e.g., Swiss rolls, Madeiras, and Victoria sandwiches. These products are often described as "whole cakes" because they are sold as single products for portioning by the end user. If the structure of the cake is too weak, so that it crumbles when cut, it will be unacceptable to the consumer. Such a fault may not be apparent when the cake is first produced but develops as moisture redistributes within the product. There are several causes for this but the most common ones are protein deficiency in the flour, insufficient egg, and/or excessive fat in the formulation and incorrect processing.

9.5.10 Moisture migration

Moisture in ambient cakes will migrate, if permitted to do so, whenever a moisture gradient exists, until equilibrium is eventually reached. This can be demonstrated by examining a freshly baked Madeira cake in which the moisture contents of the top crust and the core are typically 10% and 25%, respectively. After cooling and wrapping in a barrier film, moisture will migrate from the core to the top crust and into the air space inside the pack until equilibrium is attained. Similarly, migration will also occur between adjacent components. This is most commonly seen in fruit pies where the pastry is softened by moisture coming from the filling. Excessive moisture transfer can spoil the desired characteristics of a product by making, for example, the pastry in a fruit pie soft and soggy, and the filling dry and stale. Moisture migration can be minimized by careful formulation with a view to balancing the ERHs of the various components but this may not be feasible technically or acceptable organoleptically. An alternative solution would be to introduce a moisture barrier which inevitably incurs additional cost and may not always be practical. A useful review has been given on moisture migration and its control (Labuza and Hyman, 1998).

9.6 ERH CALC

ERH CALC™ (Cauvain and Young, 1990; Anon., 1999a) is a PC-based program for the baking industry produced by the FMBRA, which enables the technologist to run formulation simulations and rapidly calculate theoretical ERHs of the products and hence estimate their MFSL. The benefits of this computer tool are basically three-fold. First, it saves considerable time and effort in determining MFSL compared to trial and error methods. Second, it enables many more theoretical formulation options to be tried in a limited amount of time. Third, it is an excellent training aid for new development staff who are unfamiliar with cake technology.

The program, further developed and enhanced, has been relaunched recently by CCFRA (Anon., 1999a) as one of the three modules of the Cake Expert System software for flour confectionery product development. The other two modules are

- BALANCE—a cake formulation development tool that helps technologists to develop new products rapidly
- FAULT DoC—a cake fault diagnostic tool that helps technologists to enhance product quality and eliminate product faults.*

In addition to its original features (i.e., easy selection of ingredients and quantities, availability of advice and glossary sections, and, of course, automatic calculation of ERH and prediction of MFSL), the latest ERH CALC™ includes features such as the Mold Inhibitor Calculator and Composite Products and Packaging

*Courtesy of Campden & Chorleywood Food & Research Association Group, Gloucestershire, United Kingdom.

routines. Thus, additional shelf life can be estimated by choosing the type of preservative (e.g., sorbic acid) and its level. The shelf life and an indication of moisture movement in a multicomponent product can be assessed by using the Composite Product route and moisture loss through a wrapping material can be investigated by the packaging model.

To get the most out of ERH CALC™ it is necessary to understand the design assumptions and to compare predictions with actual measured results. By reconciling differences, the user will develop skills in interpreting the information produced. ERH and MFSL predictions are made on the assumption that the product is wrapped in an impermeable film and that moisture will equilibrate within the pack. In practice, this is not always the case. For example, fruit pies are not usually film-wrapped and free to lose moisture to the atmosphere. Also, assumptions are made about solubility levels of sugars whereas in practice, formulations and process conditions can greatly affect the amount of sugar which finally goes into solution. By using ERH CALC™, a newcomer to cake technology quickly understands the role and significance of humectants in increasing cake shelf life as well as the importance of accurate control of moisture loss in baking and cooling.

Considerable experience gained in the use of ERH CALC™ since it was first launched has confirmed that it is an extremely useful tool, one that can be highly recommended to anyone involved in industrial cake development and manufacture. In the UK, ERH CALC™ is now being widely used by cake technologists on a routine basis to work out "best before" dates for new cake products (Young, 1998).

9.7 THE DETERMINATION OF PRODUCT SHELF LIFE

The theoretical estimation of MFSL using ERH CALC™ is very valuable in formulation design but there are many other factors that have to be considered before a shelf life is finally set. The quantitative and qualitative data collected during concept life tests relate to a development product manufactured under tightly controlled conditions. It is necessary, however, to take a broader view and consider external factors which may put the product at risk.

Hazard Analysis Critical Control Point (HACCP) is a system designed to identify potential hazards and their associated risks, allowing appropriate controls to be introduced and monitored (Mortimore and Wallace, 1998).

9.7.1 Product data

The product data required for making shelf-life decisions are derived from the following:

- sensory analysis
- laboratory analysis (chemical analyses and physical tests)
- microbiological examination
- travel test.

9.7.1.1 Sensory analysis

The following are some of the most common quality attributes assessed in the sensory analysis of cakes:

- moistness
- crumb firmness
- color
- flavor
- texture
- mouthfeel
- odor.

Products should be evaluated by trained assessors who have been selected for flavor sensitivity, color identification, and their ability to recognize consistently a range of quality attributes and assess their intensity reliably. The analyses are usually carried out at weekly intervals until the product has achieved its desired shelf life or until it has become unacceptable. Each attribute is judged on a qualitative or numerical scale so that results can be tabulated or plotted as they become available. The information obtained can be used to make comparisons, identify trends, and set product standards.

The sensory analysis environment should ensure that assessors are not distracted or influenced by one another, or affected by any external factors, and that the correct lighting is used for color evaluation. The importance of correct experimental design and proper data analysis cannot be overemphasized. Further information on sensory analysis of foods can be found in BS 5929—Parts I–VI (BSI, 1982, 1984, 1986, 1989). If results are required to be presented graphically, spider plots, for example, can be generated using Microsoft® Excel or a specialist software such as Senpak®.

9.7.1.2 Laboratory analysis

Analyses and tests (other than microbiological examination) commonly used include the following:

- moisture content
- peroxide value (PV)/free fatty acid (FFA)
- titratable acidity
- pH value
- refractometric solids % (RS%)
- ERH%
- specific volume of cake (to identify possible shrinkage)
- product weight loss
- moisture/fat migration into packaging
- travel test and package performance.

It is necessary to carry out ERH measurement on individual components as well as on the complete product, for it is possible to have an apparently safe ERH for the comminuted product but an unacceptably high ERH for one of the components. The MFSL of a cake is usually governed by the component which has the highest ERH and hence the need to determine the ERH of individual components. Big differences between the ERHs of various components also tend to suggest that moisture/water vapor migration (and other related deterioration) is likely to be a problem.

Additional chemical analysis may be carried out for nutritional labeling purposes. If a customer has specific nutritional requirements it would be sensible to ensure that they are met at an early stage of the development program as changes that have to be made later may affect the shelf life of the product.

9.7.1.3 Microbiological examination

Molds and yeasts are the most common organisms that are likely to grow in packaged cakes, although total viable counts and presumptive coliform counts are monitored routinely. In certain intermediate products whose manufacture involves only the mixing of ingredients (e.g., synthetic filling creams), it is necessary to check for the presence of pathogenic organisms. Although the ERHs of most packaged cakes are too low to support the growth of pathogens (Adams and Moss, 1995), this procedure is useful to confirm that the formulations have been correctly weighed and processed.

9.7.2 External factors

The following external factors are usually considered during shelf-life determination of cakes:

- plant hygiene
- process control
- storage of intermediate products
- environmental and human factors
- historical data from similar products
- information from external sources
- consumer use.

9.7.2.1 Plant hygiene

It is necessary to ensure that plant and machines are capable of being cleaned to a satisfactory standard and if there are potential problems they must be identified and resolved as early as possible. Some of the more common problems encountered are

- poor access to machine
- inefficient use of cleaning chemicals, e.g., poor control of detergent and disinfectant concentrations
- inadequate supply of water at the desired temperature and pressure
- inadequate hygiene specifications
- poorly trained staff.

Regular environmental monitoring (e.g., using air sampling, swabbing of machine parts and surfaces, and so on) during factory trials can provide early warnings of such problems.

9.7.2.2 Process control

It is necessary to establish at an early stage the actual process capabilities during plant trials as over-scaling of liquids, under-scaling of powders, inaccuracy in depositing, inconsistency in baking, and so on are problems which can result in unacceptable variations in product moisture content and ERH.

9.7.2.3 Storage of intermediate products

Many cakes use intermediate products that are not heat treated (e.g., filling creams for Swiss rolls), and control of such intermediate products can be crucial to the maintenance of the assigned product shelf life. Control should include a maximum holding period, the conditions of storage, handling procedures, and if appropriate positive release based on a practical and meaningful sampling plan.

9.7.2.4 Environmental and human factors

Past experience has indicated that MFSL of cakes can be extended by 10–15% simply as a result of better hygiene, improved housekeeping, and more effective pest control in the bakery. It has been shown that MFSLs of certain cakes increased significantly, the main reason being the production of these cakes had been transferred from an old and poorly maintained factory to one which was purpose-built and had a superior environment in respect of general hygiene.

Poor personal hygiene can cause high levels of microbiological contamination at start of life, which can result in reduced product shelf life. The importance of personal hygiene cannot be overemphasized, this being especially critical in labor-intensive bakery operations.

9.7.2.5 Historical data from similar products

Information from similar products and processes can be very useful in shelf-life evaluation. It is not uncommon for similar products that are manufactured on the same plant to be grouped together, because they are similar and have the same shelf life, for production/distribution planning and scheduling purposes. Historical data and information are particularly useful to newcomers in product development as well as to the more experienced product developers who are working on shelf-life extension.

9.7.2.6 Information from external sources

Ingredient and packaging suppliers can be good sources of technical data and often have relevant application experience which can assist in the setting of product shelf life. They may also have specialist instrumentation and product testing "know-how" and be prepared to help in certain aspects of shelf life evaluation.

Research associations (RAs) (e.g., CCFRA) are an invaluable source of scientific and technological expertise. This is particularly true for medium to small sized companies that do not have a central research and development function. They can provide information/legislation services and private consultancy as well as a wide range of analytical services. It is reassuring to know when one is venturing into a new product or process area, one can turn to the RAs for informal advice or consultancy on a confidential basis.

9.7.2.7 Consumer use

It is important to give adequate thought to the product end-use environment (i.e., the climate and the way the product may be stored and consumed), particularly if the product is destined for export. Temperature has a significant effect on the shelf life of ambient packaged cakes and any shelf-life determination must take this into account. It is also important to consider at a very early stage the legal status of any additives such as preservatives (e.g., mold inhibitors) that may have to be used to achieve the target shelf life, especially when the product is being developed for the export market.

9.8 SHELF-LIFE DETERMINATION

A possible decision process for determining product shelf life is outlined in Figure 9–3. The process not only utilizes product data but also requires relevant information from the process and the environment in which the product is to be manufactured, distributed, stored, merchandised, and consumed. There may also be a need for information from specialist/external sources (e.g., RAs and suppliers) when new materials (edible and/or packaging), new product types, or new processes are involved. The decision re-

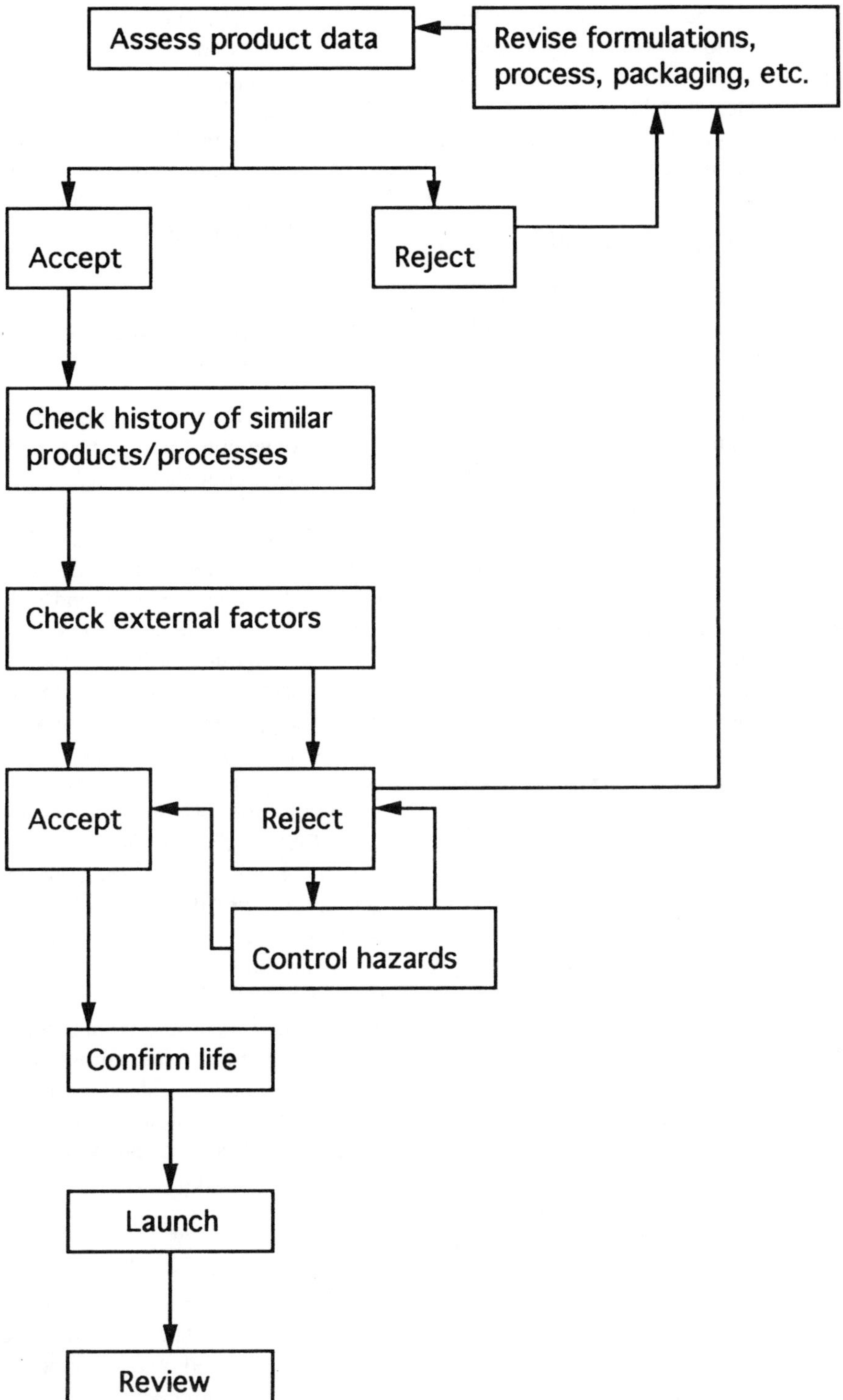

Figure 9–3 Shelf-life determination—a possible decision process.

quires judgments to be made about the robustness of the formulation, the microbiological profile, the process, and end-use conditions. The setting of a product shelf life is usually a joint effort from product development, laboratory service, and quality assurance management.

9.8.1 Product development activities

The following represents the sequence of activities for the determination of product shelf life of ambient packaged cakes.

(a) *Concept stage (development bakery):*

1. Formulation development
2. Estimation of baking and cooling losses
3. Calculation of theoretical ERH and prediction of MFSL using ERH CALC™
4. Further formulation development if necessary
5. Production of samples for shelf-life and other evaluations
6. Analysis of samples at start of life
7. Sampling for analysis at predetermined intervals
8. Comparison of preliminary shelf life with prediction
9. If necessary, further formulation adjustments and life tests
10. Setting of provisional product shelf life.

(b) *Factory stage (plant bakery):*

1. Identification of food safety hazards (HACCP)
2. Scale-up of formulations and processes, with modifications where appropriate
3. Shelf-life determination using samples made from pilot or full-scale plant
4. Comparison of shelf-life results with those obtained using concept samples
5. More full-scale factory runs (and shelf-life determinations) especially if pilot plant only was used in (3) above
6. Confirmation of product shelf life (issuing of finalized and approved complete "product-process-packaging" specifications)
7. Product launch and shelf-life validation and monitoring using samples from production runs
8. Follow-up actions and investigations if necessary.

9.8.2 An example—a white chocolate mini-sponge roll

The following information was obtained during the development of a new packaged cake which was a chocolate buttercream-filled mini-roll completely covered in white chocolate. This was the first white chocolate cake from an industrial cake manufacturer in the UK. Little was known of the performance and potential problems associated with the handling of white chocolate although the latter is more widely used in the biscuit and chocolate confectionery industries. The sponge and buttercream on the other hand were components in everyday use. White chocolate is more prone to rancidity development than normal chocolate and therefore PV and FFA determinations were carried out on both the stored chocolate and the finished product. Sensory profiles of the white chocolate and of the whole product were obtained and recorded using star diagrams or spider plots. The product was produced using a well-established process and standard plant and equipment with the exception of the chocolate enrober and its associated storage tanks. A standard pack format was used, and as anticipated the travel test gave satisfactory results and revealed no major potential problems in distribution.

9.8.2.1 Sensory data

Nine quality attributes were evaluated and the data presented (Table 9–3) were the ratings of the final product. The attributes assessed were: chocolate color and opacity, chocolate melt and mouthfeel, chocolate flavor, sponge texture, sponge crumbliness, sponge toughness, product moistness, buttercream flavor, and overall prod-

Table 9–3 White chocolate mini-sponge roll—sensory analysis results

Quality attribute	*Rating*	*Total score*	*Mean score*
Chocolate opacity	Translucent (1) Opaque (10)	31	5.2
Chocolate melt	Slow (1) Fast (10)	43	7.2
Chocolate flavor	Weak (1) Strong (10)	50	8.3
Sponge texture	Compact (1) Open (10)	46	7.6
Sponge crumbliness	Crumbly (1) Firm (10)	36	6.0
Sponge toughness	Tough (1) Soft (10)	32	5.3
Overall moistness	Very dry (1) Very moist (10)	32	5.3
Buttercream flavor	Weak (1) Strong (10)	45	7.5
Overall chocolate flavor	Weak (1) Strong (10)	30	5.0

Note: The panel consisted of six assessors.

uct flavor (see Table 9–3). Spider plots of the white chocolate and the total product were produced based on the data obtained, using a sensory panel of six assessors (Figure 9–4 a and b).

9.8.2.2 Other data

Other relevant product data obtained are summarized in Table 9–4. The dedicated white chocolate enrober and storage tanks required special control and monitoring to minimize the risk of rancidity development and color contamination. The rest of the process was well established and controlled in the usual way. The product was eventually assigned a shelf life of 3 weeks.

9.9 CURRENT DEVELOPMENTS

Two important issues in the packaged cake industry today are shelf-life extension and shorter lead time to market, both of which have implications for shelf-life evaluation.

9.9.1 Shelf-life extension

The demand for packaged cakes with a longer life is driven by the need to export as well as to remain competitive in the home market. Extending shelf life requires new technologies (e.g., new processes and techniques) which must be evaluated by the methods described, but if shelf life is to be extended to months, then the process of shelf-life evaluation will need to be speeded up. Accelerated shelf-life determinations are being successfully used in certain applications and there is now a wide variety of instruments available, which will carry out chemical and microbiological analyses faster than conventional methods.

9.9.2 Shorter lead time to market

International competition and competition within the Single European Market in foods are demanding faster entry into the marketplace

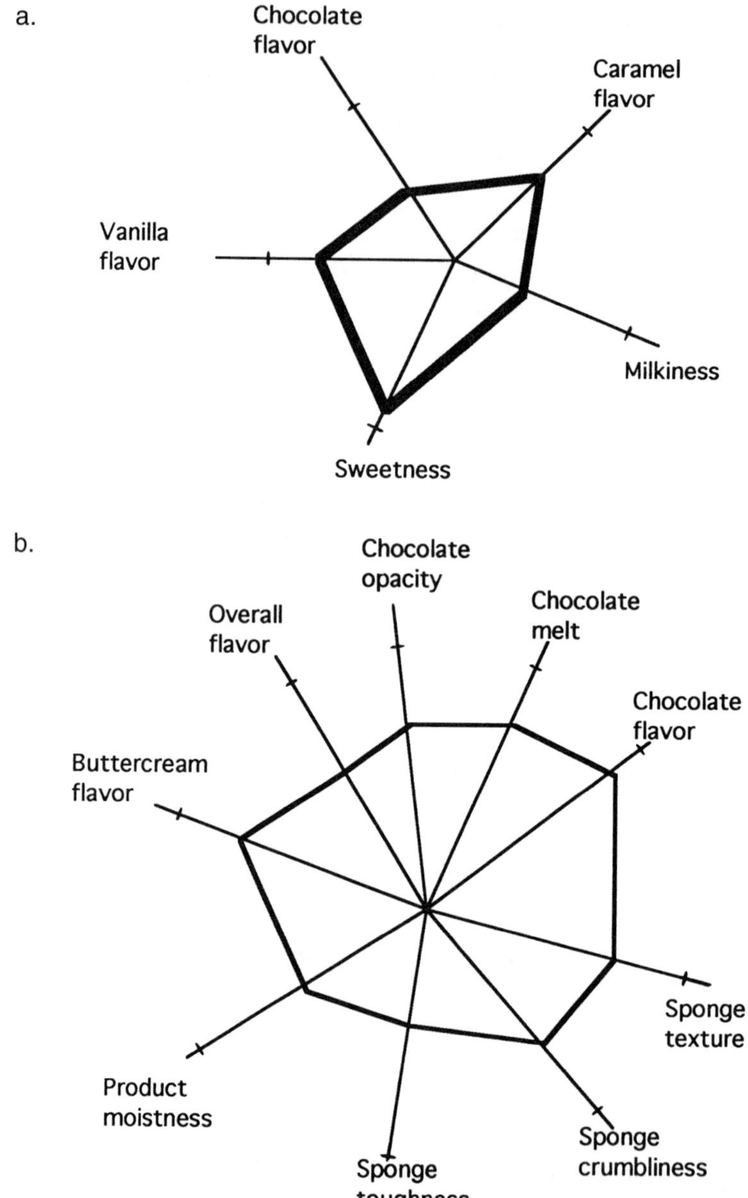

Figure 9–4 Spider plots of the white chocolate (a) and the enrobed mini-sponge roll (b).

with new products. Capital investment must begin to pay back earlier to improve profitability which in turn can be used to finance business expansion. All this has meant that the development process must be shortened without compromising product safety and quality. In addition food legislation demands that food must be safe and wholesome and, should a "due diligence" defense be used, evidence must exist to prove that "all reasonable precautions" have actually been taken. In this context, the process of shelf-life determination must not only be

Table 9–4 Shelf-life determination of white chocolate mini-sponge roll—a summary of some relevant product data

Component	*Moisture content %*	*ERH %*	*TVC[a] (cfu/g)*	*Yeasts/g[a]*	*Molds/g[a]*
Total product	17.3	78.5	280	10	<10
Golden sponge	20.3	79.2	—	—	—
Buttercream	15.4	77.4	150	20	10
White chocolate	1.6	70.1	—	—	—

[a]Results after 5 weeks' storage.

quicker but also more rigorous, comprehensive, and well documented. The route to achieving greater speed and reliability can be facilitated through the use of predictive computer tools such as the Cake Expert System and, where appropriate, the Food MicroModel (Anon., 1999b), the fruit of the UK Ministry of Agriculture, Fisheries and Food–funded microbiological predictive program for pathogens. Much benefit can also be gained by the implementation of techniques like HACCP as well as judicious introduction of modern sophisticated analytical instruments and data handling and analysis capabilities.

REFERENCES

Adams, M.R. and Moss, M.O. (1995) *Food Microbiology.* Royal Society of Chemistry, London, England.

Anon. (1960) *Odour in Packaging.* The Institute of Packaging Conference Technical Papers. United Kingdom.

Anon. (1984) *The Handbook for the Fruit Industry,* Section 14. Genu Pectins Skensued, Copenhagen Pectin Factory Limited, Denmark.

Anon. (1997) Water activity in flour confectionery product development, in *The Technology of Cake Making*, 6th edn (ed A.J. Bent), Blackie Academic and Professional, Glasgow, Scotland, pp. 386–397.

Anon. (1999a) *Cake Expert System—Software for Flour Confectionery Product Development*, CCFRA, Chipping Campden, Gloucestershire, England.

Anon. (1999b) *Food MicroModel Version 3.0*, Food MicroModel Ltd., Surrey, United Kingdom.

Bennion, M. (1972). Fats as cooking media, shortening agents and components of pastry, in *Food Theory and Applications* (eds P.C. Paul and H.H. Palmer). Wiley, London, England, pp. 213–250.

BSI (1982, 1984, 1986, 1989) *Methods For Sensory Analysis of Food,* BS 5929, Part 1–6, British Standards Institution, London, England.

Cauvain, S.P. and Seiler, D.A. (1992) *Equilibrium Relative Humidity and the Shelf Life of Cakes,* Report No. 150, Flour Milling and Baking Research Association, Chorleywood, England.

Cauvain, S.P. and Young, L.S. (1990) *ERH CALC,* Computer Software Package. Flour Milling and Baking Research Association, Chorleywood, England.

Frampton, A. (1994) Prevention of rancidity in confectionery and biscuits—a Hazard Analysis Critical Control Point (HACCP) approach, in *Rancidity in Foods,* 3rd edn (eds J.C. Allen and R.J. Hamilton), Blackie Academic and Professional, Glasgow, Scotland, pp. 160–178.

Guy, R.C.E., Hodge, D.C. and Robb, J. (1983) *An Examination of the Phenomena Associated With Cake Staling,* Report 107. Flour Milling and Baking Research Association, Chorleywood, England.

Labuza, T.P. and Hyman, C.R. (1998) Moisture migration and control in multi-domain foods. *Trends in Food Science & Technology*, **9**, 47–55.

Lees, R. (1973) *Sugar Confectionery and Chocolate Manufacture.* Leonard Hill, London, England, pp. 191–205.

Minifie, B.W. (1988) *Chocolate, Cocoa and Confectionery: Science and Technology,* 3rd edn. Van Nostrand Reinhold Inc., New York, NY.

Mortimore, S. and Wallace, C. (1998) *HACCP. A Practical Approach*, 2nd edn, Aspen Publishers, Inc., Gaithersburg, MD.

Newstubb, C.J. and Henry, B.S. (1988). Natural colours—a challenge and an opportunity. *Food Technology International Europe,* Sterling Publication Ltd, London, England.

Ooraikul, B. (1991). Modified atmosphere packaging of bakery products, in *Modified Atmosphere Packaging of Food* (eds B. Ooraikul and M.E. Stiles). Ellis Horwood, Chichester, England, pp. 49–115.

Parry, T.J. and Pawsey, R.K. (1983) *Principles of Microbiology for Students of Food Technology,* 2nd edn. Hutchinson, London, England.

Paul, P.C. (1972) Basic Scientific Principles, Sugars, and Browning Reactions. In *Food Theory and Applications* (eds P.C. Paul and H.H. Palmer). Wiley, London, England, pp. 1–76.

Seiler, D.A.L. (1976). The stability of intermediate moisture foods with respect to mould growth, in *Intermediate Moisture Foods* (eds R. Davies, G.G. Birch, and K.J. Parker). Applied Science Publishers, London, England, pp. 166–180.

Young, L. (1998). Setting the shelf life in the new product development context—for cakes and biscuits. *Ambient Stable Foods: An Essential Guide*, 27 November 1998, CCFRA, United Kingdom.

Chapter 10

Potato Chips and Savory Snacks

C.M.D. Man

10.1 INTRODUCTION

Potato chips and related savory snacks such as extruded products belong to a highly competitive, innovative, and growing sector of the food market. In the UK, they are categorized as snack products, which are generally understood as ready-to-eat ambient, "shelf stable" (6 to 12 weeks) savory products, pre-packed, and offered through a variety of outlets including supermarkets and other food retailers as well as catering establishments (Blenford, 1990). This then is the definition of snack products for the purpose of this chapter, although it is understood that the term "snack products" can mean very many different products in different countries within the European Union, and indeed in North America. In 1991, the UK consumed some 234 k tonnes of snacks, which equated to just over 4 kg per head per annum. In financial terms, the UK savory snack market is worth around £1.6 billion in 1997 and is still growing (Anon., 1998).

10.2 THE PRODUCT GROUP

Broadly speaking, snack food products may be divided into the following categories based on the process technology associated with their manufacture (Burdon, 1989):

1. ***Deep fat fried.*** This includes potato chips and potato strips/sticks/rings, and represents by far the major portion of the snack food business. In the UK (1995/96), within the potato chip market, the plain/ready-salted variety has been the most popular flavor, followed by cheese, cheese and onion, and salt and vinegar.
2. ***Quick fried.*** Pre-formed partly cooked pellets derived from potato, potato starch, and/or other cereal materials are fried at high temperature (i.e., of the order of 200°C) for a short time (10–15 seconds) giving expanded, light-textured products.
3. ***Extrusion-cooked.*** Cereal/potato powder mixes are extruded through dies at high pressure and temperature to give an expanded base which, after further drying, is enrobed with oil and flavor. A large number of products are available in this category and they display varying sizes, shapes, textures, and flavors.
4. ***Roasted.*** This is primarily represented by nuts, particularly in the form of dry roasted peanuts.
5. ***Baked.*** While products such as savory mini snack biscuits have always been a popular product, baking is increasingly being used to produce savory snacks that can have a much lower fat content.

Acknowledgments: Grateful thanks are expressed to E.D. Surman, former Development Manager of Courtaulds Films Polypropylene UK (at time of writing, part of Hoechst Trespaphan UK Ltd.), for various shelf-life reports and information on polypropylene films provided in this chapter.

Additionally, in recent years, largely as a result of consumer demand and demographic changes, low-fat and low-salt versions of some of the above categories have appeared and become popular. Premium and exotic varieties (e.g., prawn crackers, tortilla chips) (Anon., 1995; Booth, 1997) have also been developed to satisfy the growing adult specialty segment of the market.

The primary raw materials for savory snacks are therefore potato, either fresh or dehydrated, cereals (maize, rice, wheat, etc.), and nuts. Notwithstanding the important influence these raw materials have on the quality of the final products, by far the most important ingredient that can affect product shelf life is the oil used, either in frying or as a carrier for seasoning. Snack products as a group can be said to be consumed principally for pleasure, because of their unique texture and the enormous variety of flavors available. As such, well-planned and conducted consumer acceptability tests, in the form of appropriate sensory analysis, are an important part of the shelf-life evaluation of any snack product.

10.3 FACTORS AFFECTING SHELF LIFE OF POTATO CHIPS AND SAVORY SNACKS

In general, the factors that can influence the shelf life of foods are (IFST, 1993):

Intrinsic	***Extrinsic***
Raw materials	Processing
Product formulation and composition	Hygiene
Product make-up	Packaging materials and system
Water activity value	Storage, distribution, and retail display
pH value and acidity	
Availability of oxygen and redox potential	

While all the relevant factors undoubtedly have their influence on the shelf life of a snack product, some do have greater effects and deserve more attention.

10.3.1 The oil

Although the oil is primarily used as a heat transfer medium during the frying process, it does get absorbed and becomes a major part of the finished product. Ordinary potato chips contain as much as 35–40% edible oil and even in reduced-fat chips, the level of oil can still be significant (typically less than 25%). It is imperative therefore that care must be taken in the selection, transport, storage, handling, and use of frying oils since any abuse would most certainly result in reduced product shelf life (e.g., as a result of rancidity and off-flavor development) and in customer complaints.

Thus, a heavy duty frying oil, probably the most demanding use of any edible oil product, should have the following properties (Hunt, 1989):

1. Resistance of chemical breakdown such as oxidation, polymerization, or hydrolysis at temperatures up to 180°C.
2. Color stability so as not to discolor food.
3. Pourability, pumpability at ambient temperature.
4. Nutritional characteristics.
5. Oil stability not only at the high temperature of the frying process, but also during the prolonged life of the fried food when this oil may be exposed to air, for example, on the large surface area of the chips.

10.3.2 Seasoning

In the past, because of the nature and composition of savory snacks such as potato chips, microbiological spoilage during storage and distribution was not a problem. But, as demands for exotic taste and flavor continue to grow, shelf-life problems of a microbiological nature could possibly arise. Given that the microbial loads of even heat-treated spices and herbs can be significant (e.g., typical aerobic plate counts of heat-treated basil, garlic powder, ground black pepper, etc., are around 5×10^4 cfu per g), shelf-life study of products flavored with exotic spices and herbs should include, particularly during product development, relevant microbiological examination.

10.3.3 The process

Frying, be it deep fat frying or quick frying, is probably one of the most severe tests to which one can subject an oil in the manufacture of snack products. Since the quality of the frying oil is of paramount importance in determining the shelf life of fried foods, it is important to control the quality of the oil during use. In practice, the quality limits for a frying oil during use will depend on the food being fried, as different components can leach out of the food and lead to different characteristics in the oil. Although most snack product manufacturers now use fairly sophisticated continuous fryers with indirect heating systems, automatic filtration, balanced steam stacks, controlled oil flow, and efficient controls and instrumentation, all of which are aimed at keeping the oil in good condition, a few basic rules, based on the principles of good manufacturing practice (GMP), can help to maintain the quality of the oil during use (McGill, 1986; Burdon, 1989; Neilsen, 1993).

1. ***Turnover period (the time taken for the food to absorb the oil capacity of the fryer):*** Generally, as long as the recommended frying conditions are closely adhered to, the lower the turnover period, the less the oil will be degraded.
2. ***Specified frying temperatures:*** Temperatures higher than those specified promote oil degradation, and lower ones result in excessive oil uptake by the product.
3. ***Planned production:*** It is important to ensure by careful planning, a good, steady product throughput to avoid dramatic temperature changes as well as to reduce the risk of holding the oil idle in the fryer at frying temperatures.
4. ***Cleaning:*** Appropriate and efficient cleaning regimes should be employed to minimize the risk of contamination by cleaning materials or as a result of poor cleaning techniques, either of which can cause deterioration in the quality of the oil.
5. ***Plant maintenance:*** A preventive maintenance program should be operated, which can greatly reduce the risk of contamination of the oil by metals, in particular copper (e.g., from pumps, pipes, valves, fittings, or paddle shaft wear), which may act as powerful catalysts of oil oxidation.

As an indication of oil degradation one can measure the free fatty acid (FFA) level in the oil in the fryer on a regular basis (Burdon, 1989). Determination of the FFA level is a useful check, as it is an objective measurement that can be carried out easily during production runs, and can be automated. Experience has shown, however, that there can be big differences in the levels of FFA accepted in different production processes without indicating any difference in quality of the finished products. These differences more or less reflect the differences in the type of frying oil as well as fryer used, and the product being processed. Leatherhead Food Research Association in the UK has developed a rapid method based on measurement of the viscosity of the frying oil. A good correlation has been achieved between the rheological properties measured and the quality of the oil expressed by the content of the polymers, which are being formed as the oil decomposes (Gillat et al., 1991).

10.3.4 Packaging materials

One of the most important functions of food packaging is the protection of the product from environmental factors (see section 10.3.5 for further discussion) such as light, water vapor, gases, and odors during its shelf life. The exact protection required, however, depends on the nature of the product. In respect of the shelf life of snacks, the following are the main properties of flexible packaging materials (e.g., coextruded biaxially oriented polypropylene [OPP]), which can affect the storage quality of a product.

1. ***Water vapor transmission rate (WVTR):*** For a given material such as coextruded OPP, this is primarily a function of film thickness (Figure 10–1). The lower the WVTR of a film, the better is the moisture

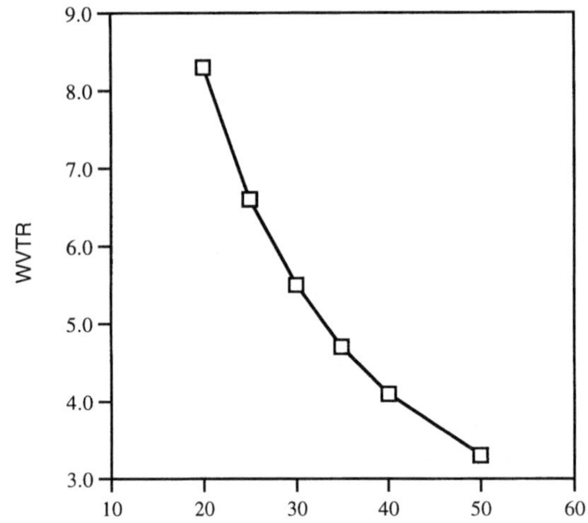

Figure 10–1 Permeability of coextruded OPP films to water vapor. (Source: CFP, UK.)

barrier. Moisture barrier of a packaging film, of course, may be further enhanced by various processes such as coating with an additional barrier material, lamination, and metallization (Table 10–1).

2. ***Oxygen transmission rate (OTR):*** Similar to WVTR, OTR of clear OPP decreases as the thickness of the film increases (Figure 10–2). Permeability to oxygen is substantially reduced by coating with a suitable barrier material (e.g., acrylic) and by metallization (Table 10–1).
3. ***Optical density (OD):*** This is defined as OD = Log_{10} (incident light / transmitted light). Determined using a suitable densitometer (at 23±2°C, 50±5% RH), it is a measure of the light barrier property of a film (Figure 10–3). The greater the OD of a film, the better is the light barrier and hence protection against light-induced changes (e.g., oxidation and bleaching).
4. ***Flavor/odor barrier property:*** The significance of flavor/odor barrier performance of a packaging film lies in its practical importance in relation to the flavor retention of packaged foods such as snacks, and also with respect to protection against tainting by extraneous sources. Snack products are usually highly flavored and contain an appreciable amount of oil, making them susceptible to contamination by taints; flavor retention and taint protection over relatively long shelf lives are therefore an important consideration. However, it has been shown by British Cellophane Limited (1982) that the odor barrier performance of coextruded OPP films should

Table 10–1 Typical water vapor transmission rates and oxygen transmission rates of some biaxially oriented polypropylene films used for snacks

Material type	*OTR (cm³/m²/24h/atm at 23±2°C and 90% RH)*	*WVTR (g/m²/24h at 38°C and 90% RH)*
clear (CFP)-30 μm	<1500	<5.5
opaque voided (CFP)-35 μm	1550	5.2
white pigmented (CFP)-30 μm	<1500	<5.5
metallized (CFP)-30 μm	<120	<0.8
PVdC coated 2 sides (ICI)-28 μm	16	5.0
PVdC coated 2 sides (ICI)-39 μm	5	2.3
PVdC/acrylic coated (Mobil)-26 μm	23	4.2
acrylic coated 2 sides (Mobil)-30 μm	650	4.0
metallized/acrylic coated (Mobil)-21 μm	1–2	1.0–1.5

Sources: Various technical data sheets, Courtaulds Films Polypropylene (CFP), ICI 'Propafilm' and Mobil Plastics Europe.

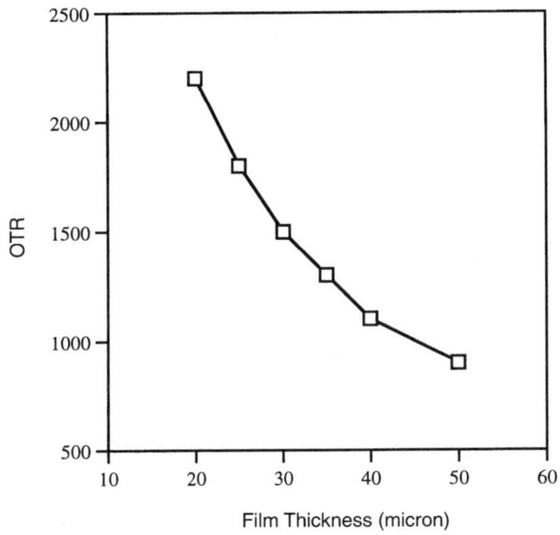

Figure 10–2 Permeability of coextruded OPP films to oxygen. (Source: CFP, UK.)

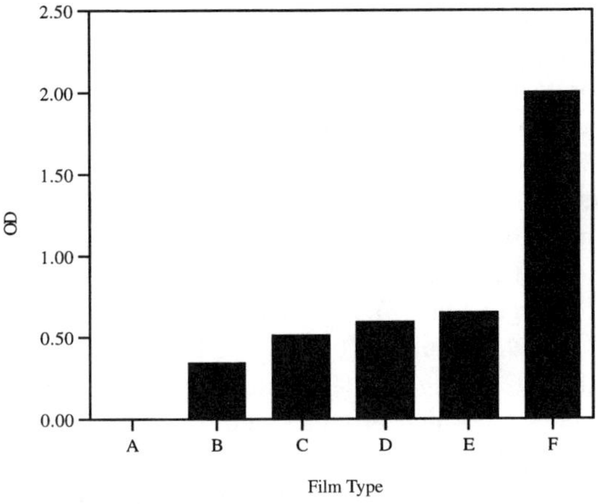

Figure 10–3 Optical densities of some coextruded OPP films. A: clear; B: opaque voided (35 micron); C: white pigmented (30 micron); D: white pigmented (35 micron); E: white pigmented (40 micron); F: metallized (20 and 30 micron). (Source: CFP, UK.)

not be assessed on the basis of oxygen permeability. Indeed because of the very many flavors that can be used and odors that may be encountered potentially, there is no general measure of the odor barrier property of a packaging film. Specific storage tests and/or calculations (Zobel, 1988) have to be carried out to assess this barrier property with respect to the flavor/odor in question (see Appendix 10–A for more information on flavor loss calculations).

During the shelf life of a snack, a number of deterioration processes take place together. They are texture loss due to moisture pick-up, rancidity development which requires oxygen and may be light-induced, and gradual loss of flavor in flavored varieties. In general, although all of these need to be controlled, relatively speaking, texture loss is more important to potato chips and related snacks, and rancidity development to nuts because of their compositional differences. Thus, currently OPP films are the most commonly used packaging materials for chips within Europe, while nuts are usually packaged in, for example, PVdC coated OPP/LDPE laminate or metallized polyester/LDPE laminate, depending on the value of the product.

OPP has qualities of toughness (against puncture and abrasion) and clarity and is rendered heat sealable by coextrusion with polyolefin copolymers (polypropylene co-ethylene) or by off-line coating (e.g., with PVdC).

Table 10–1 lists typical WVTRs and OTRs of some OPP films used for chips and snacks packaging within Europe. It must be stressed that these values are flat sheet permeability values; the protection of the product depends on the total barrier provided by the film pack. The sheet barrier properties may be altered by bad conversion techniques, or by mishandling such as scratching by poorly maintained packaging machines. Moreover, the seal integrity of the pack must be good enough to prevent leakage during the shelf life of the product. Pack barrier for snacks is thus a combination of the following:

- film barrier
- maintenance of film barrier
- seal integrity
- freedom from pack damage during transport, storage, and display.

In practice, to achieve the required level of protection of a snack, it does not necessarily follow that the highest barrier material is the best choice of packaging. The packaging material that is chosen will be a compromise between protective requirements, shelf life, aesthetics, and cost. In the chips sector, typically, a printed 30-micron white/pearlized OPP film is the most cost-effective form of packaging for commodity products with short shelf life (12 weeks or less) while metallized OPP laminates are used for larger family packs, sensitive products, and high added-value products. Certain products of the latter category may even be packaged with a 100% N_2 which provides additional protection against physical damage and oxidative changes.

Over the years, the popularity of multipacks, both of single and mixed varieties of standard packs, has been increasing. Multipacks provide variety and economy to the consumer, and a means of sales promotion to the manufacturer. The packaging of multipacks is more involved and there are many variations. In the UK, the most popular system uses a heat-sealed outside bag, usually of OPP, that takes 6, 8, 10, or 12 standard packs. A varying amount of manual operation is usually involved. The permeability of a multipack may be estimated by treating the different layers of film as resistances in series, analogous to the resistances of electrical elements in series:

$$l_t/P_t = l_1/P_1 + l_2/P_2 + \ldots\ldots\ l_n/P_n$$

where P_t = permeability of the composite
l_t = total thickness
P = permeabilities of the individual layers 1, 2, ...n
l = thicknesses of layers 1, 2, ...n

In practice, putting smaller packs into a bigger multipack bag gives a marginal laminating effect, resulting in extra protection of the products against environmental factors (Man, 1998).

In the nuts sector, commodity products with short shelf-life requirements are packed in nitrogen gas flushed, PVdC-coated OPP/LDPE laminates while higher added-value products and those with longer shelf-life requirements are packed in nitrogen gas flushed, metallized polyester/LDPE laminates. However, other materials, such as clear HDPE or PET, in the form of clear tubs, may be used for single or mixed nuts, especially during periods (e.g., festive seasons) when stock turnovers are expected to be high.

10.3.5 Environmental factors

Fatty food products such as chips can become unacceptable through the formation of rancid off-flavors due to oxidation of the oil content of the product through the process of oxidative rancidity. It is known that this reaction may be catalyzed by light and heat. Two previous investigations carried out by BCL in the 1980s looked at crisp rancidity in relation to the light barrier properties of various packaging films. The first study (1983) employed very severe test conditions where the test packs were exposed to fluorescent lighting continuously, i.e., 24 hours a day, giving a light intensity of 1000 lux on the packs which were stored flat. The second study (1985) was designed to simulate more closely the storage and display conditions to which commercial packs of chips were usually subjected. It also used a different type of fluorescent tube giving an intensity of 1700 lux, on a 12 hours on/12 hours off basis. Overall, useful conclusions have been drawn from these two studies and are summarized below:

1. Improved light barrier properties of packaging films give extended shelf life of chips in respect of rancid off-flavors. Thus, printed metallized OPP films give longer shelf life than printed OPP films which in turn give longer life than unprinted OPP films.
2. Prolonged storage under fluorescent lights (24-hour continuous exposure) at ambient humidity can cause chips to become rancid before any texture changes due to moisture uptake are detected.
3. The incorporation of a clear window in a metallized pack significantly reduces the acceptable shelf life of the chips in respect of rancid off-flavors.
4. Shingle storage of packs offers a longer shelf life than packs stored flat (Figure 10–4).

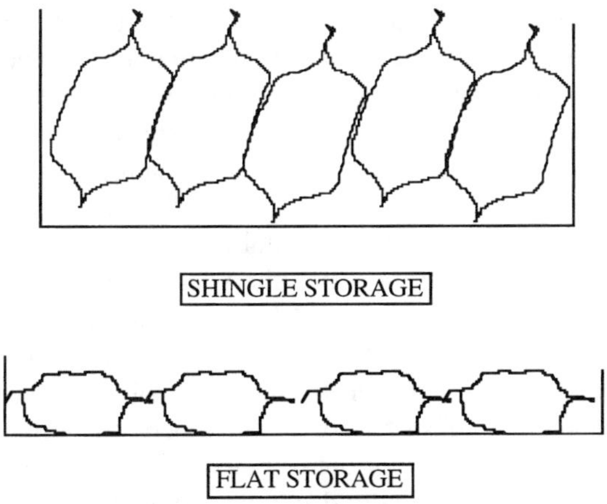

Figure 10–4 Shelf life of chips: type of pack storage. (Source: BCL, 1985; CFP, UK.)

5. Metallized films also prolong the shelf life of chips in respect of moisture uptake.

10.3.6 Legal standards

Within the European Union, the need to achieve harmonization of legislation in member states in respect of materials and articles destined for use in contact with foods and the need to protect the consumer from risks associated with the migration of harmful substances from packaging materials, have meant that food packaging migration is now under the control of the relevant legislation (Ashby et al., 1992). Today, in the UK, the Plastic Materials and Articles in Contact with Food Regulations 1998, which implement Commission Directive 97/48/EC, revoke and replace a number of earlier regulations:

- Set an overall migration limit for all food contact plastics of 10 mg/dm^2 of plastic surface area, in general, or 60 mg/kg of food.
- Establish "positive lists" of monomers and starting substances permitted for use in the manufacture of food contact plastics, many of which have a time limit on their use and some of which have specific migration limits.
- Lay down the basic rules necessary, including test times, temperatures, and simulants, for testing migration from food contact plastics, and thereby for checking compliance with the requirements of the regulations.

This piece of legislation applies to all in the chain from the plastics manufacturer, through the plastics converter, the food packer/manufacturer, to the food retailer. All have a definite role to play in the compliance of the overall migration and where appropriate, specific migration limits, though each may have a slightly different responsibility to bear. Regarding overall (global) migration values from coextruded OPP films, the following are typical results (Surman, 1993):

1. < 1 mg/dm^2 using distilled water or water of equivalent quality
2. < 1 mg/dm^2 using 3% acetic acid (w/v) in aqueous solution
3. < 1 mg/dm^2 using rectified olive oil.

The snack manufacturer obviously has a responsibility to ensure that all the information supplied to him by the packaging film supplier in respect of migration into food simulants is appropriate to his actual snack product, and this does not take into consideration any sensory analysis required additionally. Furthermore, since the snack manufacturer may also be expected to supply the retailer with compliance statements, it seems therefore some migration testing by the former cannot be totally avoided. This is particularly relevant in the case of a new/novel product and film combination, where migration testing can well be an important part of the shelf-life evaluation program.

10.4 SHELF-LIFE DETERMINATION

The nature and characteristics of snack products are such that the most common tests employed in the determination of their shelf life are

- chemical tests
- physical tests
- sensory tests
- microbiological examination where appropriate.

As an illustration, one method for ready-salted potato chips (BCL, 1985) involves storing sufficient packs of the products under controlled conditions of temperature (e.g., 25°C) and relative humidity (e.g., 75%) for a period of 12 weeks. Packs of the same products are also stored in a deep freeze to act as controls in subsequent sensory analysis. Stored products are tested at the start of the storage period, and evaluated at weekly intervals until the products are regarded as no longer acceptable by a trained taste panel of 8–10 assessors. Panel assessors are asked to give their opinion of the grading of the chips for flavor and texture directly against a control on a scale of 1–10 where:

10 = equal to control
9 = slight difference to control
8 = more distinct difference but still acceptable
7 = beginning to lose acceptability
6 = more distinct loss of acceptability
5 = very distinct loss of acceptability
4 or less = unacceptability

In addition to sensory tests, chips are used for chemical tests at every sampling occasion. To do these, the oil is extracted from the chips and the peroxide value and free fatty acid value of the extracted oil determined. Moisture uptake of the packs is monitored by weighing the packs immediately before submission to the taste panel. Figures 10–5, 10–6, and 10–7 show changes in flavor, peroxide value, and moisture uptake of chips packed in printed metallized OPP, during a shelf-life study in which the packs were stored flat and subjected to a 12-hour cycling exposure to fluorescent lights (BCL, 1985). It was found that the chips began to lose acceptability after about 8 weeks of storage. Since peroxide value gives a measure of the primary oxidation products which normally decompose quickly to secondary and tertiary oxidation products, it is apparent that changes in PV did not correlate as closely as changes in moisture uptake, with the gradual loss in product acceptability.

Besides laboratory tests, packs taken from production trials should be examined for seal integrity which can be a major factor limiting shelf life. Careful visual examination of returned packs from a travel test through the distribution and storage chain is also important to reveal transit hazards due to various mechanical

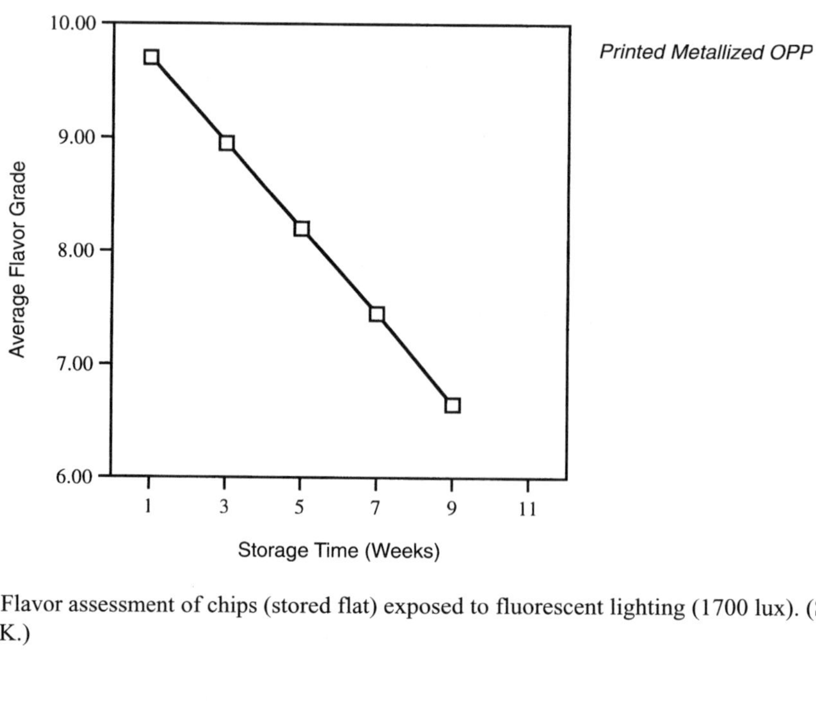

Figure 10–5 Flavor assessment of chips (stored flat) exposed to fluorescent lighting (1700 lux). (Source: BCL, 1985; CFP, UK.)

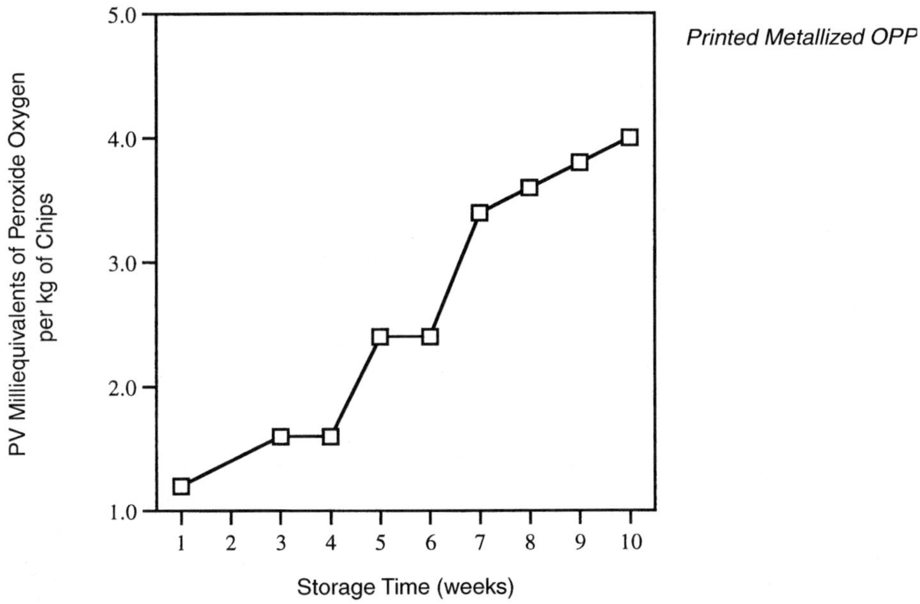

Figure 10–6 Peroxide value of chips (stored flat) exposed to fluorescent lighting (1700 lux). (Source: BCL, 1985; CFP, UK.)

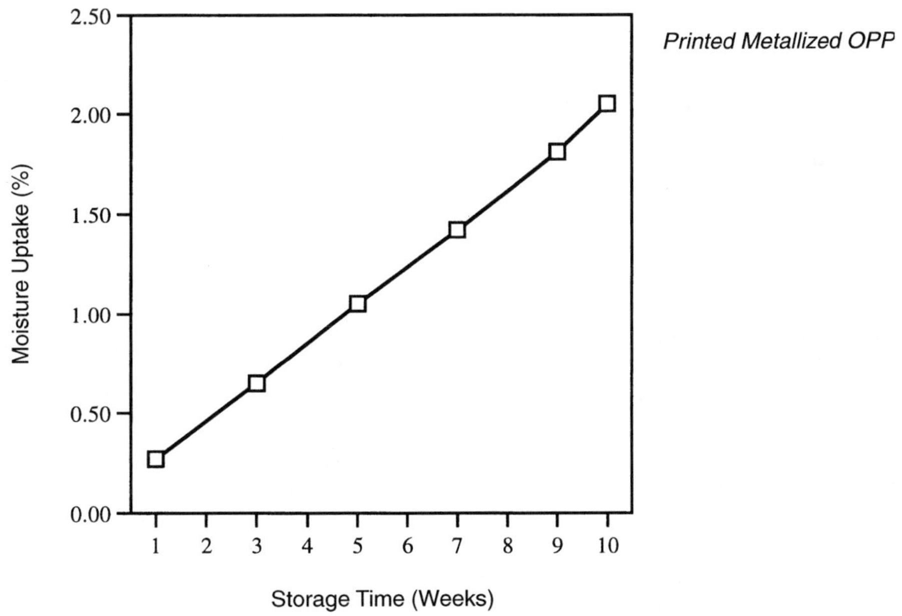

Figure 10–7 Moisture uptake of chips (stored flat) exposed to fluorescent lighting (1700 lux). (Source: BCL, 1985; CFP, UK.)

stresses and pressure, which can cause premature termination of product shelf life. Experience has shown that the level of sophistication of a distribution system (including handling, transport, and storage) can have considerable influence on the shelf life of snack products.

It is also possible to predict ambient shelf life of snacks from accelerated shelf-life tests, i.e., storage at elevated temperature and humidity. Despite obvious benefits, such tests tend to be very product specific and require experience and care in the interpretation of results. An alternative approach to accelerated tests would be to model shelf life and pack performance, using appropriate computer programs (Oswin, 1983; Hayes, 1992). For snack products, and potato chips in particular, there exist in the main, as discussed previously, two competing deterioration mechanisms, i.e., oxidative rancidity and moisture uptake. Regarding moisture uptake, the information required for modeling is well known (Labuza, 1982; Hayes, 1992):

1. the moisture-sorption isotherm data for the temperature in question
2. the packaging film properties such as WVTR
3. the storage conditions anticipated.

Armed with other film properties, e.g., OTR and permeation data, it is possible to model chemical changes such as oxidation in foods using modern computer techniques. Analytical models have been built for predicting the shelf life of oxygen-sensitive foods (Khanna and Peppas, 1982). Gnanasekharan and Floros (1993) gave a review of shelf-life prediction of packaged foods, which contains a useful scheme for modeling and predicting the shelf life of packaged food products.

REFERENCES

Anon. (1995) The UK food and drink industry—A sector by sector economic and statistical analysis. Euro PA and Associates, Cambridge, England.

Anon. (1998) *Food and Drink Statistics 1998—A Sector by Sector Guide to Agriculture, Fisheries, Food and Drink Statistics in the UK.* Euro PA and Associates, Cambridge, England.

Ashby, R., Cooper, I., Shorten, D. and Tice, P. (1992) *Food Packaging Migration and Legislation.* PIRA International, United Kingdom.

British Cellophane Limited (BCL) (1982) *An Investigation into the Odour Barrier Properties of Coextruded Polypropylene Film.* Courtaulds Films Polypropylene, UK.

BCL (1983) *A Study of the Effect of Light Barriers on the Onset of Rancidity in Crisps.* Courtaulds Films Polypropylene, United Kingdom.

BCL (1984) *Determination of the Odour Barrier Performance of Coextruded Packaging Films.* Courtaulds Films Polypropylene, United Kingdom.

BCL (1985) *Crisp Rancidity—the Effect of a Clear Window in Metallised Film on Crisp Shelf Life.* Courtaulds Films Polypropylene, United Kingdom.

Blenford, D.E. (1990) Satisfying a growing appetite for snacks, in *Food Technology International Europe*, Sterling Publications Limited, London, England, pp. 145–149.

Booth, R.G. (1997) Snack foods and breakfast cereals, in *Food Industries Manual,* 24th edn (eds M.D. Ranken, R.C. Kill, and C.G.J. Baker). Blackie Academic and Professional, Glasgow, Scotland, pp. 444–469.

Burdon, T.A. (1989) Rancidity in snack foods, in *Rancidity in Foods*, 2nd edn (eds J.C. Allen and R.J. Hamilton). Elsevier Applied Science, London, England, pp. 161–169.

Courtaulds Films. Technical Information Leaflets: SCB, SWH, SHP, SHM. Courtaulds Films Polypropylene, United Kingdom.

Gillatt, P., Kress-Rogers, E., Rossell, B. (1991) A novel sensor for the measurement of frying oil quality. *Lipid Technol.*, **3**(3), 78–82.

Gnanasekharan, V. and Floros, J.D. (1993) Shelf life prediction of packaged foods, in *Shelf Life Studies of Foods and Beverages—Chemical, Biological, Physical and Nutritional Aspects,* (ed G. Charalambous). Elsevier Science Publishers B.V., pp. 1081–1118.

Hayes, G.D. (1992) The modelling of food shelf life and package performance, using computer spreadsheets, in *Food Engineering in a Computer Climate*, Institution of Chemical Engineers, United Kingdom, pp. 215–223.

Hunt, R.W. (1989). Oils and fats, in *Food Processing—Proceedings of the Ninth British Nutrition Foundations Annual Conference* (ed R. Cottrell). The Parthenon Publishing Group, United Kingdom.

IFST. (1993) *Shelf Life of Foods—Guidelines for Its Determination and Prediction*. IFST, United Kingdom.

Khanna, R. and Peppas, N.A. (1982) Mathematical analysis of transport properties of polymer films for food packaging, *III Moisture and oxygen diffusion*. AIChE Symposium Series, **78**(128), 185–191.

Labuza, T.P. (1982) Moisture gain and loss in packaged foods. *Food Technol.,* **36**(4), 92–94, 96, 97.

Man, C.M.D. (1998) Packaging of snack foods, in *Modern Food Packaging*, Indian Institute of Packaging, Mumbai, India, pp. 522–529.

McGill, E.A. (1986) Deep frying, in *Impulse Foods*, Vol. 1 (ed D.E. Blenford). Food Trade Press, United Kingdom.

Mobil Plastics Europe. Product Information Sheets: Bicor® MB 666, MB 777, MM 778. Mobil Plastics Europe, London, England.

Neilsen, K. (1993) Frying Oils Technology. *Food Technology International Europe*, Sterling Publications Limited, London, England, pp. 127–132.

Oswin, C.R. (1983) Isotherms and package life: some advances in emballistics. *Food Chem.*, **12**, 179–188.

Propafilm. Product Information Sheets: Propafilm C, Propafilm CB. ICI, Herts, England.

Surman, E.D. (1993) Personal communication. CFP, United Kingdom.

Zobel, M.G.R. (1988) Packaging and Questions of Flavour Retention, *Food Technology International Europe*, Sterling Publications Limited, London, England, pp. 339–342.

Appendix 10–A

Flavor Loss Calculations

Work carried out by BCL (1984) resulted in the derivation of a simple equation based on Fick's law of diffusion, which can be used to estimate the likely permeation coefficients of a wide range of chemical substances through coextruded OPP as well as polyethylene, at low odorant concentrations. In the SI system, permeation coefficients have the unit – gm m^{-2} Pa^{-1} s^{-1}, which represents the weight of flavorant/odorant permeating a polymer specimen 1 meter thick, 1 square meter in area, per second, with a vapor pressure difference across the specimen of one Pascal (=10 ppm vol/vol). The permeation data, together with information relating to the pack contents and storage/display environment, provide a convenient method for estimating the transfer rates of flavors or contaminants through commercial flexible packs. Zobel (1988) has shown that the rate of transmission (R in g s^{-1}) of flavorant (e.g., vinegar flavor measured as acetic acid) through a packaging film is given by the following formula:

$$R = (PC \times VP \times A)/L$$

where PC = permeation coefficient ($g\ m\ m^{-2}\ Pa^{-1}\ s^{-1}$)
VP = vapor pressure of flavorant (Pa)
A = area of film (m^2)
L = thickness of film (m)

Strictly speaking, VP should be the difference in vapor pressure of flavorant between the inside and the outside of the pack; but for practical purposes, the inside pack vapor pressure, as measured by headspace gas chromatography, will suffice. Using this formula, it is possible, in the event that highly flavored snacks are being packaged, to conduct flavor loss calculations, thereby providing the reassurance that the chosen OPP film, for example, will give an adequate barrier.

Chapter 11

Chocolate Confectionery

A.V. Martin

11.1 INTRODUCTION

The shelf life of chocolate is the period of time during which it will retain acceptable appearance, aroma, flavor, and texture.

Chocolate is a high-energy foodstuff with a complex nutritional profile, containing 28–40% lipids (cocoa butter, possibly other vegetable fats, and milk fat), 50–70% carbohydrates, and 4–8% proteins. It is also an important source of mineral salts (Ca, Mg, Fe, Na, K) coming from cocoa liquor, and vitamins (riboflavin, vitamin A, vitamin E).

Despite this complexity, chocolate is very shelf stable due principally to the unique properties of cocoa. Cocoa solids contain a natural antioxidant in the form of tocopherol and cocoa butter breaks down to inoffensive short-chain fatty acids.

As a result, most chocolate products can be classified as "medium" or "long-life" products which, when packaged, should have a storage life under normal conditions of temperature and humidity of 9 months or longer. In the UK "normal" conditions are probably represented by a temperature of 20°C and 60–65% relative humidity.

Nevertheless, degradation of the product can occur at earlier stages whereby its physical and/or organoleptic characteristics no longer correspond to the parameters set by the manufacturer's quality standard and so are unacceptable.

It is the purpose of this chapter to examine the attributes of chocolate which provide this stability and how product degradation can occur. The various forms which chocolate confectionery can take will be considered and factors influencing their shelf lives. Methods of determining shelf lives will also be examined.

11.2 COMPOSITION OF CHOCOLATE

The precise composition of chocolate varies throughout the world due to differing national tastes and varying legislation which is mainly concerned with the percentages of cocoa and added milk solids, and amounts and types of permitted vegetable fats.

In many countries chocolate production is strictly controlled by law, in most cases restricting ingredients to sugar, cocoa butter, cocoa solids, and milk solids. Some countries such as the UK allow a percentage of cocoa butter to be replaced by vegetable fat, and others permit the use of nutmeat, whey, and so on.

11.2.1 Cocoa liquor

Cocoa beans comprise approximately 88% nibs, 11% shell, and 1% germ. Once harvested the beans are fermented, cleaned, and roasted and the nibs separated from the beans. The nib is a cellular mass containing about 55% cocoa butter locked within the cell structure. Grinding breaks down the cell walls, releasing the fat which, as it is liquefied by frictional heat, becomes a continuous phase in which the disintegrated cell particles are suspended. This fluid paste is called "cocoa liquor" or "cocoa mass."

11.2.2 Fats

The fats found in chocolate include cocoa butter, milk fat, and, in the UK, non–cocoa butter vegetable fat. This last group is commonly known as "cocoa butter equivalents" or "CBEs."

11.2.2.1 Cocoa butter

To meet legal requirements most chocolates contain at least 20% cocoa butter which has properties unmatched by any other fat. It is the cocoa butter which gives chocolate its unique flavor and exceptional shelf life through its oxidative stability. The characteristic physical properties of cocoa butter which give chocolate its special texture lie in its unique melting properties, i.e., it remains solid at room temperature up to 28°C yet melts fully and completely at 36°C (below mouth temperature). Cocoa butter is expressed from cocoa liquor leaving cocoa powder. From certain sources it must be deodorized to remove potentially harsh flavor tones and this process may reduce its antioxidant properties.

11.2.2.2 Milkfat

This is used for four reasons:

- for taste
- as a texture modifier—to soften chocolate
- as a fat bloom inhibitor
- to reduce cost.

Two main processes exist for the incorporation of milk into chocolate, which confer quite different tastes on the resultant products:

- the continental process using milk powder added to cocoa liquor
- the crumb process used in the UK and frequently in the US and certain European countries whereby milk, sugar, and cocoa liquor are condensed under vacuum and the resultant solid ground to a powder which is then added to the cocoa mass.

11.2.2.3 CBEs (Cocoa Butter Equivalents)

These are prepared from illipe, shea, and palm oils. They have a fatty acid composition almost identical to that of cocoa butter so that the physical properties of a mixture with cocoa butter in any proportion are practically the same as those of cocoa butter itself.

11.2.3 Sugar

Sugar is used to offset the bitterness of cocoa solids in plain chocolate. In milk chocolate, sugar, particularly in crumb form, has a great influence on the ultimate flavor—the combination of milk proteins and sugar through Maillard, or browning, reaction leads to the caramelized flavor characteristic of most UK milk chocolate.

11.3 MANUFACTURE OF CHOCOLATE

11.3.1 Mixing

Sugar (and crumb or milk powder for milk chocolate) is added to the cocoa liquor (or cocoa mass) and the preparation is thoroughly mixed.

11.3.2 Refining

This enables the smooth texture of the chocolate to be achieved initially by reducing the particle size. The cocoa liquor/sugar paste is passed through a series of roller refiners with extra cocoa butter being added to maintain its flow properties as the surface area of the solid particles increases.

11.3.3 Conching

Conching is the working of chocolate, milk crumb, and added fats in the form of cocoa butter, CBEs (if required), lecithin (emulsifier), and any flavors such as vanilla, into a fluid paste.

This operation further refines the texture of the mixture through continuous agitation in a large tank and coats all the particles produced in the earlier processes with fat to give the smooth mouthfeel characteristic of good chocolate. Conching also has the effect of modifying the flavor of the chocolate mass by reducing astringency and allowing the more muted and subtle pleasant flavor notes to develop.

11.3.4 Tempering

Before chocolate can be used for molding or enrobing (coating) it must be tempered so that it can set rapidly and uniformly with good permanent color, texture, and gloss. Chocolate can set in six states, only the more stable of which gives the desired gloss and snap. To achieve this stability it is necessary to cool down the chocolate with continuous mixing to produce a multitude of "seed" fat crystals and distribute them uniformly throughout the mass of liquid chocolate thus providing nuclei for subsequent crystallization. During this operation changes in viscosity can be used to obtain a measure of the condition of the chocolate.

11.3.5 Uses of chocolate

There are two principal ways in which chocolate is used:

11.3.5.1 Molding

This is the process in which tempered chocolate is deposited in warmed polished metal or plastics molds, then cooled by passing through a cooling tunnel to give rapid setting and good appearance. Good tempering will ensure that the cocoa butter will shrink in the mold and demold cleanly. In general metal molds may be more effective in that cooling can be transmitted readily through the mold whereas with plastics, cooling takes place primarily through the chocolate surface. Good mold design and molding procedure, however, should minimize this potential problem and enable the full benefits to be derived from plastics, i.e., lightness, quietness in operation, and flexibility which assists in demolding. Combination molds are now available in stainless steel with polycarbonate carriers which give the best of all worlds, but at extra cost.

11.3.5.2 Enrobing

This is the process in which centers are passed through a curtain of tempered chocolate and over a chocolate bottoming device to coat the bases. Sometimes the product is "double" enrobed to ensure extra coating if the center is especially vulnerable to external influences, or to complete covering of awkward surfaces. In these instances the first coating may not be require to be well tempered provided it is set before the second coating is applied. The property of shrinking on cooling of cocoa butter, essential for molding, must be taken into account in enrobing. If enrobed centers cool too rapidly, the chocolate could crack, but on the other hand, if it cools too slowly, it could end up discolored and soft—in either case the problem might not be apparent at the packaging stage and only emerge when the pack is opened.

11.4 TYPES OF CHOCOLATE CONFECTIONERY

11.4.1 Molded blocks

11.4.1.1 Milk or plain chocolate

This may contain:

- roasted nuts, e.g., hazelnuts, almonds, peanuts
- unroasted nuts/fruit, e.g., raisins
- fondant cream/praline
- layered wafer/biscuit
- cereal, e.g., crispies.

11.4.1.2 Molded aerated blocks

Aerated chocolate has been a very popular product for many years. The principal manufacturing processes are:

- ***vacuum aeration*** in which the tempered chocolate is subjected to a vacuum under which microbubbles of entrapped air expand leading to a cellular texture. Cooling then takes place, still under vacuum to avoid collapse of the aeration.
- ***gas injection*** where inert gas, generally CO_2, is injected into the chocolate under pressure. The product is then molded under atmospheric pressure.

11.4.2 Filled blocks

Molding of filled chocolate blocks can be done in two ways, each of which has its adherents. In straight molding the product is deposited into the mold in its entirety in one operation or a center is pushed into a mold full of chocolate. In "shell" molding the inside of the mold is in effect lined with a solid coating of chocolate to form a shell into which the center, be it a cream or a solid such as a biscuit, is filled or inserted, with additional chocolate to provide backing cover. This operation can be described as "fill-cool-tip-cool."

11.4.3 Count lines

While molded blocks or bars are generally segmented so that they can be eaten over several occasions or shared, count lines are intended to be eaten in one go and by one person. They are usually, but not exclusively, enrobed and can contain:

- wafer—layered with fondant cream or praline
- cereal, e.g., rice crispies
- biscuit
- fruit and/or nuts
- caramel (toffee)
- butterscotch cracknel
- nougat
- peanut butter.

The above list is not exclusive and it can be seen that most, if not all, of the ingredients can also be found in one form or another in molded products. It is, perhaps, worth mentioning that in molded lines caramel is a center in its own right whereas in a count line, it performs the secondary function of holding nuts, cereal, and other components together.

11.4.4 Assortments

These comprise a number of different sweets which can be enrobed or molded and may be miniature versions of the products already described. They frequently also include fondant cream centers which have been deposited in liquid form into shapes formed in starch, allowed to solidify and then brushed out and enrobed. Often the centers are formulated so that after covering they revert to a semi-liquid form. This method of producing centers has been superseded in many cases by "starchless molding" in which the fondant, for example, is deposited into plastics molds. This involves the use of release oil in the molds which must be carefully specified and controlled to avoid possible contamination of the enrobing chocolate and consequent loss of shelf life.

11.4.5 Straight lines

These comprise small sweets similar to those in an assortment but all one type in a pack. Sometimes this is done when sweets are unsuitable for mixing in an assortment where they could transfer flavor to others. This was originally the case with chocolate peppermints, but specialized "After Dinner" mints are now branded lines in their own right.

11.4.6 White chocolate

This consists mainly of deodorized cocoa butter, sugar, and milk powder and is produced in a modified version of standard milk chocolate manufacture. It can be molded or enrobed and used wherever normal chocolate would be found but as it lacks the antioxidant properties of cocoa liquor, and as the deodorizing process can reduce those of cocoa butter, it is much more susceptible to oxidative rancidity.

11.5 ATTRIBUTES OF MILK CHOCOLATE

11.5.1 Texture

1. ***Hardness:*** This characterizes the hardness of the chocolate. It is assessed by its resistance to a point and can be scored on a scale from "soft" to "hard."
2. ***Snap:*** This represents the sound intensity produced when breaking the piece of chocolate and can be scaled from "low" to "high."
3. ***Melting:*** Melting is the assessment of a solid product which quickly changes into a liquid or semi-liquid state under the action

of saliva in the mouth. It can be scored from "slow" to "quick."

4. ***Sticky or claggy:*** This describes the texture of chocolate that remains soft, sticks to the teeth, and persists in the mouth. Scoring scale is from "very sticky" to "non-sticky."
5. ***Smoothness:*** Smoothness describes chocolate particle size and texture by the impression of more or less coarseness in the mouth when eating the product. Scoring is from "very gritty" to "very smooth."

11.5.2 Flavor

1. ***Sweet*** is a basic taste whose reference is saccharosis. It is scored from "not sweet" to "very sweet."
2. ***Cocoa*** flavor is a specific complex flavor which refers to the basic note of the product. It is scored from "no cocoa flavor" to "very strong cocoa flavor."
3. ***Milky*** flavor is a general term designating the presence of a pure flavor or odor of milk. It is scored from "no milky flavor" to "very strong milky flavor."
4. ***Toffee*** flavor describes the flavor associated with specific sweets. It is scored from "no toffee flavor" to "very strong toffee flavor."
5. ***Cream/butter*** flavor designates the presence of pure flavor of cream or butter in a chocolate. It is scored from "no cream or butter flavor" to "very strong cream or butter flavor."
6. ***Acid*** represents the elementary taste caused by diluted aqueous solution of acids such as citric or lactic. It is scored from "no acid taste" to "strong acid taste."
7. ***Astringent*** refers to the ability of certain substances to cause contraction of tissues in the mouth and especially at the back of the throat. Scoring is from "no astringent sensation" to "very strong astringent sensation."
8. ***Rancid*** applies to fatty substances degraded by oxidation or hydrolysis releasing short-chain fatty acids, aldehydes, and ketones.
9. ***Butyric*** describes a piquant cheese odor deriving from fat lipolysis.
10. ***Cardboardy*** is a generic, ill-defined taste developed by oxidation of old cocoa and chocolate.

11.6 CAUSES OF DETERIORATION IN QUALITY

Chocolate, when stored under the correct conditions of temperature and humidity, is a very stable product due to its low moisture content (1.0%–1.9%) and its high fat content (28%–35% including 1.4%–1.6% of polyunsaturated fatty acids). The presence of tocopherols (vitamin E) in cocoa liquor and cocoa powder provides a high degree of protection against oxidative rancidity. Nevertheless, severe degradation may occur; fat bloom, sugar bloom, rancidity, and loss of texture can result if proper manufacturing and storage procedures are not followed.

The factors causing deterioration of chocolate products can be divided into two main classes: those which are inherent in the product itself and cannot be prevented by packaging alone and those which are dependent on the environment and may be controlled to a greater or less degree by choice and effective use of packaging.

11.6.1 Product factors

11.6.1.1 Quality of raw material

It goes without saying that the quality of the raw materials has a major bearing on the quality of the finished product and strict quality control must be exercised.

11.6.1.2 Quality of chocolate

Improperly tempered chocolate will continue to stabilize after the product is wrapped and has left the factory, leading to soft texture and the appearance of fat crystals, particularly on the surface of the product, causing the grayish, moldy appearance known as "fat bloom." This unsightly blemish can also be caused by incompatible oil or fat used as a lubricant in the manufacture of centers migrating through the chocolate or by the release of oil from high-fat centers such as roasted nuts, coconut, truffle, and so on. Injudicious use of "rework" in a product can also affect the quality of temper of the chocolate.

11.6.1.3 Oxidative deterioration

In the short term oxidative rancidity is unlikely to occur in appreciable concentration with plain or milk chocolate but it can relatively rapidly affect other ingredients such as high-fat centers or roasted nuts which do not have a full covering of chocolate. This can be the result of inadequate coating on enrobed centers or poor backing off on molded products. With milk or plain chocolate, after the lapse of a considerable period of time, staling or a "cardboardy" flavor can become apparent. White chocolate does not have the antioxidant protection of cocoa liquor and so is prone to oxidative rancidity, particularly light induced. Even with light barrier packaging its shelf life is shorter than that of milk or plain chocolate. The cost of eliminating oxygen from the packaging of chocolate would be prohibitive and not generally worthwhile in view of the relatively limited increase in shelf life which would result. It could only be justified if particularly sophisticated ingredients used in the product recipe were especially susceptible to oxygen and where the product could stand the oncost of oxygen scavengers and barrier packaging.

11.6.2 Environmental factors

11.6.2.1 Temperature

Temperature, especially high and fluctuating conditions, has a marked effect on the quality and shelf life of chocolate products with the potential to cause loss of temper, fat bloom, and accelerated staling. Although low temperatures will preserve flavors and the textural quality of the chocolate, bars containing wafers, and expanded centers, can crack under low-temperature storage. If the packaging of the bar is not moisture vapor proof the cracks will permit the entry of moisture to the center with consequent rapid deterioration.

11.6.2.2 Humidity

Storage in damp conditions, condensation due to packaging in damp conditions, or condensation of moisture in coolers or fondant centers with high equilibrium relative humidity (ERH) giving off moisture which is then trapped in the package can all cause "sugar bloom." This is similar in appearance to fat bloom but with a rough rather than greasy feel and is the result of the moisture extracting soluble sugar from the chocolate and then depositing it in the form of large granules on the chocolate after re-evaporation. Certain centers, such as cracknel or "honeycomb," are particularly hygroscopic and require extra protection against moisture pick-up compared with conventional chocolate products. This could involve the use of double chocolate coating to minimize the possibility of cracks through which the moisture could be drawn into the center. Other centers can lose moisture and dry out under some, not necessarily particularly extreme, conditions. In both these instances packaging plays a vital role in affording a reasonable shelf life to the product.

11.6.2.3 Off-odor/loss of flavor

This can be caused by external influences permeating the packaging material or, indeed, emanating from it. Storage of inadequately wrapped chocolate near strong smelling chemicals or display in shops close to strongly flavored sweets such as poorly wrapped mints can all cause unacceptable flavor pick-up. Some plastics packaging materials can cause flavor loss or "scalping" by absorbing the "high notes" but this is not likely with the materials most commonly used for packaging chocolate products.

Odor transfer from the inks used on packaging can cause taint but modern water- or alcohol-based formulations, properly applied, should minimize this risk. The increasing use of reclaimed board, for environmental reasons, on intermediate and outer packaging leads to a greater risk of taint which must be countered by enhanced barrier primary packaging.

11.6.2.4 Insect infestation

Chocolate products are very attractive to several varieties of insect and if storage or transit conditions are likely to be such as to result in infestation very careful consideration must be given to the packaging employed. It must be effectively sealed and resistant to creeping insects penetrating through folds and boring species which can drill through aluminium foil but fortunately appear to be frustrated by the flexible plastics in common use.

11.6.2.5 Light

As mentioned earlier, light-induced rancidity can be a problem with certain products such as white chocolate and chocolate containing poorly coated roast nuts. Although outer packaging should protect a product for most of its life, even short exposure on a shop counter or shelf can be sufficient to cause unacceptable deterioration.

11.7 CONTROL OF SHELF LIFE BY PACKAGING

11.7.1 Heat

Normal chocolate confectionery packaging is not designed to withstand high temperatures which are best avoided by storage and transit in controlled conditions. High temperature, approaching the melting point of cocoa butter (36°C), causes the chocolate to soften, drastically reducing its shelf life and no amount of subsequent cooling can reverse the process of deterioration.

A certain amount of insulation can be built into outer packaging by using foil lined or metallized board or expanded polystyrene, but experience shows that at best this is a short-term remedy. It is more likely that insulation only delays the effect of heat and in fact it may retain it after the pack is put into cool conditions.

11.7.2 Moisture

It is possible, by careful use of packaging material, to protect a product from external moisture almost indefinitely. The traditional foil and paper wrap used for blocks of chocolate is adequate for most products of this type stored, transported, and sold under temperate, well-regulated conditions; but if moisture protection is likely to be required and traditional appearance retained, the foil must be heat sealed. This requires the use of special machinery and the foil to be coated with either a heat seal varnish or low density polyethylene film. The latter is to be preferred as it gives a better seal and provides the foil with added protection against tearing, piercing by the corners of the product, projecting nuts, and similar kinds of physical damage, and insect penetration.

By contrast, the irregular shape of most count lines does not lend itself to the foil and paper band format while the nature of the product is generally such that more protection is required against moisture, even in temperate conditions, than is the case with block lines. For this type of product the horizontal form-fill-seal (HFFS) wrap, also known as "pillow pack," "fin seal," and "flow wrap," has been devised. This method takes a continuous web of flexible material, supplied on reel, and forms it into a tube around the product, sealing its edges together by heat and/or pressure as it progresses through the machine. This tube, containing the product, is then cut to a predetermined length and the open ends sealed, again by heat and/or pressure. This continuous flow wrap, by careful choice of the materials used, can give a very wide range of properties to the package. In particular it can provide a complete barrier to moisture vapor—indeed most materials which are capable of being heat sealed will also give an adequate barrier against moisture.

Where an unusually high degree of protection against moisture is required, either to keep it out of or to retain it within the product, it may well be necessary to use a laminate, e.g., of polyester and aluminium foil with polyethylene, as a heat sealant. The foil could be replaced by metallization of the two plastics films.

11.7.3 Oxygen

The flow wrap which provides an excellent barrier to moisture can also protect its contents against oxygen, provided the appropriate material is chosen. This would generally incorporate a PVdC (polyvinylidene chloride) coating which could also be used for heat sealing. The addition of a layer of aluminium either as foil or as a coating applied by a metallization process, will also enhance the oxygen barrier properties of a wrap.

Oxygen barrier materials are constantly being developed and PVdC can be replaced, if its chlorine content is considered to be environmentally undesirable, e.g., by silicon oxide or EVOH (ethylene vinyl alcohol), although these do not have the wide range of properties available from PVdC and are considerably more expensive.

Oxygen scavengers are available which can eliminate the oxygen contained within an oxygen barrier package. These have not been considered necessary for ordinary chocolate confectionery but have been used on occasion in Japan. Flow wrap equipment can be modified to gas flush a package, replacing air with nitrogen or carbon dioxide. This operation slows down the machine and necessitates the use of a heat seal medium, generally polyethylene, to obtain a sufficiently gas-tight closure.

11.7.4 Taint/loss of flavor

The film most used to flow wrap confectionery, coextruded, cavitated or pearlized biaxially oriented polypropylene (OPP) is a good barrier to external taint or product flavor loss. This property can be enhanced by coating the film with PVdC or acrylic instead of coextruding it with polyolefin copolymer. The latter is more common and cheaper but rather less efficient as a barrier, particularly to oxygen.

It should be noted that although materials which are good oxygen barriers are also good odor barriers it is possible to have good barrier to odor without necessarily having a corresponding oxygen barrier. Some OPP films are coated with acrylic on one side and PVdC on the other, thus giving heat sealability with a wide temperature range and protection against moisture, oxygen, and taint. Any film can have its barrier properties further improved by metallization and/or lamination, both of which, incidentally, can be said to give the wrap an extra quality appeal.

There is always the possibility that ink from the upper surface of the wrapping material can transfer to the inside in the reel and thence to the product. Correct procedures at the packaging converter should eliminate this but it is a potential source of taint which must be considered. The choice of inks and varnish must also be carefully made in order to eliminate the possibility of taint.

Residual solvents resulting from incomplete drying or oxidation of printing inks, varnishes, and sealants are the most likely offenders in this respect. Their presence can generally be detected by simple smell and taste testing but GLC procedures should be used to identify the source of the taint.

Over the years some plastics materials have caused tainting through the incomplete conversion of monomers or the use of softeners or plasticizers and others have absorbed volatile flavors from the product. These problems should have been overcome by the plastics processors and certainly have been in the case of OPP but nevertheless it is essential that packaging materials, especially those coming into contact with the chocolate, are thoroughly checked for potential taint and flavor loss before they are used.

11.7.5 Light

Pearlized OPP has been given a degree of opacity through a process of cavitation which is adequate to eliminate the "show through" or "optical staining" which was a feature of chocolate wraps made from clear film, albeit heavily printed. It is not, however, sufficiently opaque to provide much protection against light for particularly susceptible products such as white chocolate. For these products a lamination of film and aluminium foil, or at least a lamination of two metallized films plus a solid background print, is required to give sufficient protection for a reasonable shelf life to be achieved.

11.8 DETERMINATION OF SHELF LIFE

As the basic forms of chocolate confectionery have been on the market for many years it has been possible to assign reasonably accurate shelf lives for most products under varying climatic conditions. The most generally specified conditions are "Temperate" and "Tropical" although it is difficult to classify individual countries rigidly into these categories because of seasonal and regional variations. In addition, the quality of the conditions under which products are distributed and sold should be taken into account when determining shelf life.

Table 11–1 lists what may be regarded as representative shelf lives, characteristic of products adequately wrapped to withstand the effects of high relative humidities and to exclude light. These shelf lives must be regarded as the longest that can be assigned for the purpose of "best before" date on packaging materials. If there is any likelihood of the product being exposed to adverse conditions for an appreciable length of time, e.g., in the unrefrigerated hold of a ship in tropical waters, the shorter life appropriate to "Tropical" climate should be chosen.

11.8.1 Shelf life of new products

In general a shelf life can be assigned to a new product by analogy with an existing product, using the knowledge of the development confectioner and packaging technologist to choose the most appropriate reference.

Nevertheless it is essential to confirm this judgment by checking the performance of a new, or altered, product under a variety of conditions as well as normal storage.

11.8.1.1 Testing products for shelf life

A suitable range of conditioned cabinets should be available, for example:

Ambient 20°C 65% RH
Ambient 25°C 80% RH
Ambient 32°C 60% RH
Ambient 32°C 80% RH

Local circumstances may make other conditions more appropriate, e.g., the cold dry conditions of a Canadian winter, or cycling of temperature and humidity to promote the development of bloom to be expected under unregulated storage and thus the development of means to retard it.

11.8.1.2 Procedures

Samples should be kept in these cabinets together with a standard reference product of a similar nature, either an "own company" product or a competitor's product which is known to have a satisfactory shelf life and performance in the appropriate market and whose age can be established. Consideration should be given to the nature of the product and package and a decision taken as to whether additional tests, e.g., for light fastness, cold storage, or cyclical temperature or humidity changes are desirable and if they are, arrangements should be made to incorporate them into the test schedule.

Initially keeping tests are only intended to compare the keeping properties of a new or amended product with those of a standard product. With this in mind, it should be possible for product knowledgeable personnel to use their judgment and give a reasonably accurate shelf life based solely on an accelerated keeping test,

Table 11–1 Representative shelf life (in months) of some chocolate confectionery

Category of product	*Temperate conditions*	*Tropical conditions*
Milk chocolate	16	12
Plain chocolate	24	24
White chocolate	16	12
Fondant cream filled chocolate	18	12
Chocolate with nuts, etc.	12	9
Count lines and wafer/cereal centered products	12	9
Fat filled chocolate	12	9

confirmed by a long-term keeping test and long-term storage trial.

It is essential that all assessments of keeping tests should be made by expert staff. The standard product and its known response to test conditions will indicate to the expert any cabinet unreliability. Such a guide as to the reliability of storage conditions is particularly important when a new product is on test. A minimum of three assessors is required at the time of tasting.

New products may require a "fresh" sample for comparison at the time of tasting and this would be obtained by remaking.

11.8.2 Types of test

11.8.2.1 Accelerating keeping test

This test should last 6–8 weeks with the samples being examined weekly. The accelerated deterioration is caused by elevated temperature and increased humidity. A revised product deteriorating more than standard on this test should not usually proceed to production. Where a revised/new product keeps better than, or as well as, the standard it is assumed that it will have an equivalent shelf life. This, however, must be confirmed by a long-term test.

11.8.2.2 Long-term keeping test

This test would last up to 18 months and take place under ambient conditions (20°C and 60% RH) with four weekly examinations. Standards would again be used and this test should confirm the accelerated keeping test with which it initially runs concurrently. The range of keeping cabinets listed previously can be used for long-term testing to give a more detailed profile of the product's keeping characteristics.

11.8.2.3 Long-term storage test

This test would use production-made samples of the revised/new product, and would eventually supersede the long-term keeping test in confirming its shelf life.

11.8.3 Results of keeping test—scoring system

Chocolate products should be scored by the five characteristics or attributes listed below, although (4) and (5) will not always be needed.

1. appearance
2. flavor/taste
3. texture
4. color
5. others (specified, e.g., odor).

11.8.3.1 Determining the attribute score

Only the test product is scored, and the score depends on its comparison with the standard as shown in Table 11–2.

Table 11–2 Determining the attribute score in a keeping test

Test product score	*Result*	*Score criteria*
5	Acceptable	The test product has kept better than the standard product.
4	Acceptable	The test and standard products are of equal quality. They have deteriorated at the same rate.
3	Reassessment is an option	Test product is slightly inferior to standard but nothing undesirable has developed.
2	Reject	Test product is inferior to standard, but further work may be recommended.
1	Reject	Test product is much inferior to standard and further work is not recommended.

11.8.3.2 Overall product score

The product should also be given an overall product score (1–5) based on the agreed attributes and the above criteria. One attribute having a reject score means that the whole product must have a reject score. Using the keeping test scoring forms, each assessor (minimum of three) should independently taste the test and standard products, individually score the attributes and overall score and make appropriate comments on a keeping test form (Figure 11–1).

The person coordinating the keeping test should then collate the results and, in conjunction with the assessors, come to an agreed result. This agreed result, including comments, should be entered on a keeping test summary form (Figure 11–2), after each tasting session.

The test ends either at a pre-arranged time or when the product is judged to have reached the end of its acceptable shelf life.

11.9 PACKAGING MATERIALS—TAINT TESTING

As mentioned earlier, packaging materials can be a source of contamination leading to the early deterioration of chocolate. Milk chocolate in particular is very susceptible to taint resulting from inadequately controlled packaging.

Modern methods of printing and the use of, for example, water- or alcohol-based inks, together with a heightened awareness of the need for very strict quality control at all stages of the production and packaging of chocolate products, have greatly reduced the likelihood of taint from packaging.

PRODUCT: PRODUCT REFERENCE(S): REASONS FOR TEST: ADDITIONAL SAMPLE NOTES:			DATE OF TASTING CABINET DESCRIPTION: TEMPERATURE: HUMIDITY:	SAMPLE AGE WEEK/MONTHS:
ATTRIBUTE	COMMENTS STD PRODUCT	TEST PRODUCT	TEST SCORE	NOTES
APPEARANCE				
FLAVOUR/TASTE				
TEXTURE				
COLOUR (IF APPLICABLE)				
OTHER (SPECIFY)				
TOTAL PRODUCT				
RECOMMENDATION/ACTION AT THIS SAMPLING 1. Continue test as trial product: satisfactory: Y/N 2. Comments:		FINAL RECOMMENDATION - TEST COMPLETE	SCORE GUIDE 5. Better than standard 4. Same as standard 3. Slight difference, nothing undesirable 2. Inferior to standard 1. Much inferior	

Figure 11–1 A keeping test form.

Product:
Product Reference(s):
Reasons For Test:

Score Guide: 5. Better than standard
4. Same as standard
3. Slight difference, nothing undesirable
2. Inferior to standard
1. Much inferior

Attributes 1. Appearance 4. Colour (if applicable)
2. Flavour/Taste 5. Other - Specify
3. Texture 6. Overall Score

Against the attributes (1-6) enter the score (1-5) by CABINET

Date of Tasting	Age of Sample Weeks/Months (Delete as Necessary)	CABINET																													
		Description						Description						Description						Description						Description					
		°C			% RH			°C			% RH			°C			% RH			°C			% RH			°C			% RH		
		Attributes - Score																													
		1	2	3	4	5	6	1	2	3	4	5	6	1	2	3	4	5	6	1	2	3	4	5	6	1	2	3	4	5	6

COMMENTS:

Figure 11–2 A keeping test summary form.

Nevertheless incoming packaging materials should be regularly checked and three principal methods of testing are available.

11.9.1 Robinson test

This test is widely used throughout Europe. A piece of the test material, 20 cm × 22 cm, folded and perforated, not in contact with the product, is kept in a jar with 15 g of grated fresh milk chocolate, in the dark, for 48 hours at 20°C and 75% RH. A control is set up either without the material being tested or, if the material is printed, with an unprinted version of the material.

The evaluation of the test is carried out by a panel of assessors who compare blind the chocolate from the test with that from the control. The intensity of any off-flavor is evaluated on a numerical scale as follows (scored by half points):

0 No difference in flavor
1 Just perceptible difference in flavor
2 Just definable off-flavor
3 Definite off-flavor
4 Strong off-flavor.

The scores given by the assessors are added and the total divided by the number of assessors: e.g., eight assessors with scores of 2.5, 3.0, 2.0, 1.5, 2.0, 2.5, 1.5, 3.0 giving a total of 18; dividing the total by 8 gives 2.25 which on rounding up to the nearest 0.5 becomes 2.5. Acceptance levels are generally taken as 1.5 for unprinted material and 2.5 for printed.

11.9.2 Gas liquid chromatography (GLC)

By this method levels of residual solvents in printed materials can be monitored against agreed standards such as a global maximum of 20 mg/m^2 with a subtotal of hydrocarbons not to exceed 10 mg/m^2.

11.9.3 Aroma index

This method uses a Buchi Vapodest 100 or similar steam distiller to determine the presence and level of taint in packaging materials. Filtered condensate from the distiller is measured in a spectrophotometer at 280 nm and the absorbance multiplied by 100 to give the aroma index.

Packaging for chocolate products should be rejected if it has an aroma index higher than 35.

11.9.4 General

It should be mentioned that there are occasions when analytical instruments fail to pick up taints which are readily detected by a well-trained human nose or palate. For this reason the availability of a Robinson type test with experienced assessors is an essential backup to any form of scientific taint testing of packaged materials.

While it must be the responsibility of the material supplier to ensure that his product meets the agreed scientifically quantified criteria of GLC and/or aroma index testing, it is ultimately the chocolate manufacturer who must ensure that the consumer receives a product free from any contamination from the packaging designed to protect it.

Not all confectionery manufacturers can equip themselves to carry out all the appropriate scientific tests for taint, although all suppliers should, yet all must ensure that they have the trained staff for sensory analysis of the materials which they use. Indeed the same staff could well be those who carry out the analysis of keeping tests.

BIBLIOGRAPHY

Beckett, S.T. (ed.) (1993) *Industrial Chocolate Manufacture and Use,* 2nd edn, Blackie Academic and Professional, Glasgow, Scotland.

Briston, J. (1983) *Plastics in Packaging—Properties and Applications,* Practical Packaging Series, The Institute of Packaging, Melton Mowbray, United Kingdom.

Minifie, B.W. (1989) *Chocolate, Cocoa and Confectionery: Science and Technology,* 3rd edn, Van Nostrand Reinhold, New York, NY.

Oswin, C.R. (1983) *Package Life—Theory and Practice,* Practical Packaging Series, The Institute of Packaging, Melton Mowbray, United Kingdom.

Paine, F.A. and Paine, H.Y. (1992) *A Handbook of Food Packaging,* 2nd edn, Blackie Academic and Professional, Glasgow, Scotland.

Saxby, M.J. (ed) (1993) *Food Taints and Off-flavours,* Blackie Academic and Professional, Glasgow, Scotland.

Chapter 12

Ready-to-Eat Breakfast Cereals

J.A.K. Howarth

12.1 INTRODUCTION

Ready-to-eat breakfast cereals belong to the category of foods that are inherently stable and have a long shelf life. Thus, the requirements for determining shelf life are quite different from products which have a short shelf life, for example, chilled foods which may only be stable for a few days. The factors affecting the shelf life are also quite different; deterioration during life will probably not affect the safety of a breakfast cereal but will have an impact on consumer satisfaction.

Ready-to-eat breakfast cereals are designed to be eaten straight from the packet, requiring no preparation, and so do not include breakfast items such as porridge. It is therefore important for the manufacturer to identify the parameters most critical to consumer satisfaction and ensure that these are the characteristics preserved throughout the life of the product. This chapter gives a broad outline of the means of determining cereal shelf lives and includes some pointers and general guidelines for methodology which could also be applied to other "long life" products.

If determination of shelf life is new within an organization then one of the first steps is to ensure that the senior management fully understands the implications of the work. There must be clear standards laid down as to what denotes acceptable quality at the end of shelf life and everyone must be aware of these and accept them. Failure to do this at an early stage could lead to difficulties when trying to implement any new standards.

There are two main areas of interest when looking at stability of breakfast cereals—organoleptic properties and analytical values. Appearance, flavor, and texture of the product are important, as are analytical values such as vitamin/mineral content where the product has been fortified and a claim made. These areas are legally binding and it is important to have evidence to back any claims made regarding stability and analytical data (IFST, 1993).

Although it is not necessary with products like cereals to put a date of manufacture on the product (although a "lot" code is required), there is a "best before" date and consumers now expect there to be a fairly long time remaining after they have purchased the product before it becomes unacceptable. In particular, major retailers are now demanding that 75% of the shelf life is still available when the product is delivered to their depots.

Shelf-life testing and date-marking of the product can only give information on the stability of a full, sealed package—it cannot give information relating to the consumer who keeps an open carton in the cupboard for 6 months (IFST, 1993). Laboratory testing can give an indication of what could be expected but cannot be solely relied upon in setting the shelf life.

The manufacturer cannot be expected to guarantee a shelf life once the pack is opened, and when the product has been abused in some way.

12.2 THE PRODUCTS

The product group of "breakfast cereals" contains a wide range of products as can be seen on the supermarket shelves. These products can have quite different characteristics, while all being expected to have a similar shelf life.

The products can be separated into five groups which may need slightly different treatments.

12.2.1 Uncoated cereals

Examples of these are flakes, crisp rice, and shredded bran. These products are typically low in moisture content—usually below 4%—and are inherently stable if not exposed to the environment. These would traditionally be expected to have a long shelf life with very little change in product characteristics.

12.2.2 Coated products

Examples of coated products are sugar coated flakes, honey nut flakes, and coated extruded products. These products still have a relatively low moisture content, usually below 5%. They tend to be more hygroscopic than uncoated products and depending on the type of coating, a potentially unstable ingredient may be present. For example, a cereal with a nut-based coating can suffer from rancidity problems if not handled in the correct way.

12.2.3 Products with added components

Examples are flakes with raisins and flakes with mixed fruit. The moisture content of the whole cereal may be around 7%; however this could be made up of cereal at 3% and fruit at 12%. These are products which will show very different storage characteristics. Moisture transfer between cereal and fruit can be the critical factor in determining shelf life rather than the ingress of moisture from the environment. A product with a higher moisture content will generally be less stable than the one with a lower moisture content (Labuza, 1982).

12.2.4 Muesli products

These products have a fairly high moisture content, so moisture pick-up from the environment is not as critical as the stability of the ingredients themselves. A muesli product can only be acceptable for the time its least stable ingredient remains acceptable.

12.2.5 New products

These can take any form, or could be a combination of any of the particular cereal types mentioned above. If the product is a new combination then the shelf life must be determined, although previously gained experience can be used to give some guidance.

As discussed above, moisture content plays an important role in determining the shelf life of breakfast cereals. Since most products do not usually support microbial activity, microbial spoilage of the food is not a critical factor. Changes in moisture content can affect the organoleptic properties and this is the main criterion to be used.

Most cereals will reach an equilibrium moisture content naturally, which will remain stable for many years if stored in a relatively constant environment. Examples of this have been cereals which were discovered after 40 years and showed no signs of spoilage. The moisture content is usually higher than would be acceptable organoleptically. This only serves to emphasize the importance of organoleptic characteristics in storage testing.

12.3 FACTORS AFFECTING SHELF LIFE

There are a number of factors which will affect the shelf life of a breakfast cereal and these will have differing degrees of influence. All must be taken into consideration and as with most foods, the better the control in all the areas, the greater the confidence in achieving the shelf life required. The main factors which affect shelf life have been identified as:

- raw materials
- processing

- packaging materials
- climate and storage conditions.

The first three are within the control of the manufacturer and although the fourth cannot be controlled, the effects can be predicted to some degree.

12.3.1 Raw materials

The majority of raw materials used to manufacture breakfast cereals are to be processed in some way—the main exceptions to this being fruit/nut additions to products and muesli-type products. As mentioned earlier, microbial spoilage of breakfast cereals is very rare and the use of Hazard Analysis Critical Control Point (HACCP) technique can ensure that major hazards are either eliminated or controlled.

The main factor therefore concerning raw materials is that of consistently high quality and each material must have the critical factors identified which will allow manufacture of a high quality product.

Materials which may cause some concern microbiologically are ingredients such as milk powder, cocoa powder, dried fruits, nuts—particularly coconut, and any which does not undergo a sufficiently high temperature processing. The specifications for these materials should lay down stringent standards and the quality system must ensure that these requirements are met before the materials are used.

Potential causes of rancidity in a product can originate from nuts, and their quality must be carefully controlled to ensure that these materials do survive the long shelf life in an acceptable state.

Moisture content and equilibrium relative humidity (ERH) of fruit are critical, and depending on the quantity of fruit added to the cereal, they will affect the overall acceptability of the product.

High-quality consistent raw materials will ensure that the required shelf life will be met.

12.3.2 Processing

As with most foods, heat treatment extends the life of the materials and also ensures a minimum of microbial contamination.

Most breakfast cereals, after processing, have virtually no microbial count, but a low moisture content and ERH which will not support microbial growth.

Although breakfast cereals are processed, the conditions of processing are not as severe as those in the canning industry, for example. Cereals are manufactured to an image, the main criteria for acceptance being organoleptic properties, moisture content, and bulk density. That image will also have an acceptable range within which variation is acceptable. And so, while processing does have a definite influence on the shelf life, its main function is to convert the raw materials into highly palatable food products, with the production of certain distinct characteristics in the foods, such as attractive color, pleasant flavor, and crunchy texture.

12.3.3 Packaging materials

Packaging used for breakfast cereals serves two purposes—protection from spoilage and protection from physical damage. The most common form of packaging in use is that of a paperboard carton with an inner liner containing the food. The inner liner will provide most of the protection to the food while the outer carton provides protection from physical damage and also from potential tampering.

The main purpose of the inner liner is to protect from moisture transfer so as to preserve the product characteristics. The most effective type of liner will be determined during shelf-life testing. A liner which has barrier properties which are too effective can give other problems in storage, for example, the generation of unpleasant odors in the headspace. This is therefore a major consideration. Antioxidants can increase the expected shelf life and in some countries can be used in the liner material for this purpose (Labuza, 1982). As will be discussed later, compromises may have to be made on some occasions to allow for these factors.

12.3.4 Climate and storage conditions

As previously discussed, the liner type is crucial to preserving the shelf life of the product and the external conditions can have a considerable effect on the product. During determination of shelf life the climate of the market must be taken into account; for example, the requirements of products stored in Mediterranean/tropical climates are quite different from those destined for Northern Europe.

The destination country of the product is also of importance as transport facilities and hence the potential for picking up taints during transport and storage must be considered and, if possible, minimized. If there are poor standards of warehousing then the risk of infestation may be high. This knowledge may mean that a shorter shelf life is appropriate for some markets.

All the factors which may contribute to the length of acceptable shelf life in the product cannot be listed here. However, an attempt has been made to highlight the major areas. The most important factors will depend on the specific product and market, and the areas considered above will assist in identifying the most critical ones for that particular product. Methodology and useful checklists are available in other texts (IFST, 1993).

12.4 DETERMINATION OF SHELF LIFE

The initial determination of shelf life may have been carried out on a development product which will have given an indication of the expected shelf life. A product which remains on the market for a long period of time may need to have its shelf life reevaluated periodically as changes are made to raw materials, processing, or packaging materials. This emphasizes that once the life has been determined, this is not the end of the story—monitoring on a regular basis must continue to ensure confidence in the product. The methodology for determination and monitoring of shelf life of breakfast cereals can be divided into a number of areas all of which are important:

1. ***Initial determination of shelf life:***
   (a) identification of critical product characteristics
   (b) break point analysis (critical moisture content determination)
2. ***Shelf-life testing:***
   (a) samples
   (b) storage conditions
   (c) assessment of results
3. ***Monitoring activities:***
   (a) new markets
   (b) packaging materials changes
   (c) processing/raw material changes.

12.4.1 Initial determination of shelf life

Before beginning a shelf-life test there are some basic pieces of information which must be sought. These will set the ground rules for a particular product and will ensure that the correct test is set up to provide useful data.

12.4.1.1 Identification of critical product characteristics

When planning a test to determine or confirm the shelf life of a product, it is important to ensure that the most critical factors have been identified.

With breakfast cereals, each product can be quite different and the characteristics which are important for the acceptability of the product are those which must be tested.

Selection of the factors will also depend on whether the test is an initial determination or a confirmatory test when a fair amount of information about the product has been obtained.

For all breakfast cereals organoleptic quality attributes are most important as it is unlikely that any harmful substances will be generated during processing and storage. If this is to be the main guide, then the selection of assessors for sensory evaluation is critical. Details about this can be found in many relevant texts (Amerine et al., 1965; Lyon et al., 1992). One of the most important flavors/off-flavors is that of rancidity and it is well known that some people are more sensitive to it than others and the use of trained assessors may be useful.

Analytical data can also be used to evaluate changes occurring in the product and appropriate parameters should be selected—the most common of these being:

- moisture
- vitamins
- rancidity measures.

If there is no on-site laboratory, a reputable laboratory may be used. The same laboratory should be used throughout the testing period to maintain consistency.

Analytical tests together with sensory evaluation can give an accurate determination of shelf life, and it is important that both these are used (Birch et al., 1977).

12.4.1.2 Break point testing

The purpose of break point testing is to determine the moisture level at which a product is no longer acceptable.

Samples of different moisture levels are prepared and presented to selected assessors in a random order. The samples can be prepared in a laboratory or by selecting suitable samples from production runs. The target moisture level should be determined and samples close to this selected. The amount of water to be added to the product is calculated and weighed into trays. These trays are placed in a plexiglass chamber with the required quantity of product. The chamber should be sealed and allowed to stand until all the water has been absorbed—usually 4–7 days. Once the water has been absorbed the product should be allowed to equilibrate for one week in glass jars. After checking of moisture content the samples can be submitted to the sensory panel. The assessors rate the samples organoleptically and the moisture level at which the product is not acceptable is determined. This procedure should be agreed by a number of senior personnel within the organization as this will form the basis for determining the acceptable shelf life of all products.

Different products will have different break points depending on their composition and end use.

12.4.2 Shelf-life testing

The specific methodology used for a shelf-life test will depend on the type of product and the resources available. However, there are some general rules which will help to ensure that the data collected provide a realistic assessment of product shelf life.

12.4.2.1 Samples

The specific arrangements for sensory testing will depend on the available resources, but in general, a minimum of six experienced assessors are needed to give a meaningful result.

Accelerated tests require at least 18 cartons of cereal packaged in the same packaging as intended for storage. It is not usually necessary for each assessor to have a full carton, a bowl for each will be sufficient.

Ambient testing will also need at least 18 cartons of cereal which should be stored under typical conditions, e.g., in a distribution warehouse.

If enough food is envisaged to be left over after sensory evaluation for the analytical work, then one carton per assessor is sufficient. Otherwise a second carton may be needed for the analysis.

Results plotted graphically are more easily interpreted and trend lines can be drawn.

The shelf-life determination should extend beyond the proposed life of the product as this will give extra information concerning what may happen once the product is in the home of the consumer.

During testing, a reference sample available to compare with the sample under test is desirable. However, no acceptable method of preserving a cereal for this purpose is known (Stone et al., 1985). A freshly made sample will provide some reference, and the use of an experienced panel should be adequate.

12.4.2.2 Storage conditions

Storage under a number of different environmental conditions (including the use of accelerated storage) can give some predictive results.

In general, there are three types of testing conditions that can be used which have different purposes (see Table 12–1).

The most suitable conditions should be selected depending upon the product and the time available before shelf life must be established.

Table 12–1 Types of storage conditions and their uses

Test	*Suitable for*	*Not suitable*	*Time*
Ambient	All products	—	18 months
Accelerated 38°C	Flaked/puffed products	Fruit products, coated	18 weeks
Tropical 38°C 90% humidity	Investigating packaging	General shelf-life prediction	Variable

For example, in the development of a new product it may not be possible to complete an 18-month ambient test; an accelerated test could be used to give an initial prediction, followed by a full test afterwards.

12.4.2.3 Assessment of results

Data collected during a shelf-life test will consist of sensory as well as analytical data. Analyzing the data and looking at the correlation coefficient, percentage fit, and slope to assess for trends from the sensory scores and analysis results will give an effective prediction. The correlation coefficient between, for example, sensory scores and analytical results gives a good indication of whether or not there has been significant quality deterioration over time. In general, it is accepted that the higher the correlation coefficient, the more significant the observed change.

In the example shown in Table 12–2, there is little change in the appearance scores against time reflected in a slope of 0.01, a correlation coefficient of 0.12 and a percentage fit of 1.5.

Results plotted graphically are more easily interpreted and trend lines can be drawn as in Figure 12–1 (the higher the score, the higher the acceptability).

Using the results from the break point analysis, it can be determined when the product becomes unacceptable.

For analytical data, the critical point may be when a vitamin falls below the intended nutritional claim.

Addition of some vitamins, in particular vitamin C, can give some stability problems and this is affected by the method of application and storage conditions (Labuza, 1982).

Figure 12–2 shows little loss of vitamin C when applied to the product, in this case, separately from other vitamins.

This, however, changes when vitamin C is added together with other vitamins. While the vitamins are stable in solution, once on the product some changes are observed (see Figure 12–3). The graph shows a rapid initial loss and this could also result in development of off-flavors and odors.

In general, one result showing a marked adverse trend would not indicate a definite cause for concern, as this could be a one-off sample

Table 12–2 Appearance scores for a product stored under ambient conditions

Month	*Appearance*
0	7.2
1	6.2
2	5.8
3	6.5
4	5.3
6	6.2
7	6.2
8	5.8
9	5.7
10	5.7
11	5.7
12	7.3
13	6.0
14	6.0
15	6.2
16	6.0
17	7.5
18	6.3

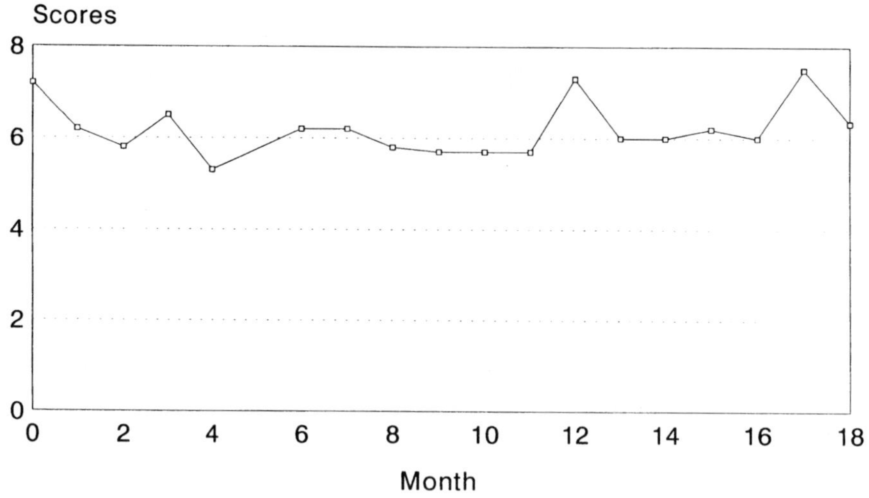

Figure 12–1 Appearance scores plotted to show trends more clearly.

with a fault in the packaging. Two consecutive results which show marked vitamin degradation would however need investigation.

The following examples illustrate the methodology of testing and how the shelf life can be determined.

12.4.3 Example 1—flaked/puffed product

The steps involved in testing this type of product, assuming it is a new product are:

1. identification of critical attributes
2. break point analysis

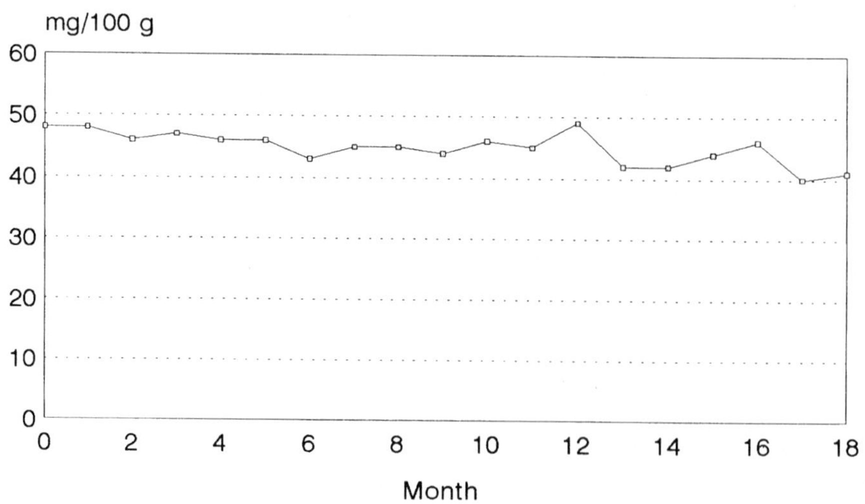

Figure 12–2 Vitamin C added as a separate spray showing little change on storage.

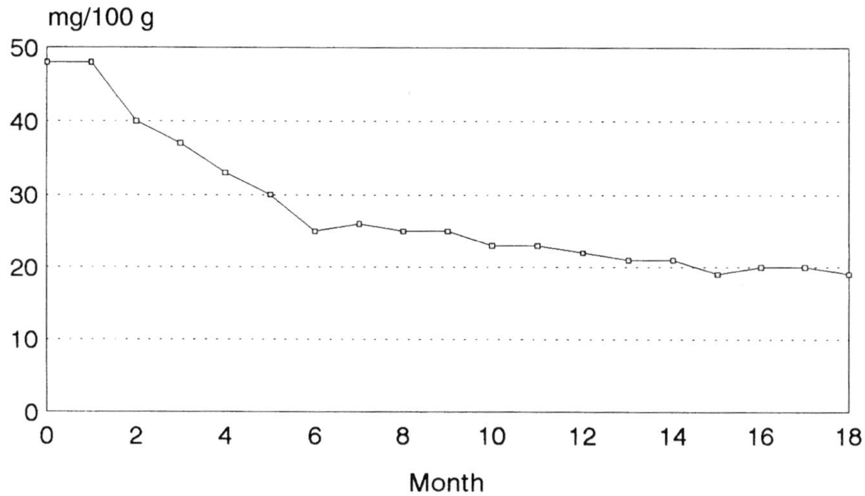

Figure 12–3 Vitamin C added as one spray with other vitamins, showing changes of vitamin C with storage.

3. test conditions
4. interpretation of results.

12.4.3.1 Critical attributes

As the product does not have any "high risk" ingredients, there are no specific factors which may need specialist assessors.

The most critical factors for this type of product are:

- appearance
- flavor/off-flavor
- texture (crisp/crunchy)
- moisture
- vitamins/minerals if fortified.

12.4.3.2 Break point analysis

This identifies the moisture content at which the product quality becomes unacceptable from an organoleptic point of view. It is assumed that for this product the break point moisture level is 6%. This will be used to confirm the final shelf life.

12.4.3.3 Test conditions

In the initial determination of shelf life, both accelerated and ambient testings are useful if run parallel to one another.

The accelerated test will give a prediction of the full shelf life and may highlight any early problems.

12.4.3.4 Interpretation of results

Figures 12–4 and 12–5 show data from storage of a product under ambient conditions. The test ran for 18 months and Figure 12–4 demonstrates that the product reached its break point in 16 months. This would suggest that the product should not have a shelf life longer than 16 months.

Figure 12–5 confirms the product acceptability by sensory evaluation. For example, the data for texture yield is

Intercept	8.25
Slope	–0.09
Correlation coefficient	0.72

It may be concluded that there were no real trends in the reduction of acceptability.

The same product tested under accelerated conditions showed little deterioration in product quality (Figure 12–6) and little increase in moisture (Figure 12–7).

From these results the predicted shelf life from accelerated testing would be 18 months (based on 1 week accelerated life being equiva-

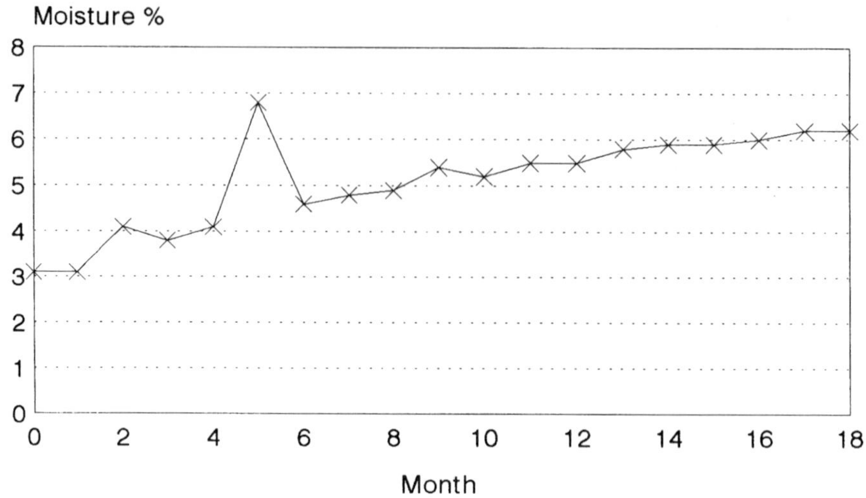

Figure 12–4 Moisture content of a flaked product stored under ambient conditions.

lent to 1 month ambient life). However, with the ambient results the acceptable recommendation would be 16 months.

The results of the test can only be used to predict the expected shelf life when using the same product formulation, manufacturing process, and packaging materials. Any major changes will invalidate the prediction based on accelerated test alone.

Break point testing/tropical conditions can be used to assess different packaging materials without doing a full test.

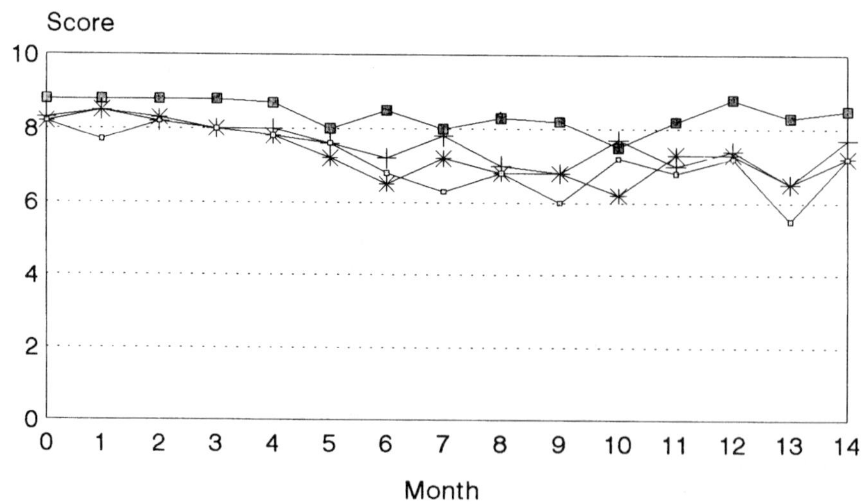

Figure 12–5 Sensory scores of a flaked product stored under ambient conditions. □, Appearance; +, texture; *, flavor; ■, off-flavor.

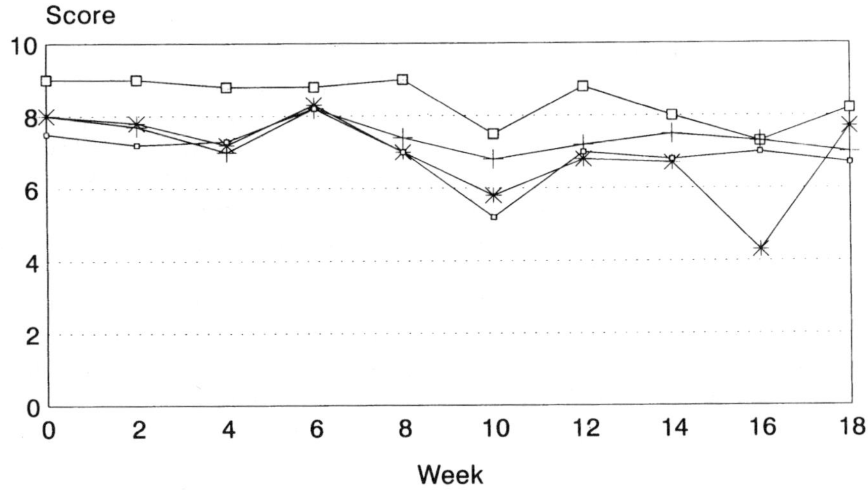

Figure 12–6 Sensory scores of a flaked product stored under accelerated conditions. □, Appearance; +, texture; *, flavor; ■, off-flavor.

12.4.4 Example 2—cereal with added components

Combining two or more materials together which have very different storage properties can give rise to complex shelf-life problems. All constituent parts must be acceptable at the end of the shelf life and the percentage combination can have a marked effect. A good example is flakes with raisins.

12.4.4.1 Identification of critical attributes

The following are identified as critical attributes:

- Appearance—plump, glossy fruit, "bright" flakes
- Texture—crisp flake/soft fruit
- Flavor—fresh, fruity flavor, fresh bran flavor

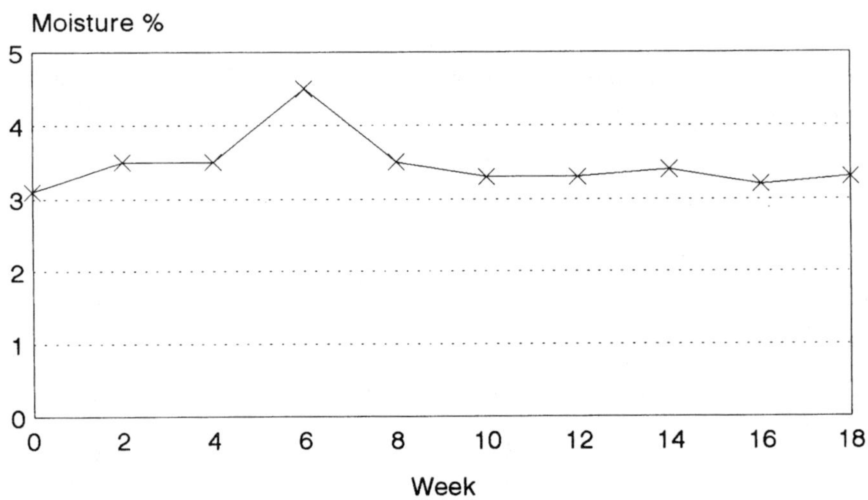

Figure 12–7 Moisture content of a flaked product stored under accelerated conditions.

- Analytical parameters will include:
  –moisture of flakes and fruit
  –vitamins/minerals.

12.4.4.2 Break point analysis

Taking a homogeneous product, the overall moisture can be determined when the product is no longer acceptable. However, in this case, the "total" moisture is not the critical parameter, but the values for the individual parts of the product are.

The break point for this type of product would have a maximum and minimum value, for example, a minimum of 2.0%, a maximum of 8.0% with a maximum for cereal at 7.0%.

12.4.4.3 Test conditions

Only ambient test conditions are suitable for products with fruit added. Accelerated tests, i.e., at high temperature, will cause premature hardening of the fruit and will not give an accurate prediction. Accelerated tests may be used to look at the stability of nutrients but not organoleptic properties.

For sensory assessment, information should be obtained separately for fruit and flake texture.

12.4.4.4 Interpretation of results

In Table 12–3 it can be seen that there is little change in the organoleptic properties of the product. However, there are clear trends in the moisture contents of the flakes and fruit. The previously determined break point of the product would, from these figures, suggest an acceptable shelf life of about 14 months, based on moisture content. The organoleptic properties confirm this. However, with this type of product experience has shown that organoleptic properties are more critical.

Another example (Table 12–4) shows a marked deterioration in fruit texture and for this product a shelf life of only 8 months would be acceptable.

In this situation, a decision then has to be made as to whether to modify the product or packaging, or offer a reduced shelf life. Any changes to the product/pack should require retesting.

Table 12–5 shows the effects of different levels of fruit addition. The higher level affects flake quality while the lower level results in deterioration of the fruit.

Table 12–3 Data from the shelf-life test of a product over 13 months showing organoleptic properties and moisture content

Month	*Appearance*	*Texture fruit*	*Texture flake*	*Flavor*	*Off-flavor*	*Moisture flake*	*Moisture fruit*
0	7.5	7.25	6.5	7.5	8.5	4.1	11.4
1	7.5	7.33	6.33	7.17	8.6	4.1	9.3
2	7.17	7.17	6.0	7.17	9.0	4.3	8.7
3	8.17	6.5	5.67	7.17	8.3	4.6	8.3
4	7.8	6.2	6.8	7.6	9.0	5.0	8.8
5	7.33	6.67	6.5	7.5	8.6	5.0	8.1
6	7.67	7.17	6.17	7.5	8.1	5.4	8.8
7	7.83	5.67	6.33	7.17	8.1	6.7	8.5
8	7.4	6.2	6.0	7.4	9.0	6.6	8.1
9	7.6	5.6	5.6	7.2	9.0	6.5	7.0
10	7.4	6.2	6.0	7.4	9.0	6.6	7.9
11	7.6	6.6	6.2	7.0	8.6	6.9	8.2
12	8.0	6.5	6.0	6.5	8.3	7.2	8.4
13	7.5	6.17	5.33	6.17	7.5	7.2	8.2

Table 12–4 Texture scores for a product stored under ambient conditions

Month	*Texture of fruit*
0	7.3
1	7.3
2	7.2
3	6.5
4	6.2
5	6.2
6	6.2
7	6.0
8	5.7
9	5.3
10	5.3
11	5.3
12	5.0
13	4.8

With other materials, e.g., nuts, it is likely that oxidative change rather than moisture transfer will be critical. In this case, either the quality of the nuts or oxygen availability, or both, will be the life-controlling factor.

12.4.5 Example 3—coated products, e.g., extruded, coated cereal

This type of product can consist of any kind of "base" material, e.g., corn, oats, wheat, or rice, and can have various coatings, from a simple sugar coating to one consisting of honey, nuts, cocoa powder, and so on.

12.4.5.1 Identification of critical attributes

An oat based product with a honey and nut coating is taken as an example.

The most critical attributes will include:

- Appearance—dullness/glossiness
- Flavor—rancidity and other potential off-flavors, loss of honey nut flavor
- Texture—stickiness, softness
- Analytical parameters—moisture PV (peroxide value), FFA (free fatty acid), vitamins/minerals.

12.4.5.2 Break point analysis

As an increase in moisture content can cause changes other than softness, the break point for this type of product is usually lower than for other cereals. For this example a maximum moisture content of 5.0% is assumed to be the critical value.

12.4.5.3 Test conditions

Ambient conditions are most suitable; accelerated tests (high temperature) may cause some unusual effects, e.g., increased stickiness, which would not be seen in ambient conditions.

Table 12–5 Effects of different levels of fruit addition on fruit and flake texture

	15% fruit		*25% fruit*		*40% fruit*	
Month	*Flake texture*	*Fruit texture*	*Flake texture*	*Fruit texture*	*Flake texture*	*Fruit texture*
0	7.5	7.7	7.2	7.2	7.9	7.7
1	7.7	7.6	7.7	7.2	7.7	7.5
2	7.8	7.3	7.3	7.5	7.5	7.5
3	7.6	7.4	7.3	7.0	7.5	7.6
4	7.5	7.4	7.6	7.2	7.2	7.6
5	7.6	7.0	7.5	7.3	7.0	7.4
6	7.7	7.0	7.4	7.2	6.8	7.4
7	7.8	6.8	7.2	7.7	6.4	7.5
8	7.3	6.8	6.8	7.3	6.4	7.3
9	7.5	6.7	7.5	7.2	6.2	7.6
10	7.7	6.4	7.4	7.2	6.1	7.7
11	7.3	6.2	7.0	6.5	6.0	7.5
12	7.6	6.0	6.7	7.3	5.6	7.5

12.4.5.4 Interpretation of results

Figures 12–8 and 12–9 show clear trends in increase in moisture content and reduction of texture quality. Using the break point analysis moisture level, the product would no longer be acceptable after 8 months. The decision must then be made as to whether this life is acceptable. If the alternative is to reformulate then the product must have a full storage test before the shelf life can be confirmed.

Figure 12–10 shows a clear trend in FFA and PV results which would suggest potential rancidity and off-flavor development.

However, in this case there has been no deterioration in the assessors' rating of flavor (Figure 12–11) which suggests that the increases in PV and FFA are not reflected in the acceptability of the product. This again emphasizes that both sensory and analytical data should be used together as they provide complementary information.

12.4.6 Monitoring activities

Once the shelf life has been initially determined, it should be monitored on a regular basis to take into consideration changes in raw material/processing which might not be major changes, but could over a period of time cause a drift in product quality.

The frequency of re-testing will depend on the facilities available and the complexity of the product. For example a product containing "high-risk" ingredients may need to be re-tested every 2–3 years whereas a relatively stable product may only need to be re-tested every 5 years.

A program should be set up to ensure all products are reexamined on a regular basis. This will also give some confidence against the risks in shelf-life testing of looking at a typical situation rather than the worst case (Paine, 1992). Mathematical models can be used to predict shelf life from currently available data which can help to minimize repeat testing (Labuza, 1982).

If there are any changes to raw material or, more critically to packaging materials, then a full test should be carried out before the change is confirmed. This will minimize the risks involved.

12.5 CONCLUSIONS

In this chapter the main criteria for consideration when evaluating the shelf life of breakfast cereals have been discussed. As indicated, there are some points for general consideration which

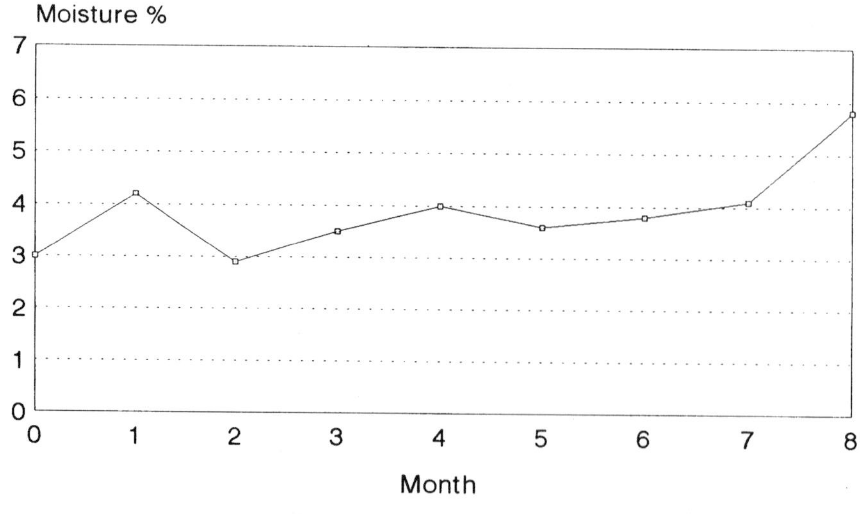

Figure 12–8 Moisture content of a coated product during storage. Break point level 5%.

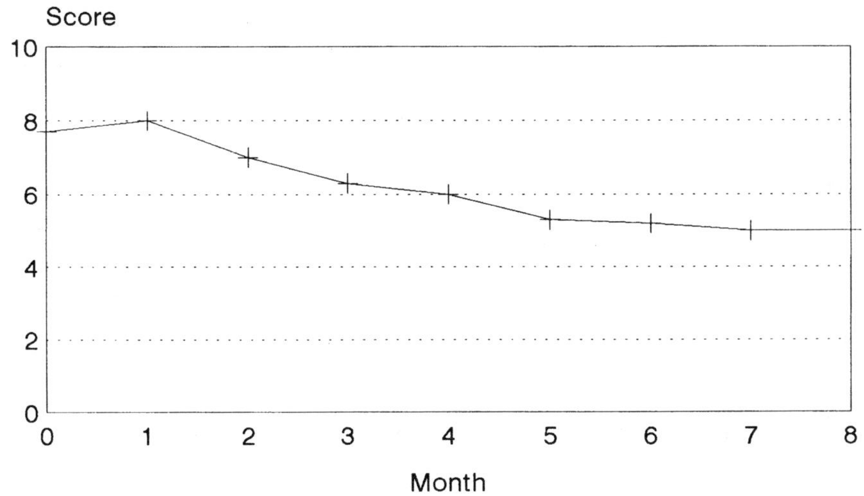

Figure 12–9 Texture scores of a coated product during storage. Break point level 5%.

relate to all types of cereals. However the wide diversity of cereal products means that each product should be treated on an individual basis. Lessons learned from other products can be used, but this information should not be treated as absolute. Small changes in processing or raw materials can have a serious effect on storage life.

Each product category and individual company facilities must determine the exact nature of the testing carried out, and the examples quoted should give an indication of how the data collected can be analyzed and interpreted to provide useful storage information. It is important to design the program to suit individual requirements.

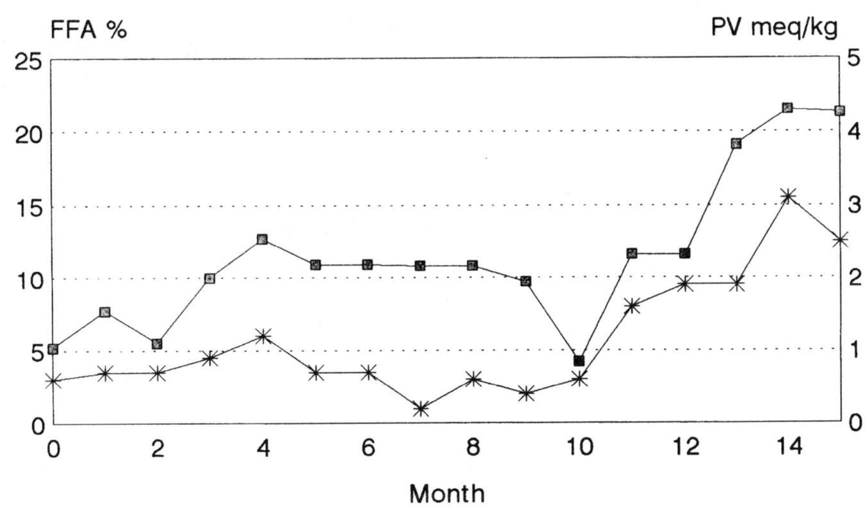

Figure 12–10 PV and FFA levels of a coated product during storage. *, FFA; ■, PV.

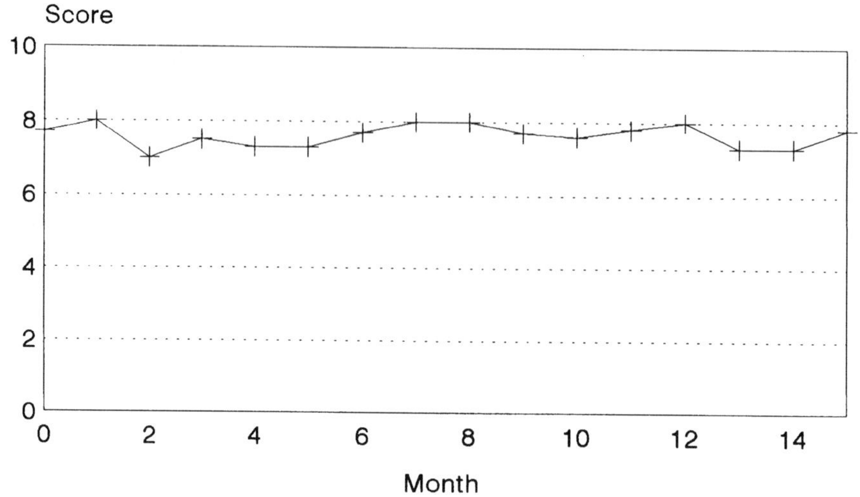

Figure 12–11 Flavor scores of a coated product during storage.

BIBLIOGRAPHY

Amerine, M.A., Pangborn, R. M. and Roessler, E. B. (1965) *Principles of Sensory Evaluation,* Academic Press, New York, NY, pp. 275–304, 321, 377, 494, 510.

Anderson, R.H., Maxwell, D.L., Mulley, A.E. and Fritsch, C.W. (1976) Effects of processing and storage on micronutrients in breakfast cereal. *Food Technology,* **30**(5), 110, 112–114.

Birch, G.G., Brennan, J.G. and Parker, K.J. (1977) *Sensory Properties of Foods,* Applied Science, Barking, Essex, England, Ch. 8, p. 113, Ch. 13, p. 213.

Borenstein, B., Caldwell, E.F., Gordon, H.T., Johnson, L. and Labuza, T.P. (1990) Fortification and preservation of cereals, in *Breakfast Cereals and How They Are Made* (eds R.B. Fast and E.F. Caldwell) American Association of Cereal Chemists, St. Paul, MN.

Fast, R.B. (1987) Breakfast cereals: processed grains for human consumption. *Cereal Foods World,* **32**(3), 241–4.

IFST (1993) *Shelf Life of Foods—Guidelines for Its Determination and Prediction,* IFST, London, England, pp. 3–5, 35.

Labuza, T.P. (1982) *Shelf Life Dating of Foods,* Food and Nutrition Press, Westport, CT, pp. 101, 107, 108, 115, 149.

Lyon, D., Francombe, M., Hasdell, T. and Lawson, K. (1992) *Guidelines for Sensory Analysis in Food Product Development and Quality Control,* Chapman and Hall, London, England.

Maxwell, D. (1990) *Prepared Foods,* **159**(2), 87–88.

Paine, F.A. (1992) *What Is Shelf Life,* Food Technology International Europe, Sterling, London, England.

Seiler, D.A.L. (1988) *Food Science and Technology Today,* **2**(1), 37–41.

Stauffer, J.E. (1991) *Cereal Foods World,* **36**(1), 11–12, 14, 16, 19–20, 23, 26.

Stone, H. and Sidel, J.L. (1985) *Sensory Evaluation Practices,* Academic Press, London, England, pp. 264, 265.

CHAPTER 13

The Storage of Thermally Processed Foods in Containers Other Than Cans

M.R. Goddard

13.1 INTRODUCTION AND BACKGROUND

The market for thermally processed foods presented in containers other than the traditional can is, in commercial terms, a relatively young one. All common packaging materials—board, glass, and plastics as well as metal—are now represented in this market, the most recent entry being plastics.

In the case of these heat-stable high-barrier plastics containers both Maskell (1991) and Seldman (1992) gave good accounts of the history of this market, showing how the first products were test marketed in 1983 and how it was not until 1987–1988, when a number of both major branded food manufacturers (General Foods, Batchelor Foods) and high street retailers (Boots, Marks & Spencer) entered the market, that the area started to grow. Growth however has been rapid, and, for the most part, sustained through the 1980s and into the 1990s. This is demonstrated quite vividly by figures on the number of retortable plastics barrier containers sold in the UK over the period 1988–1992 (Wendt, 1992), growing from 5 million units in 1988 to some 30 million units in 1992. Other figures reflect this trend with the area continuing to show a value growth of around 8% pa to an estimated total value of £26 million in 1992 (Anon., 1992c).

Other major packaging formats for ambient-stable moist/liquid products are the board/foil laminate TetraPak and Combibloc containers, dominated by the fruit juice and dairy markets, and the glass jar.

Despite the fact that glass has a considerable history as the packaging for heat processed products (Appert's original experiments in 1809 used corked jars), usage has been declining until relatively recently. Fueled by the growth in the pasta ("pour over") sauce market and more recently in the conversion of the "cook-in" sauce market from cans and board/foil laminates to glass, the format has gained favor again, partly due to its "green" image, transparency, and recyclability.

The non-pasta sauce market in the UK started with the launch of a range of canned cook-in sauces under the RHM "Homepride" label in 1974. The first launch in carton (Colmans) took place in 1987 and since then growth has been rapid (around 20% pa in terms of volume) with the market having an estimated market value in the UK of around £60 million (Anon., 1992a). The pasta sauce market is somewhat younger, having started in the UK with the launch of Campbell's "Prego" range in 1981. This market is almost entirely glass based and has also shown very rapid growth (21% volume increase in 1991) with an estimated UK market value of £65 million in 1992 (Anon., 1992b).

None of these product areas has any appreciable showing in cans. This review, therefore, focuses on these product areas and concentrates on the shelf life of heat processed foods, packed in plastics packaging. It is the case however that many of the observations and procedures will be equally applicable to other packaging formats, differences in shelf life between packaging formats being related, in the most part, to variation in permeability of the actual packaging materials and their sealing systems to external factors affecting product stability, primarily light and oxygen.

13.2 PRODUCT GROUPS AND THEIR CHARACTERISTICS

The products covered under the heading of this chapter can all be described as "meal components in sauce." The reason that such a blanket description can be applied to pasta sauces, stir-fry vegetable mixes, ready meals, and so on is that all of these products need the application of an accurate thermal process to provide shelf stability while optimizing quality. This requirement dictates that, although the ratio of sauce to components may vary and the components themselves be many and diverse, the presence of a continuous liquid phase is common to all.

In all these moist, ambient-stable, processed food systems a combination of an applied heat process, product formulation, and suitable packaging is used to confer microbiological safety and stability by either (a) producing and maintaining a microbiologically sterile system or (b) reducing microbial counts to an acceptable level and producing and maintaining a bacteriostatic system.

In the case of the product groups covered in this chapter these two categories are usually differentiated by the pH of the foods concerned. High pH products (pH ≥ 4.5) are generally cream sauces, gravy-based meals or sauces, or occasionally low water activity "dry" packed meal components such as potato products. Low pH products (pH ≤ 4.5) are generally tomato based or of the "sweet and sour" type with high levels of natural acidulants such as vinegar, fruit juice, or fermented sauces.

High pH products have to be subjected to a thermal process sufficient to destroy all microorganisms and in general terms this means a heat treatment equivalent to at least 3 minutes at 121°C ($F_0 = 3$). For low pH foods, with a pH of 4.5 or less, a less severe thermal process will achieve ambient stability as the pathogenic organisms of greatest concern—*Clostridium botulinum*—cannot grow under these conditions.

These basic microbiological rules hold regardless of whether the product is packed in glass, high-barrier plastics, or foil laminates and determine that quality change on storage due to microbial activity (i.e., spoilage) should not be a shelf-life factor in these products.

13.3 SPECIFIC FACTORS AFFECTING SHELF LIFE

In thermally processed ambient-stable foods deterioration is restricted, for the most part, to that caused by (a) enzyme activity, (b) chemical reaction, and (c) physical breakdown of components.

13.3.1 Enzymic activity

In general terms the thermal processes to which these products are subjected during manufacture are sufficient to inactivate all enzymes and it is rare for residual enzymic activity or enzyme regeneration to cause storage problems.

The possible exception to this is in the case of UHT high temperature–short time processes. Here enzyme reactivation has been noted, the most commonly quoted example being the reactivation of alkaline phosphatase, lipase, and protease in UHT processed milk (Kikakis, 1986). Possible storage effects are gelation due to protease activity and the emergence of off-notes in the flavor due to lipase action on milk fat. Such problems are however uncommon and are rarely if ever observed in retort processed systems due to the longer periods of heating involved.

13.3.2 Chemical changes

Chemical changes that take place in such systems are those related to either oxidation, nonenzymic browning (i.e., Maillard reaction), or resulting from nutrient breakdown.

13.3.2.1 Oxidation

Oxidation reactions are many and complex. They can use either headspace oxygen, trapped oxygen, or free radical oxygen released by enzymic activity or thermal treatment. They can be initiated photochemically; they are, in common with all chemical reactions, very dependent on temperature and time, and are often catalyzed by common food constituents such as trace metals. The results of these reactions are similarly varied but most commonly are observed as off-color and off-flavor development resulting from the oxidation of lipids, colors, and other reactive species. Under normal ambient storage conditions, i.e., 15–25°C, oxidation reactions are probably the most common cause of quality deterioration during storage.

Mogensen and Poulsen (1980), in a study comparing different packaging materials on the shelf life of UHT milk, showed a strong dependence of flavor quality on storage time and temperature and, most significantly, whether or not the packaging was permeable to oxygen.

Color changes however, although similarly dependent on storage temperature and time, did not show this link with oxygen and were concluded to be related to non-enzymic browning reactions (see below) rather than oxidation reactions. Trammell et al. (1986) demonstrated similar results for pasteurized orange juice where both the loss of ascorbic acid and, in this case, browning could be related to initial headspace oxygen concentration but where flavor change could not be related to oxygen concentration. Maillard-type reactions were once again implicated.

13.3.2.2 Non-enzymic browning or Maillard reaction

In general terms this reaction takes the general form of a condensation reaction between amino acids and reducing sugars and has the potential to occur in any system where these two substrates are present. Although the Maillard reaction is extremely complex the factors which control it are reasonably well understood. The application of thermal energy has a complex effect on the reaction. In general, as with all chemical reactions, heat accelerates and in many cases initiates the reaction, but due to the complexity of the sequence of the reaction steps the relationship is not a clear one (Benzig-Purdie et al., 1985). Moisture content is an important controlling factor. Very low (approaching 0%) and very high (greater than 90%) moisture systems show little or no non-enzymic browning, but few foods fall into these categories. pH has an inhibitory effect on the Maillard reaction when low and accelerates the reaction when high but there is little or no effect between pH values of 3 and 8, i.e., at the pH values of most foods. The nature of the substrate materials also affects the reaction. Proteins are less reactive than pure amino acids; polymeric carbohydrates such as starch do not take part in the reaction but pure reducing sugars such as lactose are very reactive. Many other chemical constituents of foods have either positive or negative effects on non-enzymic browning and this complex series of reactions can affect food quality in three ways. First, it can produce undesirable off-color and flavor—the brown discoloration and bitter notes in sterilized milk are usually quoted as an example of this. Second, the condensation reaction can lead to a reduction in nutritional quality of the food as it consumes carbohydrate and amino acids, and third, it has been suggested that some of the compounds produced by the latter stages of the reaction may be anti-nutritive in nature.

13.3.2.3 Nutrient breakdown

Loss of nutrients from a processed food is strongly influenced by the severity of the applied heat process and by the time and temperature of storage. The effects of thermal processing have been reviewed by Karmas and Harris (1988) who provided a detailed compilation of available kinetic data relating rates of loss of a range of nutrients to thermal processes. A general conclusion is that high temperature/short time (HTST) processes and low storage temperatures favor nutrient retention but that all processed foods exhibit some degree of nutrient loss. Similarly Kramer (1977) reviewed the available literature on the effects of storage on fresh, frozen, and canned foods and concluded that in all cases the lower the storage temperature the better it is not only for nutrient retention, but also the "aesthetic" properties of the food, i.e., appearance, taste, and so on.

13.3.3 Physical breakdown

Although there is the potential for physical change in many of the components of thermally processed ambient-stable products, e.g., texture change in meat, softening of vegetable components, these are rarely a problem. The major reason for this lies in the fact that the initial heat process to which the food has been subjected, particularly if this is a sterilization process, has already effected far more change on the system than that possible by even the most elevated storage time/temperature combination.

The two main physical factors influencing the shelf life of these foods relate therefore to the continuous aqueous sauce phase rather than to the discontinuous component phase and to two phenomena, namely starch retrogradation and emulsion breakdown.

13.3.3.1 Starch retrogradation

As previously mentioned most thermally processed foods contain an appreciable percentage of water-based sauce or gravy both to add taste to the other components and to facilitate the heat processing operation. In many cases the texture of this sauce component is controlled by the addition of thickening agents such as edible gums and starches. In actual fact careful selection of these thickeners can also improve the efficiency of the applied thermal process by providing a system with heat transfer properties optimized for each stage of the process through control of the temperature/viscosity profiles (Whistler et al., 1984). The most commonly used thickeners are those best suited to this purpose—chemically modified starches.

A great deal has been written regarding the physical and chemical nature and behavior of starch and it is not within the remit of this chapter to go into this topic in any detail. For a very complete reference work on this subject the multi-volume treatise *Starch: Chemistry and Technology* by Whistler et al. (1984) is recommended. There is one property of cooked starch systems which is of importance in the storage stability of many ambient-stable moist foods and that is the phenomenon known as "retrogradation." To understand this phenomenon a little further explanation of the physico-chemical properties of starch is necessary.

Starches are the favored thickeners for savory foods because of the short and clean texture generally desired in such systems, but, because of the relatively delicate nature of cooked native starches such as cornflour or potato starch, chemical modification is required to give the starch the necessary stability to processing.

In most applications the thickening and texture-giving properties of starch are due to its presence in the hydrated form as swollen granules entrapping concentrated aqueous biopolymer solutions. These biopolymers comprise repeating anhydroglucose units either linear in nature and connected by alpha 1–4 glycosidic links (amylose) or with a branched structure, the anhydroglucose units being again connected by alpha 1–4 links but also by an alpha 1–6 link at selected sites (amylopectin). In the presence of heat and water these glucose polymers move from a crystalline to a solution state, binding water of hydration and "swell." Starches are chemically cross-linked to "toughen" the granules and to prevent (or at least reduce) granule disintegration through the application of heat, shear, extremes of pH, or ionic change. Cross-linking is achieved by treatment of the native starch with di- or polyfunctional reagents capable of reacting with the hydroxyl groups in and between both the amylose and amylopectin molecules to form chemical bridges which restrict swelling of the hydrated polymers and hence of the starch granules.

Retrogradation is a phenomenon related to the tendency for dissolved starch molecules, particularly the linear amylose molecules, to move from a disordered to a more ordered state, but one quite different from that found in the uncooked starch granules. In solution the hydroxyl groups on parallel amylose molecules exert attractive forces (hydrogen bonding) and molecules line up into a network structure. This restructuring in solution manifests itself initially as the formation of a gel structure but with time, or when exacerbated by freeze-thaw cycling, the gel contracts and water is expelled (so-called weeping). In extreme cases this "dehydration" will result in amylose being precipitated out of solution as insoluble, high melting point granules, giving rise to a rough or grainy texture. This phenomenon is not entirely restricted to amylose. Amylopectin will also exhibit a form of retrogradation, giving rise to weak gels; but the strength of these is weak and as the melting point of crystalline amylopectin is below the serving and eating temperature of most savory products, this is not considered to be a problem.

The "toughening" of the starch granules or immobilization of polymeric material by chemical cross-linking also serves to reduce the amount of free starch molecules, in particular the amylose, which leaches out of swollen or broken granules and hence acts to reduce the chance of retrogradation. This is not a total solution however, since the percentage of starch molecules which can be cross-linked without giving rise to unacceptable reduction in thickening power of the starch is low—a maximum of one cross-link per 100 starch molecules and often as low as one link per 2,000 molecules. This still leaves a very high percentage of "free" molecules and an appreciable risk of retrogradation taking place, particularly in systems which have been subjected to extremes of heat and shear, e.g., sterilized systems. In order to overcome this, most modified starches are based on native starches low in amylose (and hence high in amylopectin)—the so-called waxy starches—and a second modification is applied to remove the unacceptable "long" or "slimy" texture associated with amylopectin.

In most cases therefore the tendency for heat processed foods to exhibit retrogradation has been formulated out but the growing need for "clean labels" and the progressive removal of chemical additives from foods is forcing a return to native starch and flour-based systems which may well result in a resurrection of the problem.

13.3.3.2 Emulsion stability

Many heat processed foods, particularly those containing cream-style sauces, contain appreciable levels of fat in the form of an oil-in-water emulsion.

In order to maintain the fat in this state the system requires the presence, at the oil/water interface, of an emulsifier—a material with both hydrophilic and lipophilic properties, which stabilizes the fat droplets against coalescence. If this material is not present at a sufficiently high level to stabilize the emulsion throughout the thermal process, or if its emulsifying properties are damaged by the applied heat, the emulsion will "break" and the oil and water phases will separate into two distinct layers.

The most common emulsifiers used in food systems are proteins—particularly milk proteins—and fat derivatives, both naturally occurring and synthetic, where hydrophilicity is derived from the presence of polar species attached to the non-polar glyceride moieties.

Fat-derived emulsifiers are in general stable to the time and temperature combinations that most conventional thermally processed foods experience. Proteins are considerably more heat labile, denaturing (changing structure), and changing functionality in the presence of heat. Proteins are also sensitive to chemical factors such as pH and free ion concentration, and under certain conditions in aqueous solutions, will, often as a function of time, denature and change in functionality. The most common visible consequence of protein denaturation is a loss of emulsifying properties, reflected in emulsion breakdown, and a textural change, such as gelation or granularity, in the aqueous phase. Any or all of these changes may be important factors in shelf-life determination.

13.3.4 Other factors influencing shelf life

Two other factors may affect the shelf life of thermally processed ambient-stable foods, both of which primarily influence the flavor of stored products. First, there is the possibility, particularly in plastics laminate packaging but also to a lesser extent in other laminate materials, of flavor migration into the packaging material or partitioning between product and packaging, and second, the much more common effect of flavor transfer between product components.

The latter effect is only really a problem in mixed component systems (such as fruit salads and ready meals) where, over a period of time, individual components start to lose their own flavor characteristics and to take on an average taste. It is rare for this phenomenon to produce truly negative quality attributes but, in order to ensure high quality in these products, it is often monitored as a guide to establishing optimum shelf life.

Flavor and color migration into packaging tends to be more of an aesthetic problem than a quality issue. The levels of loss of flavor and color into the packaging are very low and are more likely to result in discoloration of the inside of the pack rather than a significant loss of color or flavor in the product. What is of more concern is the possibility of constituents of the packaging material, e.g., plasticizers, migrating into the food. Of particular concern regarding the potential quality issues associated with the presence of these materials in processed foods are systems where the food has been subjected to high or prolonged thermal treatment while in contact with the packaging material, e.g., during a post-packaging pasteurization or sterilization process or during regeneration for consumption in either a conventional or a microwave oven.

In 1973 the EEC adopted a resolution to initiate a program for the harmonization of legislation on food-contact materials in order to remove technical barriers to intra-community trade and to protect consumers from any risk arising from the migration of packaging components into foods. The first part of the legislation on this issue related to plastics was completed in 1990 and issued as Directive 90/128/EEC, the so-called Plastics Directive, which has since been implemented across the EU.

The current legislation prescribes in some detail test methods for measuring the migration of packaging material components into foodstuffs and the maximum permitted levels of these components. It further spells out the nature of the tests required to be made with reference to the food simulants which should be used to replace actual foods, the temperature regimes to which the test materials should be subjected and the number of replicates/frequency of testing.

Although this type of testing is unlikely to form part of shelf-life evaluation, as all packaging or potential packaging materials should have already passed through this protocol, the methodologies included within it can often be usefully applied to the determination of specific storage problems, e.g., related to flavor loss or color migration into packaging.

13.4 DETERMINATION OF SHELF LIFE

From the information given above it can be seen that the storage changes which dominate in heat processed ambient-stable foods such as sauces and ready meals are mainly changes in flavor and color. Textural changes can occur in such foods but by and large these are insignificant when compared to the textural changes imposed on the system by the initial thermal process. Changes in color and flavor, therefore, are the major determinants of shelf life. This then governs the way in which changes are monitored and shelf life is set.

Textural change is limited and if it does occur, is usually restricted to the aqueous phase. Rheological measurements, such as the measurement of viscosity under specific conditions, can be used to monitor such change. In most instances however, if such change is to be a controlling factor in shelf-life determination, it will be easily recognized without the need for instrumental analysis.

Color change also tends to be relatively easy to recognize as a shelf-life determinant but in this case instrumental measurement is favored as a means of quantifying the change and finding the point of unacceptability.

Flavor change is the most common and most difficult shelf-life controlling factor to quantify in thermally processed ambient-stable products. Instrumental analysis of flavor change is very difficult. Food flavors, particularly natural food flavors, are complex and subtle, often comprising many hundreds of chemical constituents, many at extremely low levels. Many and various attempts have been made to measure flavor. Chromatographic methods such as HPLC (high performance liquid chromatography) and GC (gas chromatography) have been widely used (Barberio, 1986), both as means of producing a "fingerprint" of the flavor and, when coupled with mass spectroscopy, as means of identifying the specific aroma compounds. Experience suggests, however, that such techniques are best suited for the examination of specific problems, and are not suitable as quality control tools to be used routinely in the setting of shelf-life criteria (Fujinari, 1993).

In summary, therefore, the best approach to use in the storage testing and shelf-life determination of thermally processed ambient-stable foods is to rely on the human senses, using carefully controlled sensory analysis techniques and augmenting these, if necessary, with analytical or instrumental analyses.

The following sections describe a protocol for the storage, sensory testing, and shelf-life determination of an ambient-stable pasta product, a protocol which can be adapted to suit any of this type of product.

13.4.1 Storage protocol

The two major factors to take into consideration in setting up a storage trial for any product are the anticipated time and temperature of storage.

13.4.1.1 Time

As previously mentioned, the microbiological storage life of thermally processed products that have been processed and handled correctly is indefinite, which is why the determination of shelf life is based on the maintenance of acceptable sensory quality. A realistic assessment of the period between manufacture and consumption for this type of product gives a total life of 18 months. This figure is based largely on experience with other ambient-stable products, e.g., canned products, jams and preserves, and dry mix products, and probably errs on the side of caution for this new generation of thermally processed foods. Although it is known from in-home audits that canned and dry mix products may reach this age (or even older) before consumption, it is likely that there is less chance of this being the case with modern packaging formats, particularly with respect to the plastics laminate pack, the "technicalities" of which are less well understood by the consumer. Nevertheless, based on this assessment, storage testing is generally set to run for a minimum of 12 months. "Take offs"—the intervals at which samples are removed from storage for analysis—are usually scheduled at monthly intervals although it is common practice to sample at, for example, 0, 1, 2, 3, 6, and 12 months, only testing at the remaining take-off points if problems are becoming apparent.

13.4.1.2 Temperature

It is important to remember that an estimate of the temperature of storage of ambient-stable products is probably very different from the average ambient temperature they actually experience in practice. Temperatures within the distribution chain are likely, in Northern Europe, to range from zero in an unheated warehouse at night to in excess of 30°C inside a truck in midsummer. Because of this it is common to store at two temperatures—an average temperature, usually taken as 25°C, and an "abuse" temperature, usually taken as 35°C. It must also be noted that these are Northern European criteria. For tropical climates higher temperatures, e.g., 30°C average and 40°C abuse, are usually used.

One final point regarding temperature of storage is that if it is known that the product is likely, when in the distribution chain, to experience severe and regular temperature cycling; this should also be built into the storage protocol.

13.4.1.3 Other factors

Two other factors may need to be considered when setting up a storage protocol. First, there may be a need to consider and allow for the effect of light. This is certainly the case for products packed in glass or other transparent materials but may also be important for products packed in translucent plastics. If light is a factor, it is normal to set up two parallel exercises, with one set of samples being stored under suitable illumination, and a control set being stored in the dark.

Second, certain packaging materials are permeable to water vapor. This should not be the case for the materials used for thermally processed foods but needs to be taken into consideration if the protocol is to be extended, e.g., to dry mixes. In the case of moisture vapor transfer being suspected as a contributing factor to storage stability, a control experiment needs to be set up storing samples in both high and low relative humidity environments.

13.4.2 Sensory analysis

The definition of the shelf life of a food product in respect of sensory changes as a function of storage is based on the answering of one or more of the following questions:

1. How long can the product be stored without noticeable change in its sensory attributes?
2. How do sensory attributes change on storage?
3. How long can the product be stored before changes in sensory properties render it unacceptable?

The first two questions relate to objective analysis of the product—no qualification is required and no subjective assessment of quality is required. If question (1) is used to set shelf life a simple yes/no response to a discrimination test will suffice. There is no need for highly trained sensory assessors but a large number of people are required to ensure a statistically robust answer. Triangular tests and paired comparison tests will give a simple yes/no answer; ranking and magnitude estimation tests will add some measure of the degree of difference.

If question (2) is used then shelf life will either be estimated from determining the point at which selected sensory attributes of the product approach predetermined unacceptable levels or by the application of a secondary test which will answer either question (1) or (2). A descriptive test based on assessment of samples by a highly trained sensory panel is required to provide this sort of "magnitude" data and a profiling method such as quantitative descriptive analysis (QDA) is often used.

The third question is rather different. Changes in sensory properties are expected and a subjective assessment of the degree of change considered acceptable is required. Typical tests are based on either comparison, presenting paired samples (e.g., the test sample against a control) and seeking an absolute preference, or ranking, presenting samples either singly or in sequence and seeking a degree of preference. For this latter approach the "Hedonic" rating scale, derived by Peryam and Pilgrim (1957) is still the most widely used (Figure 13–1).

If this approach is used it is most important that the panel is composed of untrained "consumers"—to ensure a consumer-oriented rather than a product-oriented assessment—and at least 50 people to ensure statistical reliability. This approach is commonly used as an "in-house" evaluation technique and in this case it is unlikely that the panel will truly represent the consuming public. Sensory evaluations of this sort should therefore be regarded as giving guidance rather than absolute results. The only way to obtain the latter is via full consumer research.

Detailed guidelines for all of the methodologies outlined above together with a wide range of less widely used and more sophisticated tests are given in the book *Guidelines for Sensory Analysis in Food Product Development and Quality Control, Second Edition* (Carpenter et al., 2000). In the present case of an ambient-

Nine point Hedonic rating scale

Like extremely
Like very much
Like moderately
Like slightly
Neither like nor dislike
Dislike slightly
Dislike moderately
Dislike very much
Dislike extremely

Figure 13–1 Nine point Hedonic rating scale (Peryam and Pilgrim, 1957).

stable pasta in sauce product, the use of QDA to produce sensory "fingerprints" coupled with a Hedonic ranking to obtain estimates for both shelf life and the sensory characteristics which govern it will be the focus.

13.5 AN EXAMPLE—PASTA SHAPES IN SAVORY TOMATO SAUCE

13.5.1 Protocol

The test product is a simple ready meal consisting of pasta shapes in a minced meat and tomato sauce, the following being a summary of the manufacturing and distribution process.

> The product is packed, as a two shot fill (pre-cooked drained pasta followed by hot sauce) into preformed polypropylene trays, lidded with a polyethylene/aluminium foil laminate and retort sterilized to a F_0 value of 3.
>
> The product is for distribution at ambient, the individual trays being overwrapped with a printed paperboard sleeve and shrink-wrapped in dozens in cardboard trays. The desired storage life (manufacture to consumption) is 1 year.

From this information a storage protocol can be developed to determine the shelf life of the product.

1. The product is fully sterilized so that microbiological stability will not be a shelf-life determinant. Past experience suggests however that the sauce containing meat and tomato may give rise to color changes on storage and, further, it is possible due to the acidic nature of tomato, that changes in meat texture may occur. Furthermore pasta tends to soften in time due to water pick-up and starch loss, so textural changes in both pasta and sauce are possible. This combination of possible changes suggests that a QDA, coupled with a simple monadic liking test (i.e., samples are presented one at a time) is probably the best approach for this product. Although instrumental measurements may be considered to provide further information on color change, experience also tells that the heterogeneous nature of the product (coarse textured sauce, meat particles, pasta) will give poor reproducibility. Color analysis will therefore have to rely on sensory estimation.
2. The product will be stored, transported, and displayed at ambient temperatures. A Northern European environment is assumed and storage trials at +25°C and +35°C will be conducted, in order to assess the product's tolerance to abuse conditions. A control sample will be stored at +2°C, the assumption being that any change over the storage trial at this temperature will be minimal.
3. The product is packed in an opaque primary packaging and overwrapped with a paperboard sleeve. For this protocol, the packaging is assumed to be lightproof and there is no need to build a light sensitivity test into the trial.
4. The product is expected to have a storage life of 1 year. To add a margin of safety, storage trial will continue for 18 months. Take-offs however will be staggered—time zero, 1 month, 2 months, 3 months, 6 months, 12 months, and 18 months but a "safety factor" of extra samples will be included to allow for intermediate take-offs and testing if necessary.

The test protocol can therefore be summarized as in Table 13–1. This protocol allows the scheduling of the testing and gives an estimate of the total number of samples required. The 10 samples retained for the QDA at each take-off point are based on the need for 40 replicates and the assumption that each pack will provide four portions. It can be seen that a minimum of 72 samples should be stored at each temperature with a further 12 samples available for the time zero testing. Bearing in mind that these are minimum figures, the number should be increased by 50% for the storage samples and at least doubled for the time zero samples to allow for the train-

Table 13–1 Storage protocol for pasta in savory tomato sauce

Take-off point	*Date*	*Storage temperature*	*Testing required*	*Number of samples required*
Time zero	Jul 1993	n/a	QDA + Color	10 + 2
1 month	Aug 1993	+2°C, +25°C and +35°C	QDA + Color	10 + 2 for each temperature
2 months	Sep 1993	+2°C, +25°C and +35°C	QDA + Color	10 + 2 for each temperature
3 months	Oct 1993	+2°C, +25°C and +35°C	QDA + Color	10 + 2 for each temperature
6 months	Jan 1994	+2°C, +25°C and +35°C	QDA + Color	10 + 2 for each temperature
12 months	Jul 1994	+2°C, +25°C and +35°C	QDA + Color	10 + 2 for each temperature
18 months	Jan 1995	+2°C, +25°C and +35°C	QDA + Color	10 + 2 for each temperature

ing of the taste panel, i.e., a total requirement of some 350 samples.

13.5.2 QDA/liking testing

For the purpose of this example it is assumed that a suitable sensory analysis facility is available and that an experienced taste panel has been selected. If this is not the case then the reader is directed again to the book *Guidelines for Sensory Analysis in Food Product Development and Quality Control, Second Edition* (Carpenter et al., 2000), where the authors give in some detail the requirements for the design and development of such facilities and panels.

The first step in the QDA methodology is to develop a descriptive vocabulary to describe the sensory attributes of the product in question. This is done as a group exercise and after due discussion an attribute list is generated. This lists all the sensory parameters, appearance, aroma, flavor, and texture, by which the assessors can describe any given sample in terms of a set of scores. Table 13–2 gives the full attribute list for the test product and Figure 13–2 shows the scoring sheet relating to the pasta attributes.

With these attributes and a score sheet a trained assessor can produce an accurate and reproducible "fingerprint" of a sample. To ensure statistical confidence each sample is tasted blind

Table 13–2 Attribute list for pasta in savory tomato sauce

Pasta	*Code*	*Sauce*	*Code*
Color intensity	PASCOL	Thickness	THICKN
Intactness	INTACT	Color intensity	SAUCOL
Firmness	PASFIR	Glossiness	SAUGLO
Smoothness	PASSMO	Smoothness	SAUSMO
Sliminess	PASSLI	Chewiness (of meat)	CHEWIN
Starchiness	PASSTA	Overall strength of flavor	FLAVST
		Tomato flavor	TOMFLA
		Meat flavor	MEAFLA
		Herbs	HERFLA
		Spiciness	SPIFLA

Attribute	Low 1	2	3	4	5	6	7	8	9	High 10
COLOUR INTENSITY	----------									
INTACTNESS	----------									
FIRMNESS	----------									
SMOOTH-NESS	----------									
SLIMINESS	----------									
STARCHI-NESS	----------									

Figure 13–2 Scoring sheet for pasta.

in the order of 40 times. A QDA taste panel usually has 12 to 15 assessors and each assessor is presented randomly with each sample two or three times. A carefully randomized design for sample presentation coupled with internal replication and random coding of samples ensures that the sample analysis is free from order effects and other subjective influences.

As stated previously, if correctly applied, QDA is entirely objective. In order to assess changes in product quality, an acceptability question, based on Hedonic rating, is added to the end of the QDA scoring sheet.

Each storage sample is tested at each take-off and an average score calculated for every sensory attribute. With this information a sensory fingerprint can be constructed for each sample. Typically this information is represented graphically as a simple line or bar graph showing sensory scores as a function of attribute. Using this method a graphical representation of the sensory properties of the product may be produced and comparisons made between samples by simply overlaying plots. Figure 13–3 represents sensory plots for the control sample stored at +2°C compared with the sample stored at +35°C after 1 year.

Although these graphical representations show changes in sensory properties quite clearly they do not, in their own right, give any estimate of the statistical significance of these differences. It is usual therefore to couple this simple

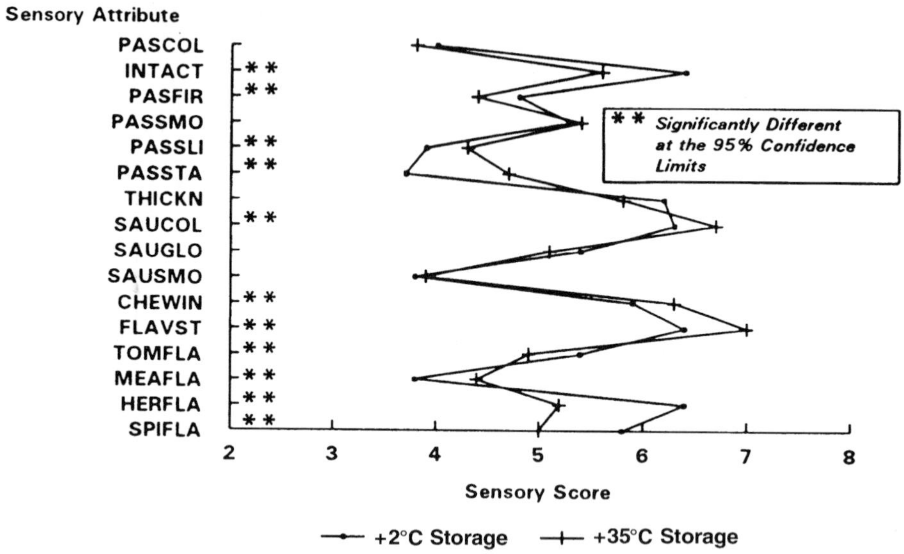

Figure 13–3 Comparison of sensory scores for "control" and "abuse" stored samples. •, +2°C storage; +, +35°C storage.

graphical approach with a statistical analysis of the raw data. Once again there are a number of methods available, which mostly rely on the analysis of variance to give a measure of confidence intervals between samples on the basis of their individual attributes. Such information may be superimposed on the plots, using asterisks to highlight those attributes which show significance at the various confidence limits.

Similarly this information gives no indication as to the effect that any changes are having on product acceptability. The liking scores, averaged in the same way from the Hedonic data, give this information (albeit in a crude form). Such information is usually expressed in the form of a simple line graph as shown in Figure 13–4.

13.5.3 Interpretation of results

13.5.3.1 Basic analysis

The simplest way of using the results produced by a storage study such as this is to use the Hedonic liking scores to determine product shelf life and the sensory analysis results to pinpoint which factors are controlling the shelf life.

To determine shelf life from acceptability data a threshold liking score is set, either by a suitable "expert" panel or from data gained from full consumer testing. The shelf life is set based on the period and conditions of storage which will reduce the acceptability score of the product to this point.

Assuming all other factors being equal, i.e., that the sensory and acceptability tests have been properly controlled and carried out, any changes in liking must be linked to changes in sensory properties. Sensory data can therefore be used to determine the factors responsible for changes in acceptability by analyzing the sensory data for statistically significant changes in attribute scores.

In the above example expert assessors have set a threshold acceptability score of 5.0 for the product. Shelf life for the product stored under normal conditions is therefore between 12 and 18 months but any temperature abuse will shorten this considerably.

The major attribute changes as a function of storage time, and temperature detected in the sensory analysis can be seen to be related to

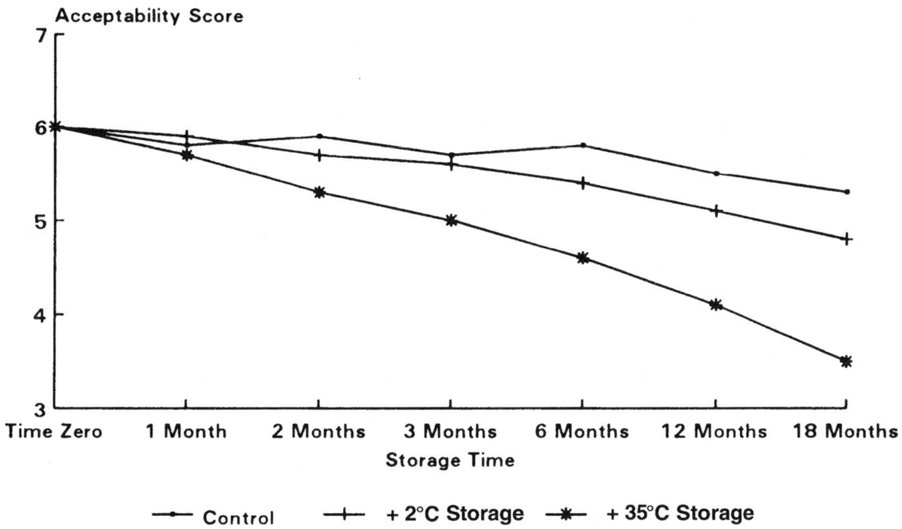

Figure 13–4 Hedonic liking scores for stored samples. •, Control; +, +25°C storage; *, +35°C storage.

pasta texture (intactness, firmness and starchiness), sauce appearance (color, glossiness) and flavor (all attributes), and meat chewiness. Although it is not possible to draw definite conclusions from this analysis it is likely that the loss of "freshness" of flavor from the sauce coupled with the loss of texture in the pasta are the prime determinants of shelf life and the first problems to be addressed if a longer shelf life is to be sought.

13.5.3.2 Predictive analysis

This simplistic approach to the determination and rationalization of shelf life from the interpretation of storage results is usually sufficient for most products. This is particularly the case where the foods, such as the example given in this chapter, have relatively short shelf lives and where storage conditions, although variable, are reasonably well controlled. In the case of products with longer shelf lives, or where for example storage temperatures may be subjected to wide variations, shelf-life prediction may be preferred to direct shelf-life determination. This approach is based on the building of quality kinetics equations, which are derived from detailed study of the effects of a carefully controlled matrix of storage parameters (time, temperature, light and so on) on the perceived quality of a specific product. This information allows the construction of a predictive equation which includes these parameters as variables and which can be used to model and predict quality performance on storage. This approach to shelf-life testing has been the subject of much and varied work and a detailed review is beyond the remit of this chapter. There may however be occasions when such an approach will need to be considered and interested readers are directed to Taoukis and Labuza (1996) for a complete review of the subject.

REFERENCES

Anon. (1992a) *The Grocer,* August 10th, p. 35.

Anon. (1992b) *The Grocer,* September 14th, p. 98.

Anon. (1992c) *Supermarketing,* September 18th, p. 22.

Barberio, G.C. (1986) in *The Shelf Life of Foods and Beverages* (ed G. Charalambous), Elsevier, London, England.

Benzig-Purdie, L.M., Ripmeester, J.A. and Ratcliffe, C.I. (1985) *Journal of Agricultural Food Chemistry,* **33**, 31–3.

Carpenter, R.P., Lyon, D.H., and Hasdell, T.A. (2000) *Guidelines for Sensory Analysis in Food Product Development and Quality Control,* Aspen Publishers, Inc., Gaithersburg, MD.

Fujinari, E.M. (1993) HPLC-CLND application to quantitate aspartame and other nitrogen containing compounds in beverages, in, *Shelf Life Studies of Foods and Beverages. Chemical, Biological, & Physical Aspects* (ed G. Charalambous). Elsevier, London, England, 1033–1048.

Karmas, E. and Harris, R.S. (1988) *Nutritional Evaluation of Food Processing,* 2nd edn, Van Nostrand Reinhold, New York, NY.

Kikakis, J.P. (1986) in *The Shelf Life of Foods and Beverages* (ed. G. Charalambous), Elsevier, London, England.

Kramer, A. (1977) *Journal of Food Quality,* **1**, 23–55.

Maskell, AJ. (1991) *Packaging Technology and Science,* **4**, 21–8.

Mogensen, G. and Poulsen, P.R. (1980) *Milchwissenschaft,* **35**(9), 552–4.

Peryam, D.R. and Pilgrim, F.J. (1957) Hedonic scale method for measuring food preferences. *Food Technology,* **11**(9), 9–14.

Seldman, J.D. (1992) *Packaging Technology and Science,* **5**, 154.

Taoukis, P. and Labuza, T.P. (1996) Summary: Integrated Concepts, in *Food Chemistry* (ed O.R. Fennema), Marcel Dekker, Inc., New York, NY, 1013–1042.

Trammell, D.J., Dalsis, D.E. and Malone, C.T. (1986) *Journal of Food Science,* **51**(4), 1021–3.

Wendt, C.H. (1992) *Packaging Technology and Science,* **5**, 159–63.

Whistler, R.L., Bemiller, J.N. and Paschall, E.F. (1984) *Starch: Chemistry and Technology,* Academic Press, London, England.

CHAPTER 14

Ambient-Stable Sauces and Pickles

A.A. Jones

14.1 INTRODUCTION

Sauces and pickles comprise a very wide range of traditional as well as more modern fruit and vegetable products. In 1998 the European retail sauces and dressings market accounted for 1.16 million tonnes of product, with a value of 2945 million Euros (LFRA, 1999). No figures are available to define the size of the catering and food service sectors of the industry.

To examine these products will necessitate subdividing the group and dealing with each subgroup in turn. Different subgroups have their own particular modes of deterioration but they all rely wholly or primarily on the use of organic acids, most notably acetic, for preservation.

In this chapter, sauces are taken to include the emulsified sauces such as mayonnaise and salad dressings, and the non-emulsified condiment sauces such as tomato ketchup, brown sauce, and mint sauce, as well as the sweet sauces such as topping syrups and fruit-based sauces. Pickles include those products traditionally regarded as pickles, such as pickled onions, pickled cabbage, pickled gherkins, and also products that are in the gray area between pickles and sauces, such as sweet pickles, piccalilli, and chutneys.

The major similarity among all these products, as pointed out earlier, is that traditionally (albeit less so these days) they belong to the category of compositionally preserved foods since they rely in the main on acetic or other similar organic acid and salt or sugar for self-preservation. In more recent times, dependence on the traditional method of preservation has been reduced, as a result of consumer demand for less acidic or less sweet products, resulting in potential vulnerability to microbiological spoilage in some of these products. Consequently the use of alternative preservation methods, such as the inclusion of chemical preservatives, thermal processing (e.g., pasteurization), and aseptic processing and packaging, has increased. Moreover, greater emphasis has been placed increasingly on preventive tools such as the one based on the Hazard Analysis Critical Control Point (HACCP) system, for the assurance of safety and quality in foods (CFDRA, 1992), as discussed in an earlier chapter in this book.

It is intended to examine briefly the scope of sauces and pickles and in more detail the main factors affecting shelf life and the way their composition can influence microbiological as well as physico-chemical changes in these products. Once the relevant shelf-life determining factors and modes of deterioration have been established, the exact methodology for shelf-life determination usually follows logically. Current practice in the shelf-life determination of these products, including the use of challenge testing and mathematical models, will be outlined together with a worked example. Reference will be made to previous chapters regarding areas which should be considered but have already been covered elsewhere in this book.

A final section will examine current developments within this product sector as well as the industry at large, which may have an impact on the shelf life of these products and the methods by which their shelf lives are determined in the future.

14.2 THE PRODUCTS

As discussed in the introduction, this product group comprises two major areas, of which sauces can be subdivided into at least three subgroups.

14.2.1 Sauces

Sauces may be defined as a range of formulated liquid or semi-solid food products, which when added to a food, alter or enhance the sensory appeal of that food, by adding richness of flavor and/or enhancing the mouthfeel.

Sauces may be used at the preparation stage of a meal. Marinades are examples of products that may be included in this category, where a certain degree of flavor needs to be introduced into the food before cooking. Sauces may be added at the cooking stage. Products in this category range from simple sauces used for their flavor, such as chilli sauce, added to spice a meat dish (e.g., chilli con carne), to more complicated savory cooking sauces such as creamy mushroom sauce, which when added to a basic commodity such as chicken, converts it into an exotic recipe dish. These sauces also include the "make-a-meal" and pasta sauces. A major subgroup is represented by the "pour-over" sauces which are added to a complete meal as an accompaniment (e.g., a salad meal or a convenience food such as pizza) in order to enhance or alter its sensory characteristics according to the consumer's preference. This subgroup of sauces includes emulsified salad dressings (e.g., mayonnaise, salad cream, French dressing), other emulsified sauces containing particulates (e.g., sauce tartare, horseradish sauce), non-emulsified sauces (e.g., tomato ketchup, brown sauce), the more exotic "ethnic"-type sauces (e.g., satay sauce) and fruit-based sauces for use with savory products (e.g., cranberry sauce and red-currant sauce). Pour-over sauces may also include the sweet (topping) and fruit-flavored sauces used to add flavor and variety to desserts such as ice cream, pudding, and yogurt. While sauces can be subdivided according to their culinary uses, from the point of view of shelf-life consideration, the approach that subdivides them into emulsified, non-emulsified, and fruit-based sauces is more useful.

14.2.1.1 Emulsified sauces

These, as their name suggests, are a group of sauces which consist of a viscous oil-in-water emulsion, often stabilized by egg yolk and normally containing vinegar. The oil content ranges from 80% plus in mayonnaise (this being very close to the maximum possible level of oil for an oil-in-water emulsion, higher levels would normally result in inversion to a water-in-oil emulsion), to the "virtually oil-free" sauces (e.g., low-calorie dressings) containing little or no fat. Emulsified sauces, therefore, have widely varying oil contents, with the mouthfeel and texture being modifiable by the inclusion of various fat mimetics or substitutes. The emulsion is usually formed by high shear mixing, which reduces the oil into minute droplets of a uniform size distribution of about a few microns in diameter and distributes them evenly in the water phase of the product. Stability of the emulsion is then maintained by the inclusion of emulsifiers and stabilizers, either natural or chemically derived. Thus, a mayonnaise emulsion may rely solely on the emulsifying properties of the egg, or include additional ingredients such as polysaccharide gums (e.g., guar gum, xanthan gum), which are often used in commercially available mayonnaise.

Some of the emulsified sauces are carefully defined, although definitions may vary from authority to authority. A mayonnaise, for example, has been defined by the European Codex as containing a minimum oil content of 78.5% w/w (from all sources), and a minimum technically pure egg yolk content of 6% (WHO/FAO, 1989). It should be mentioned, however, that at the time of writing, this standard is being considered for international rather than regional status, and as part of the consideration, the details of the standard are also under review. The Trade Association of the Mayonnaise and Condiment Sauce Industry of the EU, however, has agreed a voluntary standard for mayonnaise which requires a minimum oil content of 70% w/w and a minimum technically pure egg yolk content of 5% w/w (CIMSCEE, 1991). Technically, pure egg yolk is defined as the composition of egg yolk commercially available, and is assumed to contain about 25% of egg white. In the US, standards define mayonnaise with yet another set of criteria.

Other than mayonnaise, there is currently no further legal or informal agreement for emulsified sauces within Europe, since the repeal of the UK's Salad Cream Regulations. However, in the US, a salad dressing is defined as "an emulsified semi-solid food prepared from vegetable oils, vinegar, lemon juice and/or lime juice, egg yolk, and starchy paste" (FDA, 1999).

There are also emulsified sauces which contain added particulates. An example of these is sauce tartare, an emulsified dressing-based sauce containing particles of vegetables, usually gherkins and capers. Another sauce group which is important in the UK and Germany (but perhaps less well known in other countries) is that of the horseradish products. These again are based on an emulsified sauce base, with the inclusion of grated horseradish (and perhaps turnip), giving a distinctively flavored product for use with cooked meats and fish.

14.2.1.2 Non-emulsified sauces

These, as their name suggests, are non–fat-based sauces, and are often manufactured by a process that may include a heating stage. Of such sauces, the classic example is tomato ketchup. The European trade association has put forward proposed standards for tomato ketchup, which require a minimum tomato solids content, and limit the addition of other ingredients (CIMSCEE, 1992). Other sauces falling within this subgroup include brown sauce, various other ketchups (e.g., mushroom ketchup), and ethnic-style sauces. Like emulsified sauces, they may be smooth or contain particulates. The range of sauces in this group is again very wide, but the technological problems associated with shelf life are very similar, as discussed later in this chapter.

14.2.1.3 Fruit-based sauces/jellies

These may be further subdivided into "savory" fruit-based products such as red-currant jelly, cranberry sauce, and the sweet confectionery sauces, such as flavored topping syrups. The former group usually relies on sugar, and sometimes vinegar or citric acid for preservation. They normally have a high soluble solids content, coupled with a pH of less than 4, which inhibits most spoilage organisms. The process of manufacture often includes a heating stage that has a pasteurization effect. The sweet topping syrups also rely on a high soluble solids content and adequate thermal processing to inhibit microbial growth, with a low pH (e.g., due to the use of citric acid) also playing a part in maintaining the assigned shelf life.

14.2.2 Pickles and other related products

Pickles consist of vegetables preserved in vinegar or vinegar-based sauce. This traditional method of preservation has been practiced in many countries for centuries. Traditionally, pickling is one of the ways of ensuring that perishable fruit and vegetables are available for consumption during periods outside their normal harvesting seasons. The major varieties consumed in the UK are pickled onions, pickled beetroot, and piccalilli. This product group may be further divided into sweet and sour categories, with sour pickles being the predominant category. Chutneys and relishes are frequently included in this category.

The traditional pickling process involves a number of stages, including the brining of vegetables for storage, brine removal prior to preservation (freshening), and the preservation in vinegar or vinegar-based sauce (the pickling proper). The vegetables are often whole, but usually processed to make them suitable for prolonged storage. Thanks to efforts of J.C. Dakin and others for their work carried out at the Leatherhead Food Research Association over some 30 years, virtually all aspects of traditional pickle manufacture were scrutinized with the result that many practical problems were tackled and solutions found (Kilcast, 1984).

Chutneys and relishes are similar products, relishes often having smaller pieces of vegetables than chutneys. They basically consist of vegetable and/or fruit pieces suspended in a thick vinegar-based sauce. Their process of manufacture often includes a boiling/heating stage which reduces the moisture content of the mix, softens the fruit/vegetables, and pasteurizes the product. The preservation of these products is by acid and water activity control (using salt and sugar). Examples of these products include sweet pickles and mango chutney.

14.3 FACTORS AFFECTING SHELF LIFE

Many of the products within the area of ambient-stable sauces and pickles are well established and the specific factors that affect their shelf life are well known and documented. A very useful and comprehensive review on pickles and sauces is available (Broomfield, 1993). Nevertheless, it is useful to review some of these factors, as a good understanding of how they influence product shelf life will be important to product developers who are newcomers to this field, as well as those who are trying to develop modern versions of these traditional products.

14.3.1 Raw materials

The importance of quality and consistency of the raw materials used in the maintenance of the shelf life of sauces and pickles cannot be overemphasized. From a microbiological safety point of view, the use of pasteurized egg is vital wherever the recipe calls for egg. Indeed in many countries this is now a legal requirement. All raw materials should be handled according to good manufacturing practice (GMP) (Anderson and Blanchfield, 1991) and, if necessary, any sensitive ingredients should be decontaminated before use. Because of the long shelf life of products in this group, the quality of the raw materials is crucial if the assigned shelf life is to be met the first time and every time. For example, even the slightest signs of browning in the dehydrated horseradish flakes for use in various horseradish sauces could mean that the end product would not survive its stated shelf life as a result of color deterioration.

14.3.2 Composition and formulation

This is perhaps the most important factor that influences the shelf life of ambient-stable sauces and pickles. This is because the exact composition and formulation of a product will have profound influence on possible changes that may occur during the storage of the product. These changes can be microbiological, physical, chemical, and biochemical in nature. Such changes, of course, can have safety and/or quality implications which are intimately linked to the shelf life of the product.

In the case of emulsified sauces (i.e., oil-in-water emulsions), the oil droplets are discrete and distributed throughout the continuous aqueous phase. This structure tends to preclude the spread of lipolytic spoilage organisms in the product. Furthermore, there is little indication that industrially produced oils (and fats) have any significant role to play in foodborne diseases. The oil phase can therefore be disregarded, at least from a food safety point of view. For sauces (emulsified and non-emulsified) and pickles, therefore, the product safety determining factors are associated with the (continuous) aqueous phase. This aqueous phase contains the necessary nutrients to support microbial growth. However, it also contains the components that either kill these organisms or inhibit their growth. Factors in the aqueous phase that determine the microbiological safety and stability of the product are

1. ***Acetic acid (used as such or derived from vinegar).*** The level of acetic acid is believed to be the single most important factor in determining microbiological safety and thus shelf life in sauces and pickles (see section 14.4.2). The percentage of undissociated acetic acid has direct effect on its efficacy as a preservative agent, and this is affected by the pH of the aqueous phase and the acids dissociation constant. Acetic acid has the characteristic that at

the pH of many sauces and pickles it is mainly in the undissociated form, thereby exerting maximum antimicrobial effect. Moreover, the fact that acetic acid has a very low oil water partition coefficient (compared with sorbic and benzoic acid) means that, to all intents and purposes, all its preservative effect is retained within the aqueous phase of the emulsified sauce.

2. ***pH of the aqueous phase.*** Figure 14–1 gives the approximate pH growth ranges for some microorganisms important in food spoilage. As in other foods, *Clostridium botulinum* is unable to grow in a sauce or pickle with a pH value below 4.5.
3. ***Water activity.*** Salt and sugar contents affect the water available to microorganisms by modifying the osmotic balance. This available water is expressed as water activity (a_w) and is directly related to the equilibrium relative humidity (ERH) of the system. The water activity has a major implication in the survival of microorganisms. However too high a salt content (e.g., >4%) is unacceptable from a sensory point of view. Therefore formulation considerations must also be considered when looking at food safety aspects. The contribution of salt and sugar is indeed included in the CIMSCEE formulas (see section 14.4.2).
4. ***The presence of preservatives (either naturally present or added).*** Iso-thiocyanates in horseradish and the volatile "oil of mustard" (allyl iso-thiocyanate) in black mustard are good examples of naturally occurring preservative agents that more or less make these materials self-preserving. An example of added (permitted) preservative is potassium sorbate (e.g., in some tomato sauces and cranberry sauce), and where they are used, acidulants such as citric acid, lactic acid, and tartaric acid also have a preservative role, both by means of pH control and

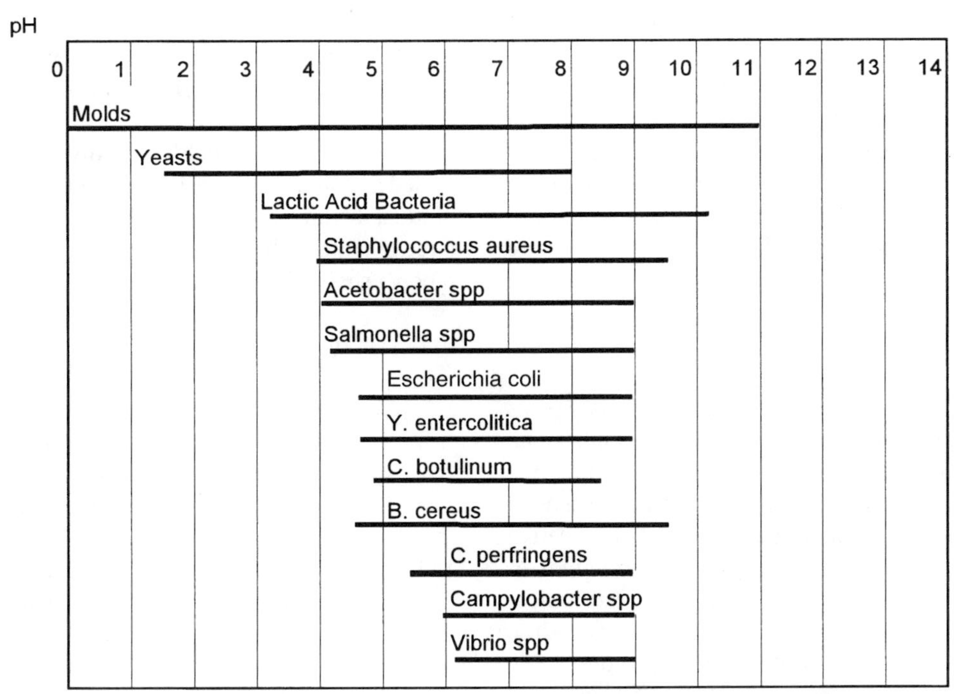

Figure 14–1 Approximate pH growth ranges for some microorganisms important in food spoilage.

by their own antimicrobial properties (although less so than acetic).

The synergistic effect of acetic acid, pH, salt, and sugar on the microbiological safety and stability of mayonnaise and similar products has been studied and defined empirically by a mathematical model which will be discussed more fully in the next section of this chapter.

Composition and formulation are also crucial to the physical stability, and hence consumer acceptance and shelf life of emulsified as well as non-emulsified sauces. In mayonnaise and salad dressings, the egg protein plays a vital role in the stability of the oil-in-water emulsion. The transient stability of the emulsion depends on factors like the efficiency of the mechanical emulsification, the size distribution of the oil droplets, the intrinsic viscosity of the mix, and so on. The long-term stability of the emulsion however depends also on the use of suitable emulsifiers and stabilizers. The egg protein remains the effective emulsifier (surfactant) and stabilizers such as guar gum and xanthan gum are commonly employed to ensure the long-term stability of the product. In some non-emulsified sauces, modified starch is often used to give a texture and to confer the long-term stability of the product. In addition to selecting the right starch (one that can withstand low pH, high shear, and heat treatment and one that is less likely to retrogradate and "weep"), correct processing is important to ensure that the texture benefits of the starch are realized (Figure 14–2).

14.3.3 Availability of oxygen

For long shelf-life products such as sauces and pickles, the presence of oxygen can have an important influence on shelf life. In emulsified sauces, the choice of vegetable oils can have a major influence on the development of oxidative rancidity. Table 14–1 lists the iodine values of some common edible vegetable oils, an indication of their susceptibility to oxidative rancidity. As can be seen from this table the more stable oils include rapeseed oil and soya oil, which is one reason these two oils tend to be the most used in the sauces industry. Sunflower oil has been marketed as being beneficial to health. While its use has been seen as desirable, its greater susceptibility to rancidity development limits the shelf life of products containing it.

Vegetables and fruit used in the manufacture of sauces and pickles often contain enzymes that have not been inactivated. In the presence of oxygen, grated fresh horseradish will brown very quickly due to its polyphenols content. Fresh horseradish should be handled and processed with the minimum of delay, and sulfur dioxide (as an antioxidant) may also be used to control browning if the shelf life of the end product is not to be reduced.

14.3.4 Processing

It is possible to produce microbiologically stable sauces and pickles based on a cold process provided certain criteria are satisfied (see section 14.4.2). Indeed for emulsified sauces, it is neither necessary nor desirable to use a hot process. In recent years, however, consumer demand for milder products has meant that a pasteurization stage (or alternatively hot filling at not less than 80°C in some cases) is included to achieve an acceptable shelf life for many pickles and some sauces.

Pasteurization of pickles involves the controlled and limited thermal processing of the final product within its container. Pasteurization has three potential advantageous effects in respect of product shelf life, namely (Broomfield, 1993):

1. Thermal destruction of spoilage microorganisms, permitting a lower acetic acid content to be used.
2. Total or partial thermal inactivation of enzymes of vegetable or of microbial origin, thus preventing deteriorative changes caused by enzymes.
3. In the case where self-venting closures are used, removal of air during pasteurizing and retention of headspace vacuum on cooling, thus minimizing oxygen-dependent and oxidative changes (e.g., darkening and softening).

	Under-cook	Optimal cook	Over cook
Appearance	Cloudy	Clear	Clear
Texture	Thin, starchy taste	Heavy bodied short textured	Cohesive long textured
Stability	Poor	Good	Fair
Viscosity	Low	Good	Viscosity drop

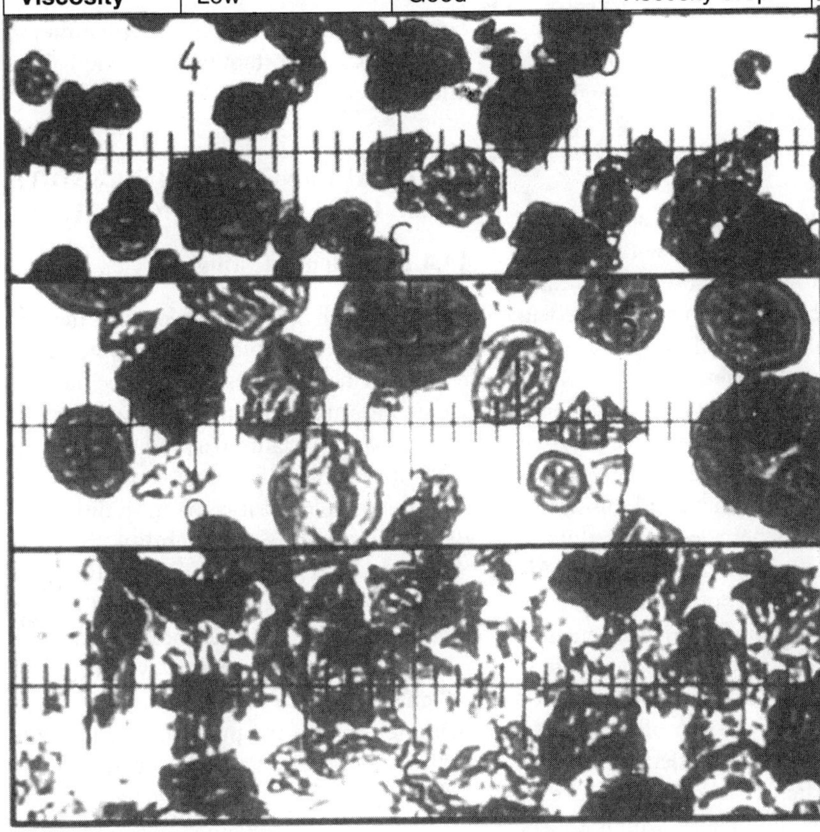

Figure 14–2 Characteristics of under-cooked, optimal, and over-cooked starch slurries. (Source: Anon (1987), *Food Starch Technology,* National Starch and Chemical, Manchester.)

Hygienic design and practices are GMP cornerstones and they are fundamental to sauces and pickles manufacture as they are to other food products, irrespective of whether the sauce or pickle is pasteurized or not.

14.3.5 Packaging

The traditional packaging material used for sauces and pickles is glass, which provides practically a perfect gas as well as water vapor barrier. In addition, where necessary, hot filling is possible with glass containers, which can be a useful alternative to in-container pasteurization. Developments in packaging materials, in particular plastics materials, have led to the introduction of the multilayer "squeezable" bottles for sauces. A typical structure for this type of bottle is: inside polypropylene (PP)/adhesive/EVOH/adhesive/PP outside, EVOH being the gas barrier layer and PP the water vapor barrier layer. While this type of container offers the

Table 14–1 Iodine values of vegetable oils—an indication of their susceptibility to oxidative rancidity

Vegetable oil	*Iodine value*
Rapeseed (Canola) oil	97–107
Soya oil	125–136
Sunflowerseed oil	125–140
Corn (Maize) oil	110–128
Olive oil	80–88

consumer convenience in use, it was not commonly filled hot although it could withstand hot filling up to 70°C. However, more recently, advances in plastics technology have resulted in variants being capable of hot filling. Because of its multilayer structure, barrier properties of the container depend on EVOH and PP being evenly present throughout the container. The oxygen barrier property of EVOH is affected by the relative humidity of the environment and this has to be taken into account when designing storage trials for products packaged in this container. Finally, both the packaging supplier and the sauce manufacturer have a responsibility to ensure, particularly for any new sauce bottle combinations, that the legislative standards in respect of food packaging migration are complied with. In the UK the requirements are detailed in the Plastics Materials and Articles in Contact with Food Regulations 1998 (as amended), which enacts the appropriate EU directive.

14.3.6 Consumer use

Consumer use of multi-use products like sauces and pickles has always been an uncertain factor in respect of product shelf life. Although it is generally believed that many mayonnaises and dressings are sensitive only to typical acetic acid–resistant organisms, which are not widespread in the consumer's home, it is important that the possibility of spoilage during use should be considered during the development of mild formulations. Spoilage may be controlled to a certain extent by limiting the shelf life of opened product at room temperature and/or recommending refrigerated storage once the product has been opened. Yeast contamination has been known to cause some tomato sauces to ferment and the pack to explode. It is, therefore, very important that precise label instructions are used for microbiologically vulnerable products. The consideration of potential consumer abuse should be an important factor of shelf-life evaluation, especially in today's litigious society.

14.4 CHALLENGE TESTING AND THE USE OF MATHEMATICAL MODELS

14.4.1 Challenge testing

To establish the microbiological safety and stability of sauces and pickles, challenge testing may be used. Challenge testing is a technique which yields information to the manufacturer as to the microbiological status of a product during its "normal" or expected life before consumption. To the food microbiologist it is the most direct evidence of product safety and stability (Shapton and Shapton, 1991). In practice, challenge testing comprises the direct inoculation of a food with microorganisms. When carried out for safety purposes, the microorganisms will be pathogens which are likely to be associated with the product being tested. For the purpose of stability testing, the spoilage organisms selected are those which are known to cause problems in the particular type of food under investigation. The choice of microorganisms for inoculation, their culture, and method of inoculation are therefore important considerations in challenge testing. The organisms chosen also require to be "trained" to be able to survive under adverse conditions (e.g., the presence of acetic acid, low pH, high sugar, and so on). Ideally, organisms likely to be used should be cultured from within the sauce plant, however, certain indicator organisms have been identified and specified in the *Code for the Production of Microbiologically Safe and Stable Emulsified and Non-Emulsified Sauces Containing Acetic Acid* produced by the Comité des Industries des Mayonnaises et Sauces Condimentaires de la Communauté Economique Europeenne

(CIMSCEE, 1991). This code states that an intrinsically stable sauce is one where acetic acid tolerant organisms will not grow. Of the organisms identified, the mold *Moniliella acetoabutans* is considered the organism most resistant to acetic acid, but is easily eliminated by satisfactory hygiene and/or pasteurization. Within the food industry contamination due to this mold is considered rare. This organism may be a useful reference, but as it is easily eliminated, it is not usually used in challenge testing.

The CIMSCEE code recommends a twofold examination, covering both safety and stability. Product safety requires that pathogens cannot survive in the sauce, the major potential hazard being due to salmonellae. However because of the risk of possible contamination where the laboratory is based at a factory, challenge testing is not carried out using *Salmonella* spp but by an indicator organism which has been shown to have mortality rates similar to that exhibited by salmonellae. The organism usually used is a strain of *Escherichia coli* (NCTC 5933 or VBEEs 01-03). Product stability on the other hand requires that the sauce does not support the growth of spoilage organisms. Several organisms considered to be indicators of resistance to the conditions likely to be present in sauces are (CIMSCEE, 1991):

Molds:	*Monascus ruber* (Van Tieghim), (VMEuMO 01-02) resistant to 1.8% acetic acid; *Penicillium roquefortiii* (VMMoPe 16-07) resistant to 1.2% acetic acid; *Penicillium verrucosum* (VMMoPe 20-07) resistant to 1.2% acetic acid.
Yeasts:	*Pichia membranaefaciens* (Hansen) (VYAPi 01-02 and 01-04) resistant to 2.5% acetic acid; *Zygosaccharomyces bailii* (VYASa 07-01 and 07-05) resistant to 3.0% acetic acid.
Lactobacilli:	*Lactobacillus buchneri* (VBLLa 18-01) resistant to 2.5% to 3.0% acetic acid.

The organisms must be adapted gradually to acetic acid before use, as detailed in the code, or as described originally by Tuynenburg Muys (1971).

Challenge testing is extremely time consuming and, more important, requires laboratory skills which may not always be available. Time is a crucial factor in today's product development programs and as an alternative approach to challenge testing, mathematical models are often used to predict microbiological safety and stability before embarking on a shelf-life determination for a new or revised product.

14.4.2 Mathematical models

The CIMSCEE code outlined in section 14.4.1 also includes a model which can be used to establish safety and stability for sauces containing acetic acid derived from vinegar or directly used. This model was developed from research into the Microbiological Composition Analysis (MCA) of edible emulsions carried out by G. Tuynenburg Muys at the Unilever Research Laboratory in the Netherlands in the 1960s. The model takes into account the acetic acid content, its degree of dissociation (this being dependent on the pH of the food), the salt content, the disaccharide content, and the hexose content of the sauce. The contents (percentages) of these ingredients are expressed as the weight in aqueous phase, i.e., water plus acetic acid (or vinegar) plus salt plus sugars. Because the partition coefficient for acetic acid in oil:water systems is very low, for practical purposes it is assumed that all the acetic acid will be available in the aqueous phase. This model which describes the relationship between the above-mentioned ingredients and the potential for microbiological growth has been expressed by the following formulae:

$$15.75\ (1-\alpha)\ (\text{total acetic acid}\ \%) + 3.08\ (\text{salt}\ \%) + (\text{hexose}\ \%) + 0.5\ (\text{disaccharide}\ \%) + 40\ (4.0 - \text{pH}) = \Sigma_s \quad (14.1)$$

For any sauce based on acetic acid, if the value of this formula (Σ_s) exceeds 63, safety from microbial pathogens is theoretically ensured.

$$15.75\ (1-\alpha)\ (\text{total acetic acid \%}) + 3.08\ (\text{salt \%}) + (\text{hexose \%}) + 0.5\ (\text{disaccharide \%}) = \Sigma \quad (14.2)$$

For any sauce based on acetic acid, if the value of this formula (Σ) exceeds 63, microbial spoilage should not occur.

(1–α) is the proportion of the total acetic acid which is undissociated and is related to the pH of the sauce and the pK_a of acetic acid.

The model has certain limitations, for example, it does not take into account the effects of any other antimicrobial ingredients, such as preservatives, or any antimicrobial agents naturally present in the foods (e.g., as in horseradish and mustard). Also care must be exercised when particulates are present, as no allowance is made for slow acid equilibration in sauces such as sauce tartare. Furthermore the code does not apply to pickles or dressed salads, and on the whole it badly needs updating.

The inclusion of preservatives can improve preservation of the sauce, allowing an apparently low stability value which might otherwise suggest instability when interpreted on its own. Work by Dakin (1968) suggests that an 800 mg/kg level of sorbic acid in a pickle or sauce has the equivalent effect of 1.0% acetic acid. Care however should be taken in the use of this finding as the partition coefficient for sorbic acid is such that the preservative will be distributed throughout a sauce which has an oil phase, and not be restricted to the aqueous phase of the sauce. This will result in a significantly reduced level of sorbic acid in the aqueous phase of an emulsion, especially in high oil systems, making the preservative action somewhat less effective than would otherwise be expected.

Another empirical model that can be used to predict the stability of unpasteurized pickles and sauces is the "preservation index" put forward by Dakin (Binstead et al., 1971). This model or rule basically states that the minimum acetic acid content necessary to achieve the satisfactory preservation of all pickles and sauces is 3.6%, calculated as a percentage of the volatile constituents of the product. Thus,

$$\text{Preservation index} = \frac{\text{Total acetic acid acidity} \times 100}{(100 - \text{total solids})} \quad (14.3)$$

Like the CIMSCEE formulas, the preservation index has to be interpreted carefully since an index lower than 3–6% does not always mean instability, as in the case of mango chutney. A comparison of the stability values of some sauces based on the CIMSCEE formulas and their corresponding values of preservation index is given in Table 14–2.

Similarly, research on adaptation to acid stress by microorganisms, including *E. coli* O157, has increased the gray area in the CIMSCEE model. Recent research shows that refrigeration can have the effect of extending the survival time of some spoilage and pathogenic bacteria. Survival of *E. coli* O157 has been seen to be as high as 35 days (although no increase in population occurred in the sauce) (Weagant et al., 1994; Tsai and Ingham, 1996). This research again indicates the need for caution when using these mathematical models in deciding on microbiological safety and stability of the product.

Work on updating the CIMSCEE code is being carried out by means of two research projects under way at the time of writing. Both Leatherhead Food Research Association (LFRA) and Campden and Chorleywood Food Research Association (CCFRA) in the UK are undertaking these two projects, funded by club membership. CCFRA has produced a revised model following microbiological tests using modern techniques such as impedance and turbidity measurements (Everis et al., 1999), which permit large numbers of tests to be carried out, enabling statistical analysis to be used. The model includes preservatives and other organic acids such as citric and lactic acids. The results obtained have been statistically modeled to produce a useful tool for the microbiologist in the pickles and sauces industry. However, at the time of writing, the results of this research are confidential to the members of the club. These models, when finalized, should help the product developer to be more certain of the microbiological stability of his products.

Table 14–2 A comparison of CIMSCEE values against preservation index values for various sauces

Sauce	*Intrinsic safety*	*Stability*	*Preservation index*
Mayonnaise	75.7	75.7	4.6
Sweet and sour	66.9	66.5	4.5
Sauce tartare	89.7	57.7	3.3
1000 Island dressing	66.7	66.7	3.7
Ranch dressing*	66.4	52.2	2.9
Hot horseradish	76.5	56.5	3.6

CIMSCEE code has a value of 63 as the cut-off point. Preservation Index has a value of 3.6 as the cut-off value.
*Preserved with addition of potassium sorbate.

14.5 SHELF-LIFE EVALUATION

The shelf life of sauces and pickles is somewhat difficult to establish predictively or by accelerated methods. Other than the microbiological mathematical models previously described, there is no known model that can be applied to the more modern varieties of sauces and pickles. In products such as sauces and pickles, composition and formulation play a cardinal role in determining the shelf life of the product and interactions between ingredients can prove very difficult to model. Accelerated tests, for example, may speed up some changes, but others may proceed at the same rate. Since accelerated tests are usually taken to mean storage at elevated temperatures, with or without cycling, they almost always tend to select for a different microflora which would not be representative of what may be present under ambient conditions. Other factors too will be of major relevance to sauces and pickles. Factors such as availability of oxygen, susceptibility to oxygen and to light, barrier properties of the packaging used, and so on may all play a part and contribute to limit the usefulness of any predictive or accelerated methods.

Fortunately, the product group of sauces and pickles is still largely a traditional one, and consequently useful information regarding shelf life is available in the literature as well as within the manufacturing companies. Similar products are often available with which to compare for an estimation of shelf life. Such comparisons must, however, be treated with caution, as subtle changes in formulation, ingredient supply, and/or packaging format could have a major effect on the shelf life of the product.

Frequently, initial shelf-life estimations may be based upon knowledge of the deterioration associated with individual ingredients. For example, in helping to decide the best oil to use in the formulation of an oil-in-water emulsion from a quality standpoint, the resistance of a vegetable oil to rancidity development can be assessed using a FIRA-Astell or Rancimat apparatus (Rossell, 1989). Similarly, the knowledge that a formulation contains materials that are prone to browning will prompt the product developer to pay particular attention to probable color changes during shelf-life determination.

Once the critical factors have been identified, storage trials will be possible with tests designed to detect the potential deteriorative changes that are likely to limit the product life. It must be said, therefore, there is yet no substitute for the direct approach to shelf-life determination whereby the sauce or pickle being studied is stored under defined and controlled conditions for the whole of the expected shelf life of the product in order that the correct life can be assigned.

14.5.1 Shelf-life determination

The first stage in a shelf-life determination must be the assessment of microbiological safety. Fortunately, past experience and historical data coupled with the use of HACCP and the models have meant that an indication of microbiological safety can be quickly established for sauces and pickles. A product failing to satisfy the CIMSCEE's criteria for safety and stability, for instance, will require further investigation (e.g., challenge testing), or else alternative methods of preservation (e.g., pasteurization) may be necessary.

Once microbiological safety (and stability) has been established, storage trials can be carried out. From past experience, it can be said that most quality deteriorations tend to manifest themselves as sensory changes so that shelf-life tests are usually selected to monitor them. Different conditions of storage may be used and the following are the more common conditions:

chill	0–4°C
ambient (UK)	15–25°C (temperate: 25°C, 75% RH)
accelerated	30–37°C (tropical: 37°C, 95% RH)

Temperature cycling (important for studying emulsion stability and/or effects of temperature abuse) may be superimposed on these conditions, depending on the purpose of the trial. Other variations in conditions may also be included such as storage in the dark or under artificial light, with or without cycling; the effect of vibration; and different packaging.

For sauces and pickles, sampling for analysis is normally carried out on a monthly basis, and where appropriate, samples stored under different conditions are compared. The following are examples of the more common tests that are carried out for shelf-life determination of sauces:

- ***Sensory analysis:*** aroma, flavor, color, texture, mouthfeel, visual appearance (signs of separation and apparent viscosity and consistency).
- ***Chemical analysis:*** pH, total volatile acidity, peroxide value and free fatty acid content of extracted oil, specific components such as the volatile and pungent isothiocyanates in horseradish.
- ***Instrumental analysis:*** viscosity measurement using a Brookfield viscometer or Bostwick consistometer and color measurement using a Minolta Chroma Meter (based on the L*a*b* color notation system).

These tests can be used to give some indication of the rate of change, and help to locate the end point of the shelf life. The aim of any shelf-life determination is to find the level of change in respect of each critical quality attribute beyond which the product is regarded as unacceptable. Some products will start to change from day one. A horseradish product, for example, will begin to brown, and lose its pungency (or "heat") immediately after packing. Shelf-life studies must inevitably include a degree of judgment regarding the level(s) at which the deteriorative changes have become unacceptable. Such decisions are usually made jointly by the technical, sales, and marketing (including the customer where appropriate) departments.

14.5.2 A worked example

An example to show the process of shelf-life determination is given in this section. The product chosen is a horseradish sauce product, with its compositional standards shown in Table 14–3. The initial study included an assessment of microbiological safety and stability of the concept product. As can be seen, the preservation values indicate the product is microbiologically safe, but that its microbiological stability is in doubt (Table 14–3). These results, produced using a popular spreadsheet program, prompted a challenge test on the product. It was believed that the product should be stable, as certain natural constituents in horseradish are bactericidal (i.e., the isothiocyanates). This was confirmed by the challenge testing, the results of which are shown in Table 14–4. A comparison with existing products also suggested that the product would remain in an acceptable condition for at least 10 months, but after that, browning and loss in "heat" could reach unacceptable levels. Storage trials were then carried out to confirm this. The important quality attributes were iden-

Table 14–3 Composition of "hot horseradish sauce" and CIMSCEE calculation of safety and stability values using a spreadsheet program

Ingredient	*% Recipe components*	*% Water phase*
Acetic acid (80%)	2.6	3.2
Hexose	0.0	0.0
Sucrose	4.6	5.7
Salt	1.6	2.0
Water (pH 3.5)	71.4	89.0
Dissociation constant (acetic acid)	0.948	
Microbiological stability	57.4	
Intrinsic safety	77.4	
Preservation index	3.6	
pH	3.5	

tified as color, heat intensity, flavor and viscosity. These parameters were examined over a period of time, by sensory as well as appropriate chemical analyses. Results of these tests are shown in Table 14–5. These results indicate that the product would remain acceptable for a period of about 11 months, and consequently a 10-month life was recommended, which satisfied the customer's requirements and gave a margin of safety of about a month.

Similar methods of determination can be applied to most, if not all of the ambient-stable sauces and pickles. The value of a mathematical model (e.g., the CIMSCEE model) as a first step in the development process is that it gives the development personnel an opportunity to fine tune the formulation before going to a full shelf-life determination. The model also enables decisions to be made early in the product development program as to whether or not a pasteurization stage is required as part of the manufacturing process.

14.6 CURRENT DEVELOPMENTS

14.6.1 Alternative ingredients and packaging

Within the sauces sector, over the past few years there have been major moves to develop reduced fat, and even fat-free products (the so-called "lite" products) as well as the use of alternative packaging formats. For example, products are being developed using various fat mimetics and replacers, based upon a number of raw materials. These are claimed by the suppliers to produce an acceptable product from a sensory point of view. Often, however, the system being sold does not completely provide all the characteristics associated with vegetable oils, when used in a sauce or salad dressing. Developers are therefore working on combining several of these systems in order to evaluate any possible synergistic effects, with a view to reproducing the quality characteristics of the product with a traditional oil content. Similarly, pickles containing less acid and/or sugar, albeit pasteurized, are becoming more popular as a result of consumer demand for milder products. Modified atmosphere packaging (MAP) (e.g., gas flushing or vacuum packing) in conjunction with plastics materials (Kilcast, 1984) has also been used to package pickles (e.g., sliced beetroot) for chill distribution, giving the products a novel and "fresher" image. These developments have major ramifications for product shelf life. Some of the systems being used have been available for many years, but their effects in a sauce or pickle system are not well documented. Others are novel in their applications, and their effects under certain conditions are not known. Combining these systems (ingredients and packaging) to effect a synergistic outcome with a view to pro-

Table 14–4 Results of the challenge testing on hot horseradish sauce

Time	*Media*	*Control*	Lactobacillus buchneri	Zygosacch-aromyces bailii *07-01*	Zygosacch-aromyces bailii *07-05*	Pichia membranae faciens *01-02*	Pichia membranae faciens *01-04*	Penicillium roqueforti	Penicillium verrucosum	Monascus ruber
24 hours	L agar	400	400	20	40	70	50	70	50	30
	L agar + 1% acetic acid	<10	<10	<10	<10	<10	<10	<10	<10	<10
8 days	L agar	200	<10	<10	<10	<10	30	20	<10	<10
	L agar + 1% acetic acid	<10	<10	<10	<10	<10	<10	<10	<10	<10
22 days	MRS	200	60	<10	<10	<10	<10	<10	<10	140
	MEA	<10	<10	<10	<10	<10	<10	<10	<10	<10
	MRS + 1% acetic acid	<10	<10	<10	<10	<10	<10	<10	<10	<10
	MEA + 1% acetic acid	<10	<10	<10	<10	<10	<10	<10	<10	<10
32 days	MRS	150	<10							
	MEA	20	<10							
	T.V.C.	400								

Table 14–5 Changes in mustard oil content (as allyl iso-thiocyanate) and assessors perceived "heat" of hot horseradish sauce over a 12-month period

Product age	*Mustard oil content (%)*	*Perceived heat**
1 month	0.81	10
2 months	0.77	9
3 months	0.74	7
4 months	0.72	7
5 months	0.74	6
6 months	0.76	6
7 months	0.74	5
8 months	0.73	4
9 months	0.73	5
10 months	0.69	4
11 months	0.69	2

*The sensory heat level is judged on a subjective 1 to 10 scale. 10 is hot, 1 lacks heat.

ducing a product of the required sensory characteristics and pack appeal may result in some beneficial effects being realized initially which cannot be sustained during the whole of the product shelf life. Consequently the product developer will need to consider carefully all aspects of potential product deterioration without the luxury previously afforded to him of comparing with any existing products.

Differing process requirements (e.g., as a result of using emerging process technologies) also add to the complexity in developing new products and establishing their shelf lives. Well-planned and well-conducted shelf-life determinations involving relevant and meaningful physical, chemical, microbiological, and sensory analyses must be carried out over time with these new products to unequivocally establish their safety and acceptability.

14.6.2 New techniques in shelf-life determinations

The pickles and sauces industry is primarily a traditional industry, and consequently there are few novel approaches to shelf-life determination of its products. Much research has been done and is still being carried out into the extension of mathematical modeling in food products in general (e.g., the UK's MAFF MicroModel). These models may in the future be more widely used by the food technologist in establishing the safe life of sauces and pickles. These models coupled with other intelligent knowledge-based (computer) systems being developed (see Chapter 3) will enable product development personnel to predict shelf life early in the development program.

14.7 CONCLUSION

Sauces and pickles traditionally have relied on a combination of acetic acid, salt, and sugar for their preservation. In more recent times the use of thermal processing (pasteurization), preservatives (e.g., sorbic acid), and other functional ingredients have enabled the food technologist to vary the composition of these products resulting in less salty and less acidic products which are increasingly being demanded by the consumer. Even more recently the introduction of fat replacement systems has additionally offered the opportunity for further new product development for sauces. For the food technologist the determination of shelf life has moved from simple monitoring of well-known deterioration processes of the basic ingredients to more complicated investigations concentrating not only on ingredient changes due to aging but also on the potential for microbiological changes as well as possible physico-chemical changes under many different conditions.

Some of the challenges for the food technologist have been highlighted in this chapter, and examples of probable degradation have been given. To be successful in this product group, the product developer must know the hazards and risks associated with the materials he is using, and be alert to any possible risks associated with changes in the product matrix brought about by the use of new formulations, novel ingredients and modern packaging formats.

REFERENCES

Anderson, K.J. and Blanchfield, J.R. (eds) (1991) *Food and Drink Good Manufacturing Practice A Guide to Its Responsible Management,* 3rd edn, IFST, London, England.

Anon. (1987) *Food Starch Technology.* National Starch & Chemical, Manchester, England.

Binstead, R., Devey, J.D. and Dakin, J.C. (1971) *Pickle and Sauce Making,* 3rd edn, Food Trade Press, London, England.

Broomfield, R. (1993) Acid, salt and sugar preserves, in *Food Industries Manual,* 23rd edn (eds M.D. Ranken and R.C. Kill), Blackie Academic and Professional, Glasgow, pp. 234-287.

Everis, L., Betts, G. and Evans, D. (1999) Final Report of the Acid Preservation Club. Campden and Chorleywood Food RA Confidential Club Member Report, Project 31777/Rep/2.

CFDRA (1992) *Hazard Analysis Critical Control Points (HACCP) A Practical Guide,* Technical Manual No 38, Campden Food & Drink Research Association.

CIMSCEE (1991) *Code for the Production of Microbiologically Safe and Stable Emulsified and Non-Emulsified Sauces Containing Acetic Acid.*

CIMSCEE (1991) *Code of Practice for the Production of Mayonnaise.*

CIMSCEE (1992) *Draft Code of Practice for the Manufacture of Tomato Ketchup.*

Dakin, J.C. (1968) *The Influence of Supplementary Preservatives on the Preservative Action of Acetic Acid,* Technical Circular No 401, Leatherhead Food RA.

FDA (1999) Section 169-150 Salad Dressing 21 CFR 169.

Kilcast, D. (1984) Preserved for the future, *Food,* November, 28, 29, 33.

Leatherhead Food Research Association. *The European Sauces and Dressings Market,* 2nd edn, August 1999.

Rossell, J.B. (1989) Measurement of rancidity, in *Rancidity in Foods,* 2nd edn (eds J.C. Alien and R.J. Hamilton), Elsevier Applied Science, Essex, England.

Shapton, D.A. and Shapton, N.F. (eds) (1991) *Principles and Practices for the Safe Processing of Foods,* Butterworth Heinemann, Oxford, England.

Tsai, Y. and Ingham, S.C. (1996) Survival of *Escherchia coli.* O157:H7 and *Salmonella* spp in acidic condiments. *J. Food Prod.,* **60**(7), 751–755.

Tuynenburg Muys, G. (1971) Microbial safety in emulsions. *Process Biochemistry,* **6**, 25–28.

Weagant S.D., Bryant J.L. and Bark, D.H. (1994) Survival of *Escherichia coli* O157:H7 in mayonnaise and mayonnaise-based sauces at room and refrigerated temperatures. *J. Food Prot.* **57**(7), 629-631.

WHO FAO (1989) Codex Alimentarius Regional Standard for Mayonnaise.

Chapter 15

Frozen Foods

H. Symons

15.1 HOW MANY INDUSTRIES?

As their name implies, all frozen foods depend for their stability and hence an adequate shelf life on product temperature being maintained colder than the freezing point. For most foods the freezing point is around –2°C (28.4°F). This superchill temperature is totally unsatisfactory for holding any frozen products, all of which demand far colder temperatures. This has led to prefixes being used in front of the adjective "frozen" (such as "quick" or "deep") to make the point that merely being hard because it is cold is an inadequate safeguard for maintaining the quality of frozen foods.

15.1.1 Speed of freezing

The term "quick" was introduced by Clarence Birdseye, the father of the retail frozen food industry. Birdseye had at least two reasons to select the adjective "quick." One was to distinguish the new industry he was struggling to develop, based on superior quality presented to the retail consumer, both from the frozen meat trade (which had adopted –10°C [14°F] as its storage temperature) and also from the "sharp" frozen industry in which the product to be frozen is placed in a chamber called the "freezer" at –7 to –10°C (19.4°F to 14°F) with no effort to separate already frozen product from ambient product introduced to be frozen in the same chamber. Not only is "sharp" freezing very slow, taking days or even weeks rather than hours, but also the product may suffer frequent large temperature fluctuations as new batches of product at ambient temperature are introduced into the chamber for freezing. Much of the frozen meat in international trade is carried today in container ships at –18°C (0°F) on account of the impracticability of providing different temperatures for various parts of the cargo. Enormous quantities of fish, fruits, vegetables, and shrimps are now also traded internationally and these products all demand frozen food temperatures.

Birdseye's second reason for calling the new industry "quick frozen" was to promote his recent invention, the multi-plate froster, which freezes product in a controlled and rapid fashion. Jul (1984) related a discussion he had, in the late 1930s, with executives in Birdseye Division of General Foods in the US, who had purchased Birdseye's patents, when he was able to obtain their admission that the speed of freezing was not as important to product quality as the maintenance of adequately low temperatures (–18°C [0°F]) or below during distribution.

Speed of freezing plays a part in determining shelf life. The first recognition of the importance of speed in freezing was probably due to experiments by Plank et al. (1916) who demonstrated superior product quality when the Ottesen method of freezing (patented in 1911) was employed and the product quality compared to "sharp" frozen material. The Ottesen method relies on intimate contact with the material to be frozen—generally fish—with cold sprays of isotonic sodium chloride. The Ottesen method was abandoned either because of the difficulty of maintaining the isotonicity of the sodium chloride solution or because, when the solution was not precisely isotonic, some salt would be picked up by the product leading to early onset of rancidity.

Rather than coin the adjective "quick" it would have been more accurate if the experimenters had stressed the need for freezing to be carried out in a controlled fashion in a device designed to freeze product, entirely separate from any storage facility. Undue emphasis on the importance of freezing speed is misleading. Speed of freezing within the range normally encountered in practice has little, if any, effect on quality. Excessive freezing speeds can ruin product. An increase in volume of around 10% is associated with freezing most foods. The build-up of internal pressure during very rapid freezing (such as obtained when product is immersed in total loss refrigerants such as liquid nitrogen or carbon dioxide) shatters the already frozen, and now supercooled and therefore brittle, external layers. Providing the speed is not excessively slow, days or weeks rather than hours, most products are comparatively insensitive to the speed of freezing. Broadly speaking faster is marginally better than slower in most products. This is particularly true for fruit and vegetable products, less so for animal tissue. Most of the advantages apparent immediately post-freezing in plant tissue are however lost during subsequent storage, distribution, and thawing. Current practice in the industry freezes product quickly but this reflects more the need economically to process a large quantity of product in as small a factory space as practicable rather than any obeisance to the demands of product quality.

The Codex Alimentarius Code of International Practice for Processing and Handling of Quick Frozen Foods (1982) requires freezing to be carried out in equipment designed to freeze product and at a speed appropriate to the product. Thawing is not the reverse of freezing, unless microwave thawing is used. It is far slower than freezing and is generally carried out by the consumer in an uncontrolled fashion.

15.1.2 Ice cream/quick frozen foods/frozen foods

If the frozen meat trade is defined as employing –10°C (14°F) as its characterizing storage temperature, two other frozen food industries can also be defined. The ice cream industry requires temperatures as cold as –20°C (–4°F) while the "quick/deep frozen food" industry (colloquially referred to as "frozen foods") demands –18°C (0°F).

A distinction between "quick frozen" and "frozen" retail products is preserved in several European countries. "Surgeé" contrasts with "Congeé" products in France; "Tiefgefroren" contrasts with "Gefroren" in Germany, and "Surgelato" with "Congelato" in Italy.

The choice of zero Fahrenheit, which translates into –17.8° on the Celsius scale, for quick frozen foods does not imply any magical properties for this temperature. Clarence Birdseye was an American who thought in Fahrenheit and advocated zero to minus five as an appropriate temperature, with the minimum of fluctuations, at which to hold and distribute quick frozen foods (Birdseye and Fitzgerald, 1932). In the 1950s a massive research program into frozen food quality was undertaken by the Western Regional Research Laboratory of USDA/ARS in Albany, California. The results of this research (Van Arsdel et al., 1969) determined the answer to the question "what happens, in microbiological, nutritional and sensory terms to the various products being frozen by the industry at different sub-freezing temperatures?" One outcome of this research was the reaffirmation of zero (forgetting the –5 advocated by Birdseye) as the industry norm by Harold Humphrey, then the General Manager of Birdseye Division of General Foods (Olson, personal communication). The concept of time–temperature tolerance (TTT) to describe frozen food stability was developed by this program.

15.2 STABILITY OF FROZEN FOODS

15.2.1 Microorganisms

Microbiological growth overwhelms physicochemical changes at around –8°C (18°F) or higher temperatures. Michener and Elliott (in Van Arsdel et al., 1969) listed the maximum recommended storage temperatures at which microbiological spoilage ceases. Most authorities put this at between –9°C (16°F) and –12°C (10°F). The discrepancies in the minimum temperatures found can probably be ascribed to the short length of some of the experiments. Molds can take months rather than weeks to develop substantially. Although microbiological spoilage can be discounted at frozen food storage temperatures, it must be borne in mind that the enzymes in any microorganisms present in the product will still play a part in spoilage, hence the point that a product produced under hygienic conditions or heat processed (blanched or cooked) will enjoy a longer shelf life than its unhygienic or raw counterpart. Many common pathogens (notably salmonellae, vegetative cells of *Clostridium perfringens,* and *Escherichia coli*) decline in numbers by several logs during prolonged cold storage but this must never be regarded as a substitute for good hygiene in processing.

15.2.2 Ice mobility

Ice, an inevitable concomitant of attaining temperatures as low as –18°C in food systems, is restless, always responding to any vapor pressure differential created by any thermal gradient, which is the reason for demanding reasonably constant temperatures during distribution of frozen foods. Fluctuations in product temperature of 2–3°C, as are likely to be found in bulk cold stores maintaining –18°C (0°F) or lower, are unlikely to cause perceptible damage even over long periods. On the other hand frequent fluctuations in temperature, particularly large fluctuations at "warm" temperatures (i.e., close to or in the zone of maximum crystal formation, i.e., between –1°C [30°F] and –5°C [23°F]), which can be experienced in retail display cabinets and during the "carry home" period, cause ice crystals to ripen or grow, coalesce, and move to the product surface or, if space allows, onto the inside of the packaging material. This leads ultimately to a freeze-dried product if the packaging is permeable to moisture allowing the sublimed or evaporated water vapor to escape. The loss of moisture results in toughening of animal tissue and greater exposure to any oxygen present.

Oxygen is the bugbear of almost all frozen foods, leading to oxidative rancidity if any unsaturated lipids are present, loss of color, and development of off-flavors. Moisture loss by sublimation from the surface of products leads to "freezer burn," an unsightly white color which can be mistaken for mold but which is resolved on rehydration during cooking unless it is severe. A whitish surface appearance, sought after in some markets, can also be induced, in products such as poultry, by very rapid surface freezing (by liquid nitrogen or carbon dioxide), which produces very small crystals leading to scattering of incident light.

15.2.3 Physico-chemical reactions

The most important quality changes during frozen storage which determine shelf life encompass color loss and color change in vegetables. In green vegetables this is due to hydrolysis of chlorophyll to pheophytin, giving a dull khaki color. Examples of the principal reactions occurring during storage are flavor loss, followed by the development of off-flavors; vitamin loss; retrogradation of starch; protein denaturation; hydrolytic (in milk) and oxidative rancidity; drip formation in animal tissues; crystalization of sucrose (on the surface this gives rise to spectacular volcanic excrescences of sucrose hydrate, the formation of which proceeds fastest at –23°C [–10°F]); and crystalization of lactose (in ice cream this causes a grainy texture). Good accounts of these changes in the various product categories can be found in Jul (1984), Van Arsdel et al. (1969), and Mallet (1993).

Nucleation, crystal ripening, changes on thawing, and associated quality defects peculiar to frozen products (Fennema, 1991; Bald, 1991; Jul, 1984; Van Arsdel et al., 1969; Reid, in Mallet, 1993) are all important considerations in assessing shelf life and can affect formulation considerations in selecting ingredients and processes which minimize these induced changes.

15.2.4 Time–temperature tolerance (TTT)

The time–temperature tolerance concept, developed as a result of the massive program of research into frozen food stability at Albany, postulated that quality loss in frozen foods is proportional to the reciprocal of the storage temperature. This program was concerned with evaluating the quality stability of frozen products then being marketed in the US and paid relatively little attention to the suitability of different vegetable varieties for freezing, growing conditions, processing, or packaging but concentrated on evaluating product stability taking for granted the current industry practices.

A survey of frozen food quality conducted in 1983/1984 (Earl et al., 1985) found frozen products in the distribution system as old as 840 days! Arrhenius plots on the assumption that quality loss is a zero order reaction fitted the data best. The reaction order of quality loss can in fact lie between zero and two but the data fitted best to zero order kinetics.

15.2.5 Product quality, processing, and packaging (PPP) factors

Jul (1984) drew attention to the very profound effects that the level of intrinsic product quality (freshness, suitability of variety for freezing, and so on), the processing (whether blanched or cooked; whole, sliced, or comminuted; addition of antioxidants; etc.), and the packaging (presence of glaze, breading, batter, or gravy; oxygen and moisture permeability of primary packaging; how close fitting to the contours of the product; whether evacuated or gas flushed) all exert on the final product quality. These can overwhelm TTT considerations in determining shelf life. Figure 15–1 illustrates the values for shelf life at different temperatures for various frozen products, including the effects of different packaging on bacon.

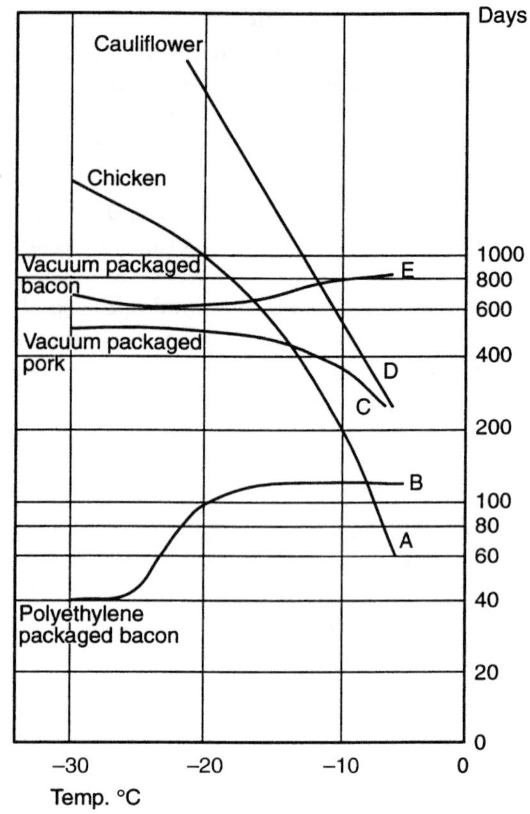

Figure 15–1 Estimated shapes of shelf-life curves for various frozen foods. (Source: M. Jul, 1984. In: Thermal processing and quality of foods, Zeuthen et al. (eds), 1984.)

15.2.6 Glassy state

The classic TTT concept, which has long served the frozen food industry well, especially after taking account of the modifications proposed by consideration of the PPP factors, has recently been refined by the recognition of the importance of the glassy state in frozen foods (Reid, in Mallett, 1993; Blond and Colas, in Bald, 1991). As an increasing proportion of water is converted to ice with decreasing temperature, the remaining solutes become progressively more concentrated and exceedingly viscous until, in a maximally freeze-concentrated solution, the glass transition temperature ($T_{g'}$) peculiar to the food is reached. At this temperature, the material between the ice crystals becomes a glass: an amorphous metastable solid, but retaining the disordered structure of a liquid, with a high viscosity (> 10^{13} Pa s). This change from a visco-elastic liquid (rubber) to a glass is accompanied by a dramatic decrease in mobility. All reactions slowed down as a consequence and hence quality loss greatly retarded, almost to the point of being suspended, over the few years of interest to frozen foods. The $T_{g'}$ of ice cream varies between –23° and –43°C depending on formulation (Goff et al., 1993).

Moisture content has little influence on the $T_{g'}$; the presence of solutes of low molecular weight (such as sugars) lowers the $T_{g'}$; however substances of high molecular weight (such as starches) exert little effect. This has the consequence that increasing maturity in many vegetables (maturity being commonly associated with a decrease in sugars and an increase in starch) will result in raising the $T_{g'}$ as also will increasing blanching time in products such as French fries in which the cut surfaces allow the leaching of sugars. Reid (in Mallet, 1993) further suggested that, at frozen food temperatures higher than the $T_{g'}$, the rate of quality loss may be dependent on the difference between the storage temperature and $T_{g'}$ rather than with the reciprocal of the storage temperature. Developing knowledge of the glass transition temperatures of materials commonly frozen and the factors which alter these temperatures will enable biotechnologists to manipulate plant genomes to achieve a higher $T_{g'}$ and product developers to formulate products and adjust processes to give as high a $T_{g'}$ as feasible. It is clearly advantageous to plan storage temperatures to take account of the $T_{g'}$ of the stored products, remembering that it is a waste of energy and money to store at temperatures lower than the $T_{g'}$ while, in the case of long-term bulk storage, the quality/cost equation is likely to favor attaining temperatures close to the $T_{g'}$.

15.2.7 Know your product

A prerequisite in planning any shelf-life determination of a frozen product is to select the most thermo-labile important quality attribute, whether this is flavor loss, color change, toughening, development of rancidity, and so on. If experience is lacking in the likely stabilities of the various quality attributes, help can be sought in the literature. A list of anticipated shelf lives for a number of products commonly frozen can be found in the literature, notably in Table 15–1 of practical storage life (PSL) taken from the International Institute of Refrigeration (IIR) (1986) where expected PSLs are tabulated for three temperatures: –12°C (10°F), –18°C (0°F) and –24°C (–12°F). The shelf lives given for –24°C (–12°F) are conservative, none more than 24 months because of a lack of published data since few investigators are concerned with longer storage periods for food products. Guadagni (in Van Arsdel, 1969) found no measurable changes over 5 years at –29°C (–20°F) in most frozen fruits. Jul (1984) quoting Bengtsson et al. (1972) made the same point. Many meat products probably have a shelf life of over 2 years if PPP factors are well managed and the product is held at frozen food temperatures. Winger and Pope (1982) found no loss in eating quality of grass fed lamb, hygienically processed and well packaged, after 2 years at –10°C (14°F). Anecdotal evidence tells us that the flesh of mammoths, after several thousand years in the permafrost in Siberia, was still recognizable and edible if not actually delicious! Help in determining what is the most thermo-labile attribute in a new product may be obtained by a judicious assessment of similar competitors" products. If date marking is practiced it is a simple matter to search out both "old" and recent production and assess the changes found. In the absence of date marking, production codes can often be broken since these are seldom cryptic.

Once the decision has been made as to which quality attributes are critical, a taste panel needs to be assembled and trained, selecting those who show reasonable sensitivity to the particular attribute and who can demonstrate consistent scores.

15.2.8 Objective tests

Objective tests are available for a number of quality attributes of frozen foods. These seldom faithfully reflect consumer preferences and are little used in day-to-day industry testing in the

Table 15–1 Practical storage life (PSL) in months at several storage temperatures

Product	–12°C (10°F)	–18°C (0°F)	–24°C (–12°F)
Fruits			
Raspberries/strawberries (raw)	5	24	>24
Raspberries/strawberries in sugar	3	24	>24
Peaches, apricots, cherries (raw)	4	18	>24
Peaches, apricots, cherries in sugar	3	18	>24
Fruit juice concentrate	—	24	>24
Vegetables			
Asparagus (with green spears)	3	12	>24
Beans, green	4	15	>24
Beans, lima	—	18	>24
Broccoli	—	15	24
Brussels sprouts	6	15	>24
Carrots	10	18	>24
Cauliflower	4	12	24
Corn-on-the-cob	—	12	18
Cut corn	4	15	>24
Mushrooms (cultivated)	2	8	>24
Peas, green	6	24	>24
Peppers, red and green	—	6	12
Potatoes, French fried	9	24	>24
Spinach (chopped)	4	18	>24
Onions	—	10	15
Leeks (blanched)	—	18	—
Meats and poultry			
Beef carcass (unpackaged)[a]	8	15	24
Beef steaks/cuts	8	18	24
Ground beef	6	10	15
Veal carcass (unpackaged)[a]	6	12	15
Veal steaks/cuts	6	12	15
Lamb carcass, grass fed (unpackaged)[a]	18	24	>24
Lamb steaks	12	18	24
Pork carcass (unpackaged)[a]	6	10	15
Pork steaks/cuts	6	10	15
Sliced bacon (vacuum packed)	12	12	12
Chicken, whole	9	18	>24
Chicken, parts/cuts	9	18	>24
Turkey, whole	8	15	>24
Ducks, geese, whole	6	12	18
Liver	4	12	18
Seafood			
Fatty fish, glazed	3	5	>9
Lean fish[b]	4	9	>12
Lobster, crab, shrimps in shell (cooked)	4	5	>12
Clams and oysters	4	6	>9
Shrimps (cooked/peeled)	2	5	>9

continues

Table 15–1 continued

Product	*–12°C (10°F)*	*–18°C (0°F)*	*–24°C (–12°F)*
Eggs			
Whole egg magma	—	12	>24
Milk and milk products			
Butter, lactic, unsalted pH 4.7	15	18	20
Butter, lactic, salted pH 4.7	8	12	14
Butter, sweet cream, unsalted pH 6.6	—	>24	>24
Butter, sweet cream, salted (2%) pH 6.6	20	>24	>24
Cream	—	12	15
Ice cream	1	6	24
Bakery and confectionery products			
Cakes (cheese, sponge, chocolate, fruit, etc.)	—	15	24
Breads	—	3	—
Raw dough	—	12	18

[a]Carcass may be wrapped in stockinette.
[b]The PSL for single fillets of lean fish would be 6, 9 and 12 months at –18°C (0°F), –24°C (–12°F), and –30°C (22°F), respectively.
(Source: IIR, 1986.)

course of shelf-life predictions or quality assurance. If a nutritional claim is made, then of course how that nutrient withstands prolonged cold storage and simulated thermal abuse is important. The loss of vitamin C and the ratio of chlorophyll to pheophytin in green vegetables can be a measure of thermal abuse (Olson and Dietrich, in Van Arsdel et al., 1969) as can the decrease in optical density of a muscle homogenate in cod (Love, 1962), the development of thiobarbituric acid (TBA) or peroxide value in products containing unsaturated lipids (Dalhoff and Jul, 1965), the development of cold store flavor in fish (McGill et al., 1974) and a host of other indicators in various products. These objective tests may be useful to determine alongside sensory changes and may also be necessary in any claim or litigation alleging unsatisfactory product handling or failure to meet a quality specification by a third party. They have seldom, however, been found to reflect accurately, and with sufficient sensitivity, consumer preferences. In general they only confirm what is already abundantly apparent from taste panel assessments.

15.2.9 Accelerated testing

Seldom can the luxury be permitted of long-term assessment of product quality over the months or years of anticipated shelf life at frozen food temperatures. Some form of accelerated testing is necessary which is likely to result in sacrificing a degree of accuracy in the results obtained. This presupposes that all frozen products lose quality, at higher temperatures than lower, i.e., possess a positive $Q_{10}(q_{10})$. $Q_{10}(q_{10})$ is the ratio of quality loss at two temperatures in the range of frozen food temperatures that are 10° apart: 10°C for Q_{10} and 10°F for q_{10}. Not all frozen products exhibit a positive $Q_{10}(q_{10})$. Neutral stability, or a $Q_{10}(q_{10})$ of 1, is found in a few products, such as salami, which may be a component in a frozen product. The quality of this cured, fermented, and dried product is temperature independent and it enjoys an equally long shelf life at frozen food temperatures as at chill or ambient. A negative Q_{10} is rare and was first described by Boegh-Sorensen (1968) who found reverse stability in frozen vacuum-packaged cured sliced pork products. Shelf life was almost independent of storage temperatures between –6 and –23°C (21 and –9°F) while one product, smoked streaky bacon, exhibited a longer shelf life at the highest temperature. Lindeloev (1978) showed that at a storage temperature as low as –25°C (–13°F) the shelf life of bacon is shorter than at –5°C (23°F). This is an example of the effect of concentration of catalysts of oxidative rancidity (sodium chloride, nitrate, and nitrite) by the subtraction of ice from the solution overwhelming the effect of temperature reduction. Christensen and Jensen (in Zeuthen et al., 1984) did not find reverse stability in cured fish.

Accelerated testing can usefully be carried out by holding control product at –40°C (–40°F) and exposing test samples to three or four different storage temperatures between –30°C (–22°F) and –10°C (14°F). For many products the reference samples may be held at –30°C (–22°F) since little change occurs in most products at lower temperatures.

15.2.10 Just noticeable difference

The goal of shelf-life determination is to estimate the length of time during which the product under test will still be appreciated by the ultimate consumer as being of the level of quality expected for the product, that is, for how long it will maintain practical storage life (PSL). Loss of quality in frozen foods is a gradual process, the changes being slow or very slow, cumulative, and irreversible. PSL is a highly subjective concept and difficult to measure by taste panel procedures. In the rarefied atmosphere of a taste panel, as opposed to a meal situation, the trained assessors are likely to discriminate between nuances of quality unlikely to be appreciated by the consumer. A way out of this difficulty is to measure the time to "just noticeable difference" (JND) or "high quality life" (HQL) by a trained taste panel and then multiply this by an arbitrary figure, generally between 2 and 5, to arrive at PSL. In products peculiarly sensitive to color loss (peaches, cauliflower, red pigmented fish) PSL may be close to JND.

JND is conveniently defined as the moment during storage at the expected temperatures of the cold chain at which a proportion (generally 70%) of the trained assessors can correctly distinguish the product under test from the control (held at –30°C [–22°F] or –40°C) using a triangular or duo–trio test. It can be argued that the proportion of assessors able to discriminate successfully should be much lower, say 25%, on the basis that a small proportion of consumers is likely to be highly discriminatory concerning the particular quality attribute deemed to be critical. Anyone with experience of taste panels knows how very sensitive individuals can be to particular organoleptic attributes and sometimes surprisingly insensitive to others.

Dalhoff and Jul (1965) preferred a sensory scoring system (–5 equals dislike extremely, 0 neither like nor dislike, + 5 like extremely) to determining JND by triangular or duo–trio testing. A decrease of 1 point on the scale would equate to JND. An advantage of this procedure is that fewer replicates are required than the four replicates for each time/temperature stipulated by Guadagni (in Van Arsdel et al., 1969). An objection is that the preference scoring system is likely to be population specific. For instance, quality expectations for fish flavor would be very different between a Danish population (who expect their fish to be alive when purchased) and an English population accustomed to fish held in ice for several days. The English taste panel would probably score the Danish fish as lacking in fish flavor whereas the Danes would score the English fish as partly rotten at best!

15.2.11 $Q_{10}(q_{10})$

Most frozen products enjoy shelf life measured in many months or even years. During product development answers as to the likely shelf life of a new or modified product are required in weeks or a few months to be of any use. Hence there is the need for accelerated shelf-life determinations. Accelerated shelf-life tests aim to establish the *rate* of quality loss with temperature change and hence to enable PSL at conventional distribution temperatures to be assessed far more rapidly than the long-term tests at the anticipated storage temperature. This rate can be expressed as the ratio of the shelf lives found at two different temperatures, generally 10°C (Q_{10}) or 10°F (q_{10}) apart. For instance, Guadagni (in Van Arsdel et al., 1969) found stabilities of strawberries, assessing both color and flavor, held at three different temperatures, −18°C (0°F), −12°C (10°F), and −7°C (20°F), to be 360, 60, and 10 days, respectively. The q_{10} is therefore 360/60 or 6 and 60/10 or 6 again for the respective 10°F differences measured. Not all rate determinations work out as neatly as strawberries and it is probably futile to try to fit quality loss determinations to a particular mathematical formula. Foods are complicated biological systems and do not lend themselves to convenient mathematical expressions which can be applied to a wide variety of products.

15.3 PRODUCT CHARACTERISTICS

15.3.1 Fruits and vegetables

Most fruits freeze well; that is, the thawed product is barely distinguishable from the raw fruit before freezing. Although strawberries do not fall into this category, losing texture during freezing and thawing, the early days of the frozen food industry saw strawberries dominating the market for frozen fruit. The fruit was mixed with sugar in wooden barrels placed in a frozen food warehouse and rolled periodically to mix the fruit and sugar. The frozen product was used in preserves and ice creams. Heat inactivation of the enzymes that cause rapid discoloration by enzymic browning leads to a cooked flavor which is unacceptable and the addition of syrups with ascorbic acid as an antioxidant is used instead. q_{10}s for frozen fruit, measured on flavor and color, have been found to range from 2.2 for blueberries to 6.0 for strawberries in dry sugar (Guadagni, in Van Arsdel et al., 1969).

Frozen vegetables have enjoyed enormous success in many different parts of the world. Since many vegetables do not can well, the introduction of frozen vegetables has meant populations living in high latitudes can enjoy summer vegetables all year round instead of having to rely largely on root crops in the winter. Properly blanched vegetables have a long shelf life at frozen food temperatures, enabling them to be exported all over the world to remedy crop shortfalls and to span the seasons. During storage chlorophyll is converted to pheophytin, with loss of green color and introduction of a khaki hue, and this can be used as an objective test for quality loss in green vegetables. Vitamin C can also be used as an objective indicator of quality loss. The most sensitive quality attribute is a diminution of flavor accompanied by a loss of color.

15.3.2 Seafood

Microbiological spoilage is the principal cause of quality loss in seafood at temperatures above chill; around 0°C (32°F) enzymic breakdown of protein becomes the principal cause of quality loss; below −8°C (18°F) microbiological spoilage ceases and protein denaturation coupled with oxidative rancidity in fatty species become the chief factors affecting quality.

Figure 15–2 illustrates the stabilities at various temperatures of several species of commonly consumed finfish. Lane (1984) pointed out that sensory characteristics of frozen fish change in intensity during storage rather than in character. The development of rancid flavors and progressive toughening accompanied by the development of cold store flavor are the principal sensory changes. In the case of young fish (e.g. codling) the initial stages of cold store damage may give rise to an increase in acceptability scores. The "firming" of the soft texture characteristic of young fish by the early onset of protein denaturation is preferred by most taste panelists (Kelly, 1969). Assessing codling in storage trials may therefore result in an initial rise in preference for texture before the expected decline sets in. Flavor is probably more critical than texture (Connell and Howgate, 1971) since this tends to be lost before protein denaturation is appreciable. Formaldehyde is formed during cold storage by enzymic decomposition of trimethylamine oxide (TMAO). Formaldehyde content is a good objective criterion of time/temperature exposure in frozen gadoid species

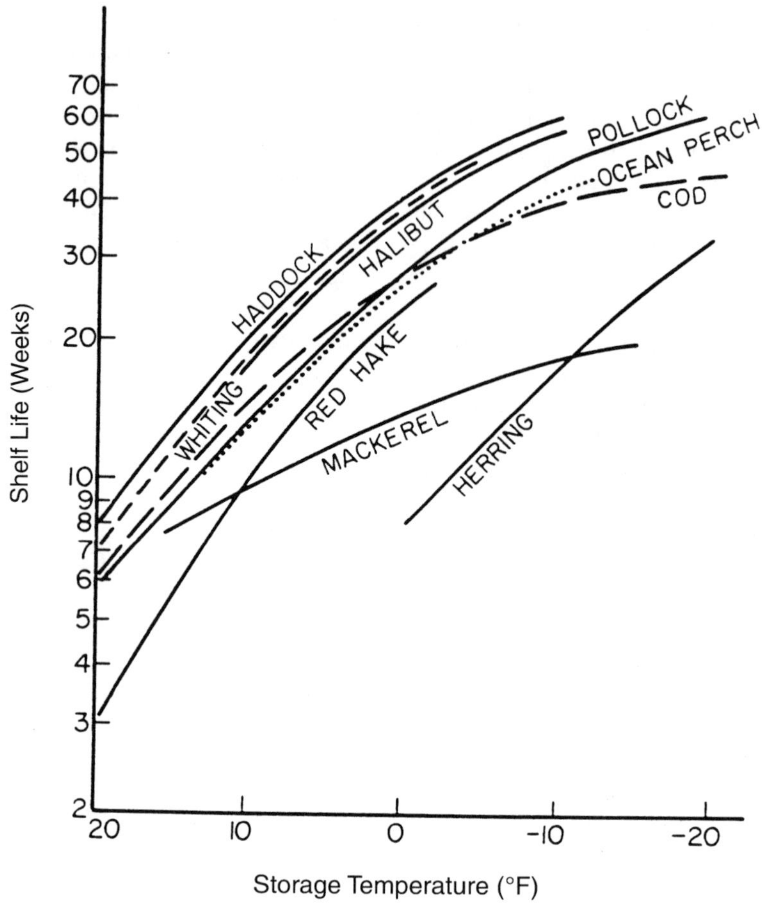

Figure 15–2 Average shelf-life values for selected finfish based on literature reported values. (Source: IIR, Proceedings of "Storage lives of chilled and frozen fish and fish products," 1985.)

(Boeri et al., 1993). This formaldehyde reacts with proteins, thereby decreasing their solubility in salt and buffer solutions (Rehbein, 1971). Quality loss during cold storage of fish is characterized by increasing loss of water-holding capacity, decrease in protein extractability, and a slight loss in ATPase activity (Reid et al., 1985). Much lower temperatures than –18°C (0°F) are useful for bulk storage of seafood. In Japan the retention of a pristine white color in surimi demands storage temperatures as low as –60°C (–6°F). Magnussen (1992) reviewed the use of such temperatures on mackerel and salmon. Steynor (1992) described a 2,000 tonnes chamber holding –60°C (–6°F) in South Africa.

Of all seafood species, squid *(Loligo duvaucelli)* probably is most sensitive to lack of product freshness leading to reduced stability in subsequent storage. Squids held for two days in ice lose flavor and the cold storage life is dramatically shortened (Jose Joseph et al., 1985).

15.3.3 Meat

Meat, whether it be lamb, venison, beef, or poultry, all freeze well. Consumer preference is for chilled meat. Some 80% of the chilled beef purchased at supermarkets in the US is subsequently frozen by the housewife at home. This reaction against purchasing frozen beef probably reflects a desire on the part of the purchaser to assess quality (ratio of bone to muscle, degree of marbling, presence of excess fat and so on) rather than any prejudice against the frozen beef processor. Stability of frozen meats can be affected by the diet of the animal. Pigs reflect in their depot fat the degree of unsaturation of the fat component of their diet. Figure 15–3 gives examples of the influence of the degree of unsaturated fat on shelf life of pig meat. Grass fed frozen lamb is recognized as having a greater stability in the frozen state than its grain-fed counterpart. Rapidly chilled muscles become tough on freezing and thawing, a phenomenon known as "cold shortening," which scarcely affects poultry. This applies with especial severity to small animals such as lamb, as the sheer bulk of the beef carcass precludes all but superficial muscles from chilling with sufficient rapidity to undergo "cold shortening." Electrical stimulation post-slaughter can reduce the time the carcass needs to spend above 10°C (50°F) to avoid "cold shortening." Eating quality of electrically stimulated grass-fed lamb, processed to strict hygiene standards and wrapped in plastics, is essentially unchanged for at least 2 years storage at –10°C (14°F) (Winger, 1984). In ground beefburgers the critical quality attribute may be a lack of balance in the spices present. Pepper flavor persists well in cold store whereas other spices diminish in intensity. A high pepper note may indicate this lack of balance long before rancidity sets in. The lack of clear and unambiguous data on the shelf lives of meats, which is ascribed to differing definitions of the end of storage life, is clearly seen in a review by Evans and James (1993).

15.3.4 Bakery products

If seafood loses quality within days of capture bakery products lose freshness within a few hours of being removed from the oven. Freezing is an obvious method to preserve "fresh from the oven" characteristics were it not for the need to traverse the chill temperature zone twice, on freezing and again when thawing. Quality loss, through starch retrogradation, occurs most rapidly at chill temperatures in baked goods. Slow freezing is to be avoided in order to reduce the time spent at chill temperatures. Despite this constraint a well-formulated, frozen, stored, and

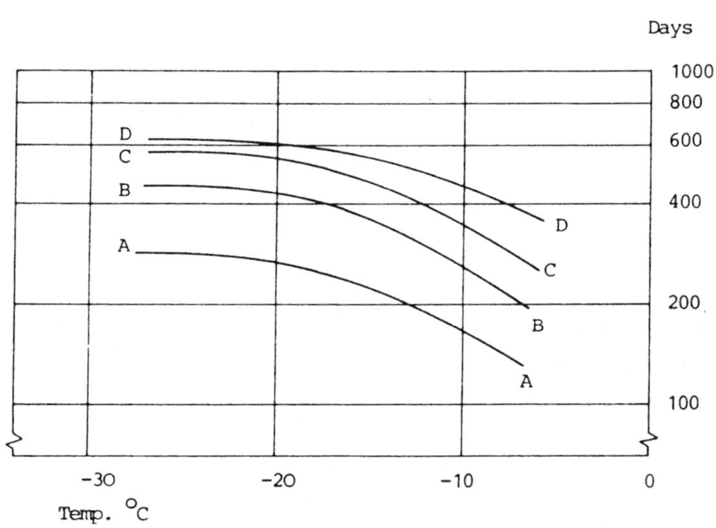

Figure 15–3 Examples of the influence of PPP factors on shelf lives of frozen foods. **A**. Pig meat with high iodine number in fat; **B**. Pig meat with low iodine number in fat; **C**. Same meat as B, but cooked; **D**. Same as C, but vacuum packaged. (Source: M. Jul, 1984. In: Thermal processing and quality of foods, Zeuthen et al. (eds), 1984.)

refreshened bakery product (regardless of the time spent in frozen storage) has sensory properties comparable with the same (unfrozen) product less than 1 day after baking (Stauffer, 1992). Amylase is a useful antidote to bread staling. Moisture migration during frozen storage is the principal cause of staling in general. With careful formulation, strict regard for the first "P" coupled with modifying production and handling techniques, the second "P" can guard against these defects. Cooling before freezing is important for stability: a hot bakery product placed in a blast freezer will result in an ice rind formed just below the crust surface, which will become detached on slicing. Some bakery products, notably croissants, are difficult to formulate to enable them to be frozen without becoming leathery and losing their fresh aroma. The open interior of the product mitigates against rapid freezing.

Yeast cells do not withstand freezing well. Yeast-raised products have some of the lowest stabilities of any frozen foods. This can be partially compensated for by increasing the amount of yeast used in the formulation; new and improved yeast strains exhibit a better survival rate in freezing.

15.3.5 Prepared products

Complex multicomponent products not only bring all the problems of the stability characteristics of each component but a few extra, such as aroma migration between the components and the likelihood of spaces being present in the package, leading to easy access for oxygen. If fluctuating temperatures are encountered moisture migration from a high water activity component to a low one may result in grave quality defects. A common complaint concerning frozen multi-component prepared foods is a lack of, or "sameness" of, flavor. This is probably due to the beginning of aroma migration between the components, the more highly flavored component transmitting flavor to the bland ones so that a depressing uniformity in flavor results. "Sameness" of flavor should be noted as possibly the first perceptible difference in many products. Otherwise the rule of assessing stability of multi-component products is to identify the most thermo-labile component and its most sensitive quality attribute and concentrate on assessing these as an indicator of JND. Figure 15–4 illustrates the enormous degree of variability reported by various authors in the shelf life of ready-to-eat frozen meat dishes.

15.4 THE COLD CHAIN

The final decision as to the likely shelf life of a given frozen product must take account of the fact that the product is not going to be stored at a steady temperature of –18°C (0°F) from the end of freez-

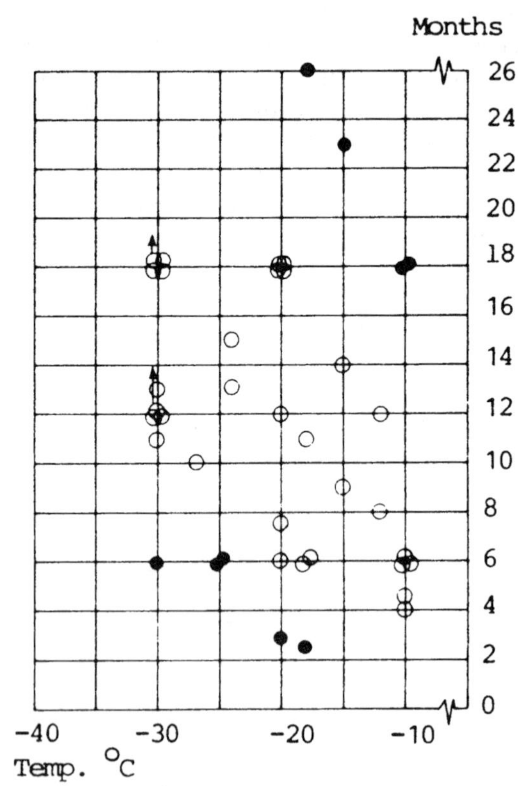

Figure 15–4 Shelf-life data for ready-to-eat meat dishes as reported by various authors, as collected by Spiess, and quoted by Jul (1984). (Source: M. Jul, 1984. In: Thermal processing and quality of foods, Zeuthen et al. (eds), 1984.)

ing until thawing for consumption or during end cooking. A typical frozen product will spend part of its shelf life in a bulk cold store, a refrigerated vehicle or container, a distribution store, a retail display cabinet or institutional frozen food storage cabinet, a period out of refrigeration during the journey from the retail outlet to home, and time in a home freezer or "star marked" frozen food storage compartment, before being consumed in the frozen state, after thawing or cooking. Figure 15–5 illustrates the various components of a typical cold chain in Europe where "star marked" frozen food storage compartments are common in household refrigerators.

An essential part of any shelf-life determination in markets where "star marked" frozen food compartments are in use is to assess the shelf life at the relevant temperatures for 1-Star, –6°C (21°F), and 2-Star, –12°C (10°F) storage.

15.4.1 Temperature monitoring

There is a plethora of indicating and recording devices to enable product temperatures to be monitored throughout distribution; from simple indicating thermometers to computerized recorders which can be buried in loads where they record temperature every 15 minutes, the complete record being read out on a computer within a minute or so at the end of the journey or period in storage. Refrigerated vehicles can be fitted with transmitters which enable the dispatcher to call up the vehicle at any time and discover its location and load temperature without any need for intervention by the driver.

15.4.2 Time–temperature indicators

The literature is replete with more than 500 patents for freeze-thaw indicators and temperature limit indicators as well as time–temperature integrators. Few have been brought to market and none has achieved an enduring success. Some are unwieldy or expensive; the best are inexpensive, costing a few pennies each, but, despite numerous trials in products as diverse as cut flowers, vaccines, ice cream, frozen military rations, or frozen sous vide products, no sustained successful use has yet been announced. Trials seem always to be abandoned even if the device itself performs its task adequately.

More attention is being paid to frozen food temperatures in Europe in response to the EU directive that mandates specific temperatures for various commodities throughout the cold chain. The need for this law was long disputed on the grounds that

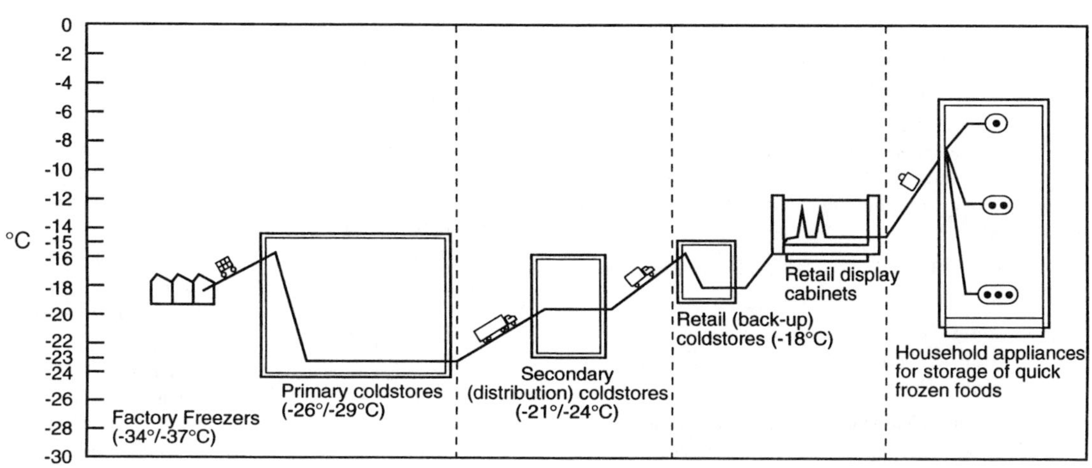

Figure 15–5 Temperature profile of the cold chain. (Source: Sanderson-Walker, 1984. In: Thermal processing and quality of foods, Zeuthen et al. (eds), 1984.)

such temperatures have little, if anything, to do with health, are only concerned with quality, and that temperature at any particular moment is an imperfect indicator even of quality.

15.4.3 Control of time in storage

Frozen foods that are well formulated, correctly processed, and adequately packaged have a very long cold storage life at sufficiently cold temperatures but nothing is forever. Some system of control is required over the time spent in the various segments of the cold chain. This calls for control over pallets, over the cases of product, and, finally, over the individual packages.

Control over pallets is generally achieved by the computerized recording system of the producer in concert with the bulk storage facility. Control over cases is achieved in Europe by printing the mandatory open shelf-life date which is demanded on individual packets on the outer case in standard format. In the US no such assistance is provided to the supermarket store clerk who may have to choose between two identical cases to be placed in the display cabinet. Production codes in cryptic form do not help much. In countries where open shelf-life dating is mandated, such as in the EU, a minimum durability, "best before" or final consumption date is printed on the consumer package together with the relevant "star marking" storage instructions. In the US some producers give a shelf-life date on individual packets but, over the years, there seems to have been little consumer reaction. Labuza (1982) gives the rationale for open shelf-life dating on consumer packets as an aid to engendering consumer confidence and supplying the consumer with useful information. In the absence of a temperature history this information can be misleading. Opponents to mandatory open shelf-life dating on consumer packets allege that this will give undue prominence to time to the detriment of maintaining adequate temperatures which is immeasurably more important to frozen food quality. Also when more than one date is on display very few, if any, prospective customers will select the earlier date; most will ignore the date while a few will "rummage" for the latest date, upsetting the "FIFO" (first in, first out) principle of stock rotation. Placing a date on the consumer package also gives the erroneous impression that frozen food quality expires suddenly whereas it is more likely to decline very slowly over long periods if held at the required temperature.

REFERENCES

Bald, W.B. (1991) *Food Freezing Today and Tomorrow,* Springer-Verlag, London, England.

Bengtsson, N., Liljemark, A., Olsson, P. and Nilsson, B. (1972) An attempt to systemize time–temperature tolerance (T-TT) data as a basis for the development of time–temperature indicators. *Bull. Int. Inst. Refrig.* (Annexe 1972–2), 303–11.

Birdseye, C. and Fitzgerald, G.A. (1932) *Ind. Eng. Chem.,* **24**, 676.

Boegh-Sorensen, L. (1968) *Keeping Times for Frozen Pork Cold Cuts,* Danish Meat Products Laboratory, Copenhagen, Denmark.

Boeri, R.I., Alamandos, M.E., Ciarlo, A.S. and Gianni, D.H. (1993) Formaldehyde instead of dimethylamine as a measure of total formaldehyde formed in frozen Argentine hake *(Merluccius hubbsi). Int. J. Food Sci. Tech.,* **28**, 289–92.

Codex (1982) *International Code of Practice for Processing and Handling of Quick Frozen Foods,* Codex Alimentarius, FAO, Rome, Italy.

Connell, J.J. and Howgate, P. (1971) Consumer evaluation of fresh and frozen fish, in *Fish Inspection and Quality* (ed R. Kreuzer), Fishing News (Books), Surrey, England, pp. 155–159.

Dalhoff, E. and Jul, M. (1965) Factors affecting the keeping quality of frozen foods, in *Progress in Refrigeration Science and Technology, vol. 1,* Pergamon Press, New York, NY, pp. 57–66.

Earl, W.H., Allen, D.H. and Kindleysides, L. (1985) Simulation modelling of frozen food distribution: retail operations and their effect on quality, in *Refrigeration Science and Technology. Storage Lives of Chilled and Frozen Fish Products,* Proceedings of Meetings of Commissions C2 and D3 of IIR, October 1985, International Institute of Refrigeration, Paris, France.

Evans, J. and James, S. (1993) Freezing and meat quality, in *Food Technology International Europe,* Sterling, London, England.

Fennema, O. (1991) *Instability of Nonequilibrium States of Water in Frozen Foods,* 8th World Congress of Food Science and Technology, Toronto, Canada.

Goff, H.D., Caldwell, K.B. and Stanley, D.W. (1993) Influence of polysaccharides on the glass transition in frozen sucrose solutions and ice cream. *J. Dairy Sci.,* **76**(5), 1268–77.

Jose Joseph, P.A., Perigreen, P.A. and Nair, M.R. (1985) Effect of raw material quality on the shelf-life of frozen squid *(Loligo duvaucelli)* mantles, in *Refrigeration Science and Technology. Storage Lives of Chilled and Frozen Fish Products,* Proceedings of meetings of Commission C2 and D3 of IIR, October 1985, International Institute of Refrigeration, Paris, France.

Jul, M. (1984) *The Quality of Frozen Foods,* Academic Press, London, England.

Kelly, T.R. (1969) Quality in frozen cod and limiting factors on its shelf life. *J. Food Technol.,* **4**, 95–103.

IIR (1986) *Recommendations for the Processing and Handling of Frozen Foods,* International Institute of Refrigeration, Paris, France.

Labuza, T.P. (1982) *Shelf Life Dating of Foods,* Food and Nutrition Press, Westport, CT.

Lane, J.P. (1984) Time–temperature tolerance of frozen seafoods. 1. Review of some recent literature on the storage life of frozen fishery products. *Food Technol.,* **18**, 1100–6.

Lindeloev, F. (1978) Freezer storage of salted products. Laboratoriet fur Levnedsmiddelindustri, Danmarks Tekniske Hojskole, Copenhagen, Denmark.

Love, R.M. (1962) Protein denaturation in frozen food. VI. Cold-storage studies on cod using the cell fragility method. *J. Sci. Food Agric.,* **13**, 269–78.

Magnussen, O.M. (1992) "Superfreezing" low temperature technology. *Scan. Ref.,* **21**(4), 3441.

Mallett, C. (1993) *Frozen Food Technology,* Blackie Academic and Professional, Glasgow, Scotland.

McGill, A.S., Hardy, R. and Burt, J.R. (1974) Hept-cis-4-enal and its contribution to off-flavour in cold stored cod. *J. Sci. Food Agric.,* **25**, 1477–89.

Plank, R., Ehrenbaum, E. and Reuter, K. (1916) *Die Konservierung von Fischen durch das Gefrierberfahrn (Fish Freezing),* Zentral-Einkaufgesellschaft, Berlin, Germany.

Rehbein, H. (1971) Does formaldehyde form cross-links between myofibrillar proteins during frozen storage of fish muscle? In *Fish Inspection and Quality* (ed R. Kreuzer), Fishing News, Surrey, England.

Reid, D.S., Doong, N.F., Foin, A. and Snider, M. (1985) Studies on the frozen storage of fish, in *Refrigeration Science and Technology. Storage Lives of Chilled and Frozen Fish Products,* Proceedings of meetings of Commissions C2 and D3 of IIR, October 1985, International Institute of Refrigeration, Paris, France.

Stauffer, C.E. (1992) Frozen bakery products, in *Frozen Food Technology* (ed C. Mallett), Blackie Academic and Professional, Glasgow, Scotland.

Steynor, P. (1992) Cold storage reaches lowest temperature. *S. Afr. Refrig. Air Cond., ZA,* Vol. B., (5), 20–22.

Van Arsdel, W.B., Copley, M.J. and Olson, R.L. (1969) *Quality and Stability in Frozen Foods,* Wiley Inter-Science, New York, NY.

Winger, R.J. and Pope, C.G. (1982) The effect of processing variables on the storage stability of frozen lamb. *Bull. Int. Inst. Refrig.,* **4**(6), 335–9.

Winger, R.J. (1984) Storage life and eating quality of NZ frozen lamb: A compendium of impressive longevity, in *Thermal Processing and Quality of Foods* (eds P. Zeuthen, J.C. Cheftel, C. Eriksson, M. Jul, H. Leniger, P. Linko, G. Varela and G. Vos), Elsevier Applied Science, London, England.

Zeuthen, P., Cheftel, J.C., Eriksson, C., Jul, M., Leniger, H. and Linko, P. (eds) (1984) *Thermal Processing and Quality of Foods,* Elsevier Applied Science, London, England.

Chapter 16

Minimally Processed, Ready-to-Eat, and Ambient-Stable Meat Products

L. Leistner

16.1 INTRODUCTION

Minimally processed, ready-to-eat, and ambient-stable meat products are in great demand. This chapter will focus on this group of food products. Meat products that are chilled, frozen, or canned or require cooking before consumption will not be discussed here.

A common feature of all minimally processed, ready-to-eat, and ambient-stable foods is that their microbiological stability and safety as well as their organoleptic quality are based on empirical rules. Increasingly, however, combined methods, i.e., the application of intentional hurdle technology, are being used to preserve these products. Therefore, in the first section of this chapter, the principles of hurdle technology will be briefly outlined. The basic aspects of hurdle technology foods and their total quality will also be described. The latter depends on a balance between preservation and quality hurdles. The next sections of this chapter cover the properties and shelf lives of minimally processed, ready-to-eat, and ambient-stable fermented as well as heat-processed meat products. A further section is devoted to the design and shelf-life evaluation of minimally processed, ready-to-eat, and ambient-stable meats using an example based on an extensive study of provisions for the German army. From this study a user guide to product design has been derived and will be discussed, which combines hurdle technology, predictive microbiology, and the HACCP concept. This guide has proved useful in solving real development problems in the food industry. The final section of this chapter provides a summary of current trends in hurdle technology foods in industrialized as well as developing countries.

16.2 HURDLE TECHNOLOGY FOODS

Many traditional methods of food preservation, such as heating, chilling, freezing, drying, salting, curing, sugar addition, acidification, fermentation, smoking, and oxygen removal, are used for making foods stable and safe. However, these procedures are based on relatively few preservation factors or hurdles, i.e., high temperature (F value), low temperature (t value), water activity (a_w), acidification (pH), redox potential (E_h), preservatives, and the competitive flora (c.f.). In some of the preservation methods mentioned, these factors are of major importance, in others they are only secondary hurdles (Leistner et al., 1981; Leistner, 1992a). The use of inhibiting factors in combination is advantageous by allowing the less extreme use of a single treatment (Gould and Jones, 1989).

The critical values of many important parameters for the death, survival, or growth of microorganisms in foods have been determined in recent decades and have formed the basis of modern food preservation. However, the critical value of a particular parameter changes if additional preservation factors are present in the food. For instance, it is well known that the heat resistance of bacteria increases at low a_w and decreases at low pH or in the presence of preservatives, whereas a low E_h increases the inhibition of microorganisms caused by reduced a_w. The simultaneous effects of different preservation factors (hurdles) could be additive or even synergistic. The microbiological stability and safety of many foods are based on the combined effects of hurdles. For instance, mildly heated canned meats ("half-preserved" or "three-quarter-preserved") need refrigeration during storage, and fermented sausages are only stable and safe if both the a_w and the pH are in the appropriate range. Therefore, in food preservation the combined effect of different preservation factors must be taken into account, and most foods are indeed hurdle technology foods.

The most important hurdles commonly used in food preservation are temperature (high or low), water activity (a_w), acidity (pH), redox potential (E_h), preservatives (e.g., nitrite, sorbate, sulfite), and competitive microorganisms (e.g., lactic acid bacteria). However, more than 60 potential hurdles for foods of animal or plant origin, which can improve the microbiological stability and/or the organoleptic quality of these products, have already been described, and the list of possible hurdles for food preservation is by no means complete (Leistner, 1999). At present, some physical, non-thermal processes (using high hydrostatic pressure, heat, pressure, and ultrasound, oscillating magnetic fields, pulsed electric fields, or light pulses, etc.) are receiving considerable attention, since in combination with other conventional hurdles they are of potential use for the preservation of food products with minimal loss of nutritional and eating qualities (Barbosa-Cánovas et al., 1998). Another group of hurdles which are at present of special interest in industrialized as well as in developing countries are the "natural preservatives" (spices and their extracts, hop extracts, lysozyme, chitosan, protamin, antibiotic peptides, etc.). In most countries these "green preservatives" are preferred because they are not synthetic, and in some developing countries that are short of foreign currency they are given preference because spices are readily available and cheaper than imported chemicals.

16.2.1 Hurdle effect

The microbiological stability and safety of most traditional and novel foods are based on a combination of several preservation factors (called hurdles) which microorganisms present in them are unable to overcome. This is illustrated by the so-called hurdle effect, first introduced by Leistner (1978). The hurdle effect is of fundamental importance to the preservation of foods, since the hurdles in a stable product control microbial spoilage and food poisoning, as well as desired fermentation processes (Leistner et al., 1981). Leistner and co-workers acknowledged that the hurdle concept illustrates only the well-known fact that complex interactions of temperature, water activity, acidity, redox potential, preservatives, etc., are significant to the microbiological stability and safety of foods. For each stable and safe food a certain set of hurdles is inherent, which differs in quality and intensity depending on the product. In any case the hurdles must be able to keep the "normal" population of microorganisms in the food under control. The microorganisms present (at the start) in a food should not be able to overcome ("leap over") the hurdles present during storage of the product, otherwise the food will spoil or even cause food poisoning.

In previous publications (Leistner, 1987, 1992a, 1995a) a diagram with several examples illustrating the hurdle effect in foods has been presented, which will not be repeated here. However, it should be mentioned that the hurdle effect is, for instance, important in the ultraclean or aseptic packaging of foods, because if there are very few microorganisms present at the start, then fewer or lower hurdles may be sufficient to ensure the stability of the product. The same is true if the initial microbial load of a food (e.g., bacteria on carcass meat) has been substantially reduced by some decontamination procedures (e.g., by the application of steam), the remaining microorganisms can be more easily inhibited. The number and intensity of hurdles needed for microbiological stability are also lower if the microorganisms present are sublethally injured, because then they lack "vitality" and are thus easier to inhibit. On the other hand, a food product rich in nutrients and vitamins will favor the growth of microorganisms (this is called the "booster" or "trampoline" effect), and the hurdles in such a product must therefore be reinforced, otherwise they will be overcome. The latter also happens if due to poor hygiene, too many undesirable microorganisms are present so that the usual hurdles inherent in a product may be unable to prevent spoilage or even food poisoning. In fermented foods (salami, cheese, pickled vegetables, etc.) a sequence of hurdles comes into play with hurdles arising at different stages of the ripening process, leading eventually to a microbiologically stable and safe finished product.

16.2.2 Hurdle technology

A better understanding of the impact and interactions of different preservation factors (hurdles) in foods is the basis for improvements in food preservation. This is because if the hurdles in a food are known and their interactions visualized, the microbiological stability and safety of this food might be optimized by changing the intensity or quality of these hurdles (Leistner, 1999). Understanding the hurdle effect has been the key to an understanding of the effectiveness of traditional methods of food preservation. The ensuing development has been the optimization of traditional foods from a preservation standpoint as well as the development of novel products by the intelligent combinations of hurdles. Thus, from an understanding of the hurdle effect, hurdle technology has been derived (Leistner, 1985a), which is based upon the deliberate and intentional combination of hurdles in the preservation of traditional and novel food products.

Intelligent application of hurdle technology is advancing worldwide as evidenced by various expressions used for the same concept in different languages (e.g., Hürden-Technologie in German, hurdle technology in English, technologie des barrières in French, barjernaja technologija in Russian, technologija prepreka in Serbian, tecnologia degli ostacoli in Italian, tecnologia de obstaculos in Spanish, shogai gijutsu in Japanese, and zanglangishu in Chinese). A number of groups of researchers are working at present in different countries on the application of hurdle technology to a variety of foods; a treasure of results which will foster the advancement of food preservation is anticipated.

16.2.3 Basic aspects

Food preservation implies exposing pathogenic and spoilage microorganisms to a hostile environment in order to inhibit their growth or shorten their survival or cause their immediate death. The feasible response of these microorganisms to such a hostile environment determines whether they would grow or die. In order to understand this response more research is needed. Recent advances have been made by considering the homeostasis, metabolic exhaustion, and stress reactions of microorganisms in relation to hurdle technology, as well as by introducing the novel concept of multitarget preservation for the gentle and yet effective preservation of hurdle technology foods (Leistner, 1995a, 1995b).

16.2.3.1 Homeostasis

Homeostasis is the tendency toward uniformity and stability in the internal status of a living organism. For instance, the maintenance of a defined pH within narrow limits is a prerequisite and feature of all living cells, and this applies to higher organisms as well as microorganisms (Häussinger, 1988). In food preservation the homeostasis of microorganisms is a key phenomenon which deserves attention. This is because if the homeostasis of these organisms is disturbed by some preservation factors in foods, they will not multiply, i.e., they will remain in the lag-phase or may even die before their homeostasis is reestablished ("repaired"). Thus, in effect, food preservation is achieved by disturbing the homeostasis of microorganisms in a food temporarily or permanently (Leistner, 1995a).

Gould (1988) drew attention to the interference by the food itself with the homeostasis of the microorganisms present in food. During their evolution a wide range of more or less rapidly acting mechanisms (e.g., osmoregulation to counterbalance a hostile water activity of foods) have developed in microorganisms that act to keep important physiological systems in balance and settled even when the environment around them is significantly perturbed (Gould, 1995). In most foods microorganisms are able to operate homeostatically in order to react to the environmental stresses imposed by the preservation procedures applied. The most useful procedures employed to preserve foods are effective in overcoming the various homeostatic mechanisms that microorganisms have evolved in order to survive extreme environmental stresses (Gould, 1995). As repair of disturbed homeostasis demands energy from the microorganism concerned, a restriction of the energy supply will inhibit repair mechanisms of the microbial cells and lead to a synergistic effect of preservation factors. Energy restrictions in microorganisms can be caused by anaerobic conditions, such as created in vacuum or modified atmosphere packaging of foods (Gould, 1995).

16.2.3.2 Metabolic exhaustion

Another phenomenon of practical importance is the metabolic exhaustion of microorganisms which could lead to "autosterilization" of foods. The latter was first observed in experiments with mildly heated (95°C core temperature) liver sausages adjusted to different water activities by the addition of salt and fat. The product was inoculated with *Clostridium sporogenes* and stored at 37°C. Spores that had survived the heat treatment vanished in the product which remained stable during storage (Leistner and Karan-Djurdjic, 1970). Later this behavior of *Clostridium* and *Bacillus* spores was regularly observed during storage of shelf-stable meat products (SSP) stored at ambient temperatures (Leistner, 1994b). The most likely explanation is that bacterial spores which survive the heat treatment are able to germinate in these foods under less favorable conditions than those under which vegetative bacteria are able to multiply (Leistner, 1992a). Then, during storage of these products some viable spores start to germinate, but the germinating spores or the vegetative cells derived from them die. Thus, the spore counts in these stable hurdle technology foods, especially the unrefrigerated ones, actually decrease during storage. A similar behavior has been observed in vegetative microorganisms in studies with high-moisture fruit products (Alzamora et al., 1993, 1995; Tapia de Daza et al., 1996). Counts of a variety of bacteria, yeasts, and molds which had survived the mild heat treatment decreased quite rapidly in these products during ambient storage, since the hurdles applied (pH, a_w, sorbate, sulfite) did not allow growth.

A possible explanation for this behavior might be that vegetative microorganisms which cannot grow will die, and they will die more quickly if the state of stability of the food is close to the threshold for growth, the storage temperature is elevated, antimicrobials are present, and they are sublethally injured (Leistner, 1995a). Apparently, microorganisms in stable hurdle technology foods strain every possible repair mechanism to maintain their homeostasis in an attempt to overcome the hostile environment. In so doing they completely use up their energy supply, become metabolically exhausted and die, leading to autosterilization of the foods (Leistner, 1995b). As a result of autosterilization, hurdle technology foods, which are microbiologically stable, become even more safe during storage, particularly at ambient temperature.

16.2.3.3 Stress reactions

A limitation of hurdle technology foods could be microbial adaptation or the development of stress reactions by the microorganisms. Some bacteria can become more resistant or even more virulent under stress, since they generate stress shock proteins. The synthesis of protective stress shock proteins is known to be induced by heat, pH, a_w, ethanol, oxidant compounds, etc., as well as starvation. These stress reactions can cause an increased tolerance toward the stress which has induced them, e.g., sublethal heat treatment causes the microorganisms to tolerate higher heat treatment. Besides, stress reactions might also have a nonspecific effect so that in response to a particular stress microorganisms may become more tolerant to other stresses, i.e., they acquire "cross-tolerance" against other stresses. Such stress responses might hamper food preservation and could turn out to be problematic for the application of hurdle technology. On the other hand, the activation of genes for the synthesis of stress shock proteins, which help microorganisms to cope with stress situations, should be more difficult if different stresses are applied to them at the same time. Simultaneous exposure to different stresses will trigger the energy-intensive synthesis of several or at least more protective stress shock proteins, which in turn may cause the microorganism in question to become metabolically exhausted. Therefore, multitarget preservation of foods could be an answer to prevent the synthesis of stress shock proteins, which would otherwise jeopardize the microbiological stability and safety of hurdle technology foods (Leistner, 1995b).

16.2.3.4 Multitarget preservation

The concept of multitarget preservation of foods was introduced by Leistner (1995a, 1995b). Multitarget food preservation should be an ambitious goal for the gentle but most effective preservation of foods. For foods preserved using hurdle technology, it has been suspected for some time that different hurdles in a food might not only have an additive effect on microbiological stability, but also a synergistic effect (Leistner, 1978). A synergistic effect could be realized if the hurdles in a food hit different targets at the same time (e.g., cell membrane, DNA, enzyme systems, pH buffering systems, etc.) within the microbial cells thereby disturbing the homeostasis of the microorganisms concerned in several respects. If this happens, the repair of homeostasis as well as the activation of stress shock proteins could become more difficult. Therefore, employing simultaneously different hurdles in the preservation of food should lead to optimal microbiological stability. In practical terms, this could mean that it would be more effective to employ different preservation factors of small intensity than one preservation factor of large intensity, because different preservation factors might give rise to a synergistic effect (Leistner, 1994a).

It is anticipated that the targets in microorganisms for different preservation factors (hurdles) will be elucidated, and that the hurdles could then be classified according to their targets. Mild and yet effective preservation of foods, i.e., based on hurdles that work synergistically, is likely if the preservation procedure is based on an intelligent selection and combination of hurdles taken from different target classes (Leistner, 1995a). Food microbiologists could learn from pharmacologists, because the mechanisms of action of biocides have been studied extensively in the medical field. At least 12 classes of biocides are already known, which have different targets, often more than one within the microbial cell (Denyer and Hugo, 1991). Multidrug attack has proved successful in the fight against bacterial infections (e.g., tuberculosis) as well as viral infections (e.g., AIDS). A multitarget attack of microorganisms should be a promising approach too, in food microbiology (Leistner, 1995b).

16.2.4 Total quality

It is has been proposed (Stanley, 1991) that the hurdle technology approach could be applicable to a wider concept of food preservation than just microbiological stability of foods, but in order for it to work, a precise knowledge of the effectiveness of each hurdle for a given food would be required. Furthermore, it was suggested that a distinction should be made between positive and negative hurdles for the quality of foods. It would seem that hurdle technology is applicable not only to microbiological stability and safety, but also to the quality aspects of foods, although the latter possibility has been much less explored. By an intelligent combination of hurdles the microbiological stability and safety as well as the organoleptic, nutritional, and economic properties of a food can be secured. Economically, it is important to determine the amount of water in a product that is compatible with its microbiological stability. It may well be that an increased a_w can be compensated for by other hurdles (pH, E_h, preservatives, etc.) resulting in a more economical product. McKenna (1994) emphasized that while hurdle technology is appropriate for securing the microbiological stability and safety of foods, total quality of foods is a much broader concept encompassing a wide range of physical, biological, and chemical attributes. The concept of combined processes should work toward the total quality of foods rather than the narrow albeit very important aspects of microbiological stability and safety. At present the tools for applying hurdle technology to total quality are still not adequate; this is equally true for predicting food quality by modeling. However, researchers should appreciate the wider power of the hurdle technology concept, and the food industry should use the available tools of combined processes as often as possible for the purpose of quality enhancement (McKenna, 1994).

Some hurdles, e.g., Maillard reaction products, are known to influence the safety as well as the quality of foods because they have antimicrobial properties and can at the same time improve the flavor of many products. This is true also of nitrite used in the curing of meat. Hurdles in foods may be negative or positive in respect of the desired total quality of a food product. Moreover, the same hurdle could have either positive or negative effects on foods, depending on its intensity. For instance, chilling to an unsuitably low temperature is detrimental to fruit quality ("chill injury"), whereas moderate chilling is beneficial. Likewise the pH of fermented sausages, which should be low enough to inhibit pathogenic bacteria, should not be so low as to impair taste. If the intensity of a particular hurdle in a food is too small it should be strengthened; on the other hand, if it is detrimental to quality it should be weakened. By careful adjustment, the hurdles in foods should be kept in the optimal range, ensuring both safety and quality, and hence the total quality of food products (Leistner, 1994a; 1995a).

16.3 FERMENTED MEAT PRODUCTS

When discussing minimally processed, ready-to-eat, and ambient-stable meat products a distinction should be made between raw and heat-processed meat products, because the hurdles responsible for the microbiological stability and safety of these two groups differ in quality and intensity. For most raw meat products a competitive flora, mainly of lactic acid bacteria, is significant, whereas for heat-processed products the presence of microorganisms is generally undesirable. Furthermore, the intensity of the water activity hurdle in these two groups is also different. In minimally processed raw meat products the a_w is a major hurdle for microbiological stability, whereas it is only a secondary hurdle for many of the heat-processed ones. In the following sections, raw sausages, mini salami, and raw hams will be discussed.

16.3.1 Raw sausages

No microbiologist would have invented the raw sausage, because the process involved is in fact a monstrous one: raw meat and fat are stuffed into a casing; then the product is stored until consumption. That this is practicable and the sausage does not become putrid are due to the lactic acid bacteria and, particularly, lactobacilli. But the fact that these lactobacilli, which are usually only present in the raw material in relatively small counts, almost always prevail over the countless other microorganisms in a raw sausage mix, is attributable to the addition of salt and the exclusion of oxygen, as well as the drying of the product (Leistner, 1995c). However, it is not only the lactobacilli that make the raw sausage into a meat product which keeps well, but also there are several other hurdles, which take effect one after the other in a specific sequence, leading to a stable and safe product. Together, they inhibit food poisoning and spoilage bacteria, and select the desired competitive flora (Table 16–1). Important hurdles in the early stages of the ripening process of fermented sausages are salt and nitrite which inhibit many bacteria in the batter. Nitrite (125 ppm should be added) is important especially for inhibiting any salmonellae that may be present in the raw sausage mix, whereas the addition of nitrate might foster the growth of these food poisoning bacte-

Table 16–1 Sequence of hurdles in the ripening of fermented sausages

Nitrite → E_h → c.f. → pH → a_w

Secures inhibition of:

Salmonella spp., *Listeria monocytogenes*, *Clostridium botulinum*, *Staphylococcus aureus*, and spoilage organisms

Leads to selection of:

Lactic acid bacteria and other desirable organisms

Nitrite: nitrite-curing-salt; E_h: redox potential; c.f.: competitive flora; pH: acidity; a_w: water activity.

ria. However, nitrite is broken down during the course of the fermentation, thus, the nitrite hurdle is short-lived. The next hurdle is the redox potential (E_h). Because several bacteria still grow well in the presence of nitrite (i.e., curing salt containing nitrite), they use up oxygen causing the E_h to decrease. A low redox potential inhibits aerophilic bacteria (particularly Pseudomonadaceae) and favors the selection of lactic acid bacteria. These are the competitive flora (c.f.) which inhibit undesirable bacteria (e.g., *Listeria*) and flourish by metabolizing the added sugars, causing an increase in acidity, i.e., strengthening of the pH hurdle. In long-ripened salami, as the nitrite is depleted, lactic acid bacteria die, while the E_h and pH rise again, i.e., all these hurdles become weak in long-ripened products. Only the water activity (a_w) hurdle is strengthened with time, and it is this hurdle that is mainly responsible for the microbiological stability of long-ripened raw sausages (Leistner, 1986a, 1987, 1995a). Since the revelation of this sequence of hurdles, the production of fermented sausages has become more advanced (Leistner, 1995c).

Another feature peculiar to fermented sausages is their microstructure, which influences the desired ripening process as well as the survival of pathogenic bacteria. Therefore, the microstructure, too, is an important hurdle for the stability of these products. Electron microscopy studies (Katsaras and Leistner, 1991) have revealed that the natural flora as well as the added starter cultures are not evenly distributed in fermented sausages, but are arrested in tiny cavities of the product, i.e., the ripening flora only grows in nests. These nests are 100–5,000 μm apart, and large areas of the sausage must be influenced by metabolites (e.g., nitrate reductase, catalase, lactic acid, bacteriocins) accumulated in such nests or cavities. In short, the ripening raw sausage is a "solid-state fermenter." Within these nests the bacteria are in keen competition for nutrients and impair one another with their metabolic products. In nests of mixed cultures, generally, the lactic acid bacteria prevail because they are tolerant of low E_h, pH, and a_w. At the beginning of the fermentation process, the lactobacilli appear vigorous and metabolically active, whereas at the end of the ripening process, many will have degenerated and died (Katsaras and Leistner, 1991). In the sausage mix small and equal distances between nests of desirable bacteria should be advantageous, since this would facilitate the necessary ripening process. Thorough mixing of the meat and fat particles of the sausage batter prior to stuffing the mix into casings would bring about the desired distribution of bacteria within the product. Furthermore, if starter cultures are used, they should be added in a fashion which favors an even distribution. This is usually achieved by using starter cultures in a liquid rather than a powder form (Leistner, 1995a).

The production of fermented sausages could be controlled by Hazard Analysis Critical Control Point (HACCP). Leistner (1985b) proposed 19 control points for the process, taking into account safety as well as quality criteria. Leistner's scheme should really be called good manufacturing practice (GMP) guidelines for the production of fermented sausages, because in the strictest sense HACCP concerns only biological, chemical, and physical safety hazards, and should not include quality aspects. The proposed guidelines (Table 16–2) will control *Salmonella* spp., *Listeria monocytogenes*, *Staphylococcus aureus*, and *Clostridium botulinum* as well as the spoilage flora in fermented sausages. These

Table 16–2 Guidelines for the production of fermented sausages

1. Colony count of raw material	Aerobic colony count < 5 x 10^6 cfu/g; Enterobacteriaceae count < 10^5 cfu/g
2. Temperature of raw material	Meat < 7°C, fat –20°C; sausage batter during chopping –5°C and stuffing –3°C
3. pH of raw material	Do not use pork with pH > 5.8 and beef with pH > 6.0
4. a_w and pH of sausage batter	a_w should be in the range 0.97 to 0.96; pH not more than 5.9
5. Sodium chloride addition	Dependent on sausage type, but not > 3.0%, and never < 2.5%
6. Sodium nitrite addition	In quick-fermented sausages at least 125 ppm
7. Potassium nitrate addition	Up to 300 ppm acceptable; does not improve stability, but possibly aroma
8. Sucrose/glucose addition	0.3% for slow-fermented products; 0.5–0.7% for fast-fermented products
9. Glucono-delta-lactone addition	0.3%, only for fast-fermented products
10. Starter cultures	Beneficial for fast- and normal-ripened but not for slow-ripened products
11. Fermentation: temperature	Not > 25°C at the start of fermentation; not > 18°C for nitrate-ripened products
12. Fermentation: relative humidity	At the start of fermentation 90%, gradually decreasing to 75% RH
13. Fermentation: air velocity	At the start 0.8 m/s, gradually reduced to 0.5 m/s, and finally to 0.05 m/s
14. Microflora inside sausage	Predominantly lactobacilli, pediococci; desirable also Micrococcacea, yeasts
15. Microflora on the sausage	Strains of *Penicillium nalgiovense* and/or of *Debaryomyces hansenii*
16. Undesirable surface flora	Prevention by light smoke and/or immersion in 20% sorbate solution
17. Smoking of sausage	Depending on sausage type; never > 1 ppb benzo-a-pyrene residue
18. pH of the products	Decrease > 5.2 in 2 (fast-fermentation) or in 4–6 days in normal fermentation; after extended fermentation 5.8–6.0
19. a_w of finished products	Fast-fermented sausage 0.95–0.90; slow-fermented sausage 0.92–0.85

guidelines are open to further improvements and apply primarily to fermented sausages commonly available in Germany (> 100 different types). For other types of sausages (e.g., Italian or Chinese raw sausages) appropriate modifications must be introduced.

16.3.2 Mini-salami

A popular snack item is the mini-salami. These small diameter (18–20 mm) sausages are ready-to-eat and ambient-stable. Two types of mini-salami are on the market: raw sausage with a shelf life of ~7 months at 25°C if packaged with an $a_w < 0.82$, and heat-processed sausage with a shelf life of ~9 months at 25°C if packaged with an $a_w < 0.85$ (Tändler and Rödel, 1982, 1983). The longer shelf life of the heat-processed product is due to the inactivation of enzymes (e.g., lipases) at a product core temperature of ~75°C. However, in both product types shelf life is limited by rancidity.

Here only the processing and storage of raw mini-salami will be briefly discussed. A shelf life of several months at ambient temperatures is feasible if, for this product, meat with a low microbial load and fresh back fat of hogs (e.g., which has not been stored for an extended period in frozen condition) are used. In a typical formulation, often equal proportions of beef, pork, and fat are used. To these meats, nitrite-curing-salt (2.6%), sugar (0.5%), ascorbate, antioxidant spices (rosemary, sage, garlic), and starter cultures are added. The batter should have an a_w of 0.97 to 0.96 and a pH of 5.7 to 5.5. It is then stuffed into artificial casings. The sausages are carefully dried and lightly smoked. With an a_w of < 0.82, the products are packaged (only one sausage per pack) in sealed laminated aluminium pouches and flushed with 100% nitrogen to exclude light and oxygen because they would otherwise promote rancidity (Tändler and Rödel, 1982, 1983).

Thus, raw mini-salami is a typical hurdle technology food, not only because of its fermentation-based manufacturing process but also because of its elaborate packaging which provides an additional hurdle. In Germany the raw mini-salami is produced in large quantities (e.g., one company alone manufactures 250 million packs per year) and is always stored without refrigeration. This is because storage of mini-salami at 10°C does little to improve its organoleptic quality compared to storage at 20°C and, ambient storage is often more convenient for a snack item. Food poisoning attributed to the consumption of raw mini-salami has never been experienced in Germany and, therefore, it is puzzling that such a product exported to the United Kingdom and Sweden should cause salmonellosis there. The reason for this surprising fact has been difficult to find but could be simple. Apparently, the importing countries stored the German mini-salami under refrigeration and this was wrong (Leistner, 1995a). Salmonellae could survive the ripening process in raw mini-salami and would sometimes be viable in the finished product. However, since salmonellae are not able to multiply at $a_w < 0.95$, they cannot grow in mini-salami although they may have survived. In raw sausages (as in mayonnaise, margarine, etc.) pathogenic bacteria tend to survive much longer at chill than ambient temperatures, because at ambient temperatures viable pathogens live homeostatically, use up their energy quickly, and die as a result of metabolic exhaustion. Manufacturers of raw mini-salami take advantage of this metabolic exhaustion by keeping the finished products prior to distribution for about 10 days at 20°C. During this conditioning period the surviving pathogens (*Salmonella* spp., *Escherichia coli* O157, *Listeria monocytogenes*, etc.) perish. Thus, it could be concluded that refrigerated storage is not always beneficial to the microbiological safety and stability of foods. Chill storage is beneficial if the hurdles present in a food are not fully effective in inhibiting the growth of relevant food poisoning and spoilage microorganisms and a chill chain is required to achieve that inhibition; refrigerated storage, in the case of raw mini-salami, could make the product unsafe.

16.3.3 Raw hams

A variety of raw hams (more than 50 different types in Germany alone) are made of discrete pieces of meat from pork, beef, or lamb, cured with salt (sodium chloride) and nitrite or nitrate, or, in the case of long-ripened bone-in hams, with salt alone. The hams are carefully dried in order to remove water and reduce the water activity, leading to an ambient-stable, ready-to-eat product with delicious taste and flavor. The typical flavor and color of raw hams are mainly caused by the activities of indigenous meat enzymes, without much participation of desirable bacteria, yeasts, or molds. However, undesirable bacteria might lead to spoilage (which is most often caused by psychrotolerant Enterobacteriaceae) or even food poisoning (caused by nonproteolytic *Clostridium botulinum* type B). Spoilage and food poisoning problems usually occur in hams with a high initial pH (> 5.8) and as a result of a curing temperature that is too high to begin with and is then not maintained at below 5°C for sufficient time.

Raw hams are definitely hurdle technology foods (Leistner, 1986b). The primary hurdle in these hams is the low water activity (a_w), which is caused by the addition of salt and the removal of water. However, this hurdle only becomes effective after salt penetration and drying, both of which take time. Therefore, the initial microbial count in the hams must be low and growth of bacteria must be prevented by an initially low pH in the ham, and by the application of strict refrigeration until a critical water activity has been reached in all parts of the ham after the curing and equilibration phases, which ensures microbiological stability of the product. Only then is it possible to raise the temperature in order to speed up enzymic ripening, i.e., flavor and color development of the product. The finished hams should have an increased pH (~6.0) due to advantageous proteolysis, but the a_w is at this stage sufficiently low to have secured a stable and safe product. Microbial growth on the surface of the ham is generally not wanted, and, therefore, cold smoke and/or a potassium sorbate dip are applied. Another option is the inoculation of the surface with mold starter cultures, which inhibit undesirable mold growth by occupying the available space on the surface of the ham. For mold-fermented hams (speck and coppa of Italy, bündnerfleisch of Switzerland, etc.) safe starter cultures (selected strains of *Penicillium nalgiovense*) must be employed, since the usual "house flora" might include many toxigenic mold species. The final products should contain sufficient salt to secure microbiological stability but not too much as to impair taste; they are generally stored without refrigeration and consumed in the raw state. If occurrence of *Trichinella spiralis* in the raw pork is not controlled, the ham should be heated before consumption, but this diminishes the flavor of the product and a salty taste dominates. Guidelines for the proper processing of raw hams are outlined in Table 16–3.

16.4 HEAT-PROCESSED MEAT PRODUCTS

Heat-processed high-moisture foods based on hurdle technology, which are minimally processed, ambient-stable, and ready-to-eat, have been named shelf-stable products (SSP) (Leistner, 1985a). They offer the following advantages: the mild heat treatment (70–110°C) improves the organoleptic and nutritional properties of the food, and their ambient stability simplifies distribution and requires less energy during storage. SSP are heat-processed in sealed containers (casings, pouches, but rarely in cans), which prevent recontamination after processing. However, because of the mild heat treatment these products still contain viable spores of bacilli and clostridia, which can be inhibited by adjusting the a_w, pH, and E_h, and, in the case of autoclaved sausages, by causing sublethal injury to the spores. At present four different types of SSP foods are recognized: F-SSP, a_w-SSP, pH-SSP, and Combi-SSP, based on their primary hurdles, although additional hurdles will enhance the safety and stability of these products. All these mildly heat-processed products are stable without refrigeration, and have fresh product characteristics (Leistner, 1992a, 1994b, 1995a).

16.4.1 F-SSP

In F-SSP (Leistner, 1985a, 1987, 1992a) the mild heat treatment that results in sublethal damage of bacterial spores is the primary hurdle. The

Table 16–3 Guidelines for the production of raw hams

1. Treatment of animal	Careful transport and slaughter of animals, secures in hams a total count of < 100 cfu/g
2. Temperature of raw material	Chilling of hams in 24 h to < 4°C in the core, inhibits growth of bacteria inside the hams
3. pH of raw material	Hams with pH > 5.8 shall be omitted, because of poor curing and thus high spoilage potential
4. Sodium chloride addition	Sufficient but not too high, because finished products shall contain > 4.5% but < 6% salt
5. Sodium nitrite addition	Important for color and flavor but less for preservation; avoided in long-ripened hams
6. Potassium nitrate addition	Max. 300 ppm used for long-ripened hams, beneficial for color, not for microbial stability
7. Curing temperature	Below 5°C, until in all parts of the ham the a_w is < 0.960 (corresponds to > 4.5% salt)
8. Ripening of hams	After stabilization, 2–8 weeks for short-ripened hams and 9–18 months for long-ripened products
9. Smoking of hams	After stabilization, for short-ripened hams at 25°C and long-ripened products at 15–18°C
10. Undesirable mold growth	Undesirable mold growth inhibited by smoke, or a dip into 20% potassium sorbate solution
11. Desirable mold growth	Typical for some products of Switzerland and Italy; starters (*Penicillium nalgiovense*) must be used
12. Bacteria in finished hams	The total count inside finished hams of high quality shall be < 10^4 cfu/g
13. a_w of finished hams	All products shall have an a_w of < 0.92, long-ripened hams might have an a_w of ~ 0.80

sausages, with an adjusted a_w (bologna-type sausages < 0.97; liver and blood sausages < 0.96) and packed in PVdC-coated casings (impermeable to water vapor and air), are processed in counterpressure autoclaves to give $F_0 > 0.4$. For bologna-type sausages, reduction of the a_w to < 0.97 is sufficient because nitrite is an additional hurdle, whereas liver and blood sausages have to be adjusted to an a_w < 0.96 as the high iron content of such products inactivates the nitrite hurdle. F-SSP can be stored unrefrigerated for several weeks or even months and have no problems with regard to food poisoning or spoilage because the critical control points in their manufacture have been identified and closely monitored (Hechelmann and Leistner, 1984). F-SSP autosterilize during storage as a result of metabolic exhaustion of the microorganisms (see section 16.2.3). In general, casings are more advisable than cans for F-SSP, because during cooling of the cans after autoclaving, some water condensation may accumulate underneath the lid. If drops of water fall back on the surface of the sausage mix, the critical a_w could be breached and clostridia (and some bacilli) may start to grow. Since flexible casings are filled tightly, condensation is unlikely to occur. Therefore, F-SSP in casings are preferred to cans with a headspace (Hechelmann et al., 1985). Guidelines for the processing of F-SSP are outlined in Table 16–4.

16.4.2 a_w-SSP

The stability of a_w-SSP (Leistner *et al.*, 1979; Leistner, 1985a, 1987, 1992a, 1994b) is primarily achieved by a reduced water activity of < 0.95. Examples of traditional a_w-SSP meat products include Italian mortadella and German brühdauerwurst. Both product groups, preserved through empirical adjustment of a_w, can be stored uncut without refrigeration. More recently, a large variety of a_w-SSP with an intentionally adjusted water activity have entered the market, most of which are snack items. The shelf life of a_w-SSP at ambient temperatures is even longer than that of fermented sausages. Lipases in a_w-SSP are inactivated by the mild heat treatment (product internal temperature > 75°C) with the result that these products are less prone to the development of lipolytic/hydrolytic rancidity (see section 16.3.2). Guidelines for the processing of a_w-SSP are outlined in Table 16–5.

16.4.3 pH-SSP

In pH-SSP (Leistner, 1985a, 1987, 1992a) increased acidity is the primary hurdle. This is the principle used in Gelderse rookworst in The Netherlands, in which designed hurdle technology has been employed for a number of years. In this product the pH is adjusted to 5.4–5.6 by the addition of 0.5% glucono-delta-lactone. This sausage is stable for several weeks at ambient temperatures if vacuum-packaged and heated at 80°C for about 1 hour in its packaging. The heat treatment eliminates vegetative organisms, and bacterial spores are apparently of little concern as their numbers decrease during the heat process and any surviving spores are inhibited by the low pH and other hurdles present (e.g., nitrite). Gelderse rookworst is exported from The

Table 16–4 Guidelines for the production of F-SSP

1. Bacterial spore counts	Only very few spores should be in sausage mix; use spice extracts instead of natural spices
2. Heat treatment of sausages	Sausages must be heated to F_0 > 0.4 (103–108°C core) in counterpressure autoclaves
3. Water activity adjustment	a_w must be < 0.97 in bologna-type sausages and < 0.96 in liver and blood sausages
4. Redox potential adjustment	E_h in the sausage mix should be low, thus casings impermeable to oxygen are used
5. pH adjustment	pH of the sausage mix should be < 6.2, but this is only critical in blood sausages
6. Required packaging	Casings or pouches are suitable, cans may be used if water condensation can be avoided

Table 16–5 Guidelines for the production of a_w-SSP

1. Heat treatment	Sausages should be heated to > 75°C (core), in sealed containers, preferably in casings
2. Water activity adjustment	a_w must be adjusted to < 0.95 by formulation and/or by drying of the products
3. Redox potential adjustment	E_h in the sausage should by low, in order to facilitate inhibition of aerobic bacilli
4. Surface of products	Since permeable casings are used molds may grow, but are inhibited by a sorbate dip

Netherlands in large quantities to Britain, and with a pH > 5.4, it is acceptable from the organoleptic point of view because the taste is only slightly sour. Despite the relatively low pH, the binding of water and fat in rookworst is not a problem because pork rinds and/or phosphates are added to the product (Leistner, 1985a, 1994b). Another traditional meat product of the pH-SSP type is brawns (e.g., jelly sausages). The product is adjusted to an appropriate pH by the addition of acetic acid. The sausage mix usually consists of a brine (pH < 4.8) made up of water, gelatine, salt, sugar, agar-agar (2%), spice, and acetic acid and a solid phase of bologna-type sausage in cubes with an a_w of < 0.98. The two components are mixed in the proportions of two parts brine to three parts meat and filled into casings. The product is heated to an internal temperature of > 72°C but not higher than 80°C because the sausage must solidify after chilling. At equilibrium, the product should have a final pH < 5.2; it can be stored for several days at ambient and is ready-to-eat. It is significant to note that while the heat resistance of vegetative bacteria and spores increases with decreasing a_w, the heat resistance of bacteria and spores diminishes with decreasing pH. It follows that pH-SSP need less heat treatment for the inactivation of microorganisms than do a_w-SSP.

16.4.4 Combi-SSP

In Combi-SSP (Leistner, 1992a; Leistner and Hechelmann, 1993) a combination of rather equal hurdles is applied. Experimental work led by Leistner suggested that even small enhancements of individual hurdles had a distinct effect on the microbiological safety and stability of a food product. For instance, in terms of the stability and safety of a product, it is a matter of great significance whether its F_0 is 0.3 or 0.4, its a_w is 0.975 or 0.970, its pH is 6.5 or 6.3, and its E_h value is somewhat higher or lower. Every small strengthening or reinforcement of a hurdle brings some weight to the overall balance, and the sum of these weights determines whether a food is microbiologically unstable, uncertain, or stable (Figure 16–1). In other words, many small steps in the direction of stability can swing the balance from an unstable into a stable state (Leistner, 1992a). This principle was utilized in the product design of bologna-type sausages as Combi-SSP. Different types of brühwurst (wieners, bockwurst, fleischwurst, fleischkäse, etc.) have been developed which prove to be microbiologically stable and safe for at least one week at 30°C. The initial spore load of the sausage mix is low because spice extracts instead of natural spices are used. Also, nitrite (100 ppm) with the curing salt must be added to the mix. The products, adjusted to an a_w and pH of < 0.965 and < 5.7, respectively, are heated to a core temperature of > 72°C. After vacuum packaging, they are then repasteurized for 45–60 minutes (depending on the diameter of the product) at 82–85°C (Hechelmann et al., 1991; Leistner, 1994b). Combi-SSP are amazing products. They are ready-to-eat meats which, once

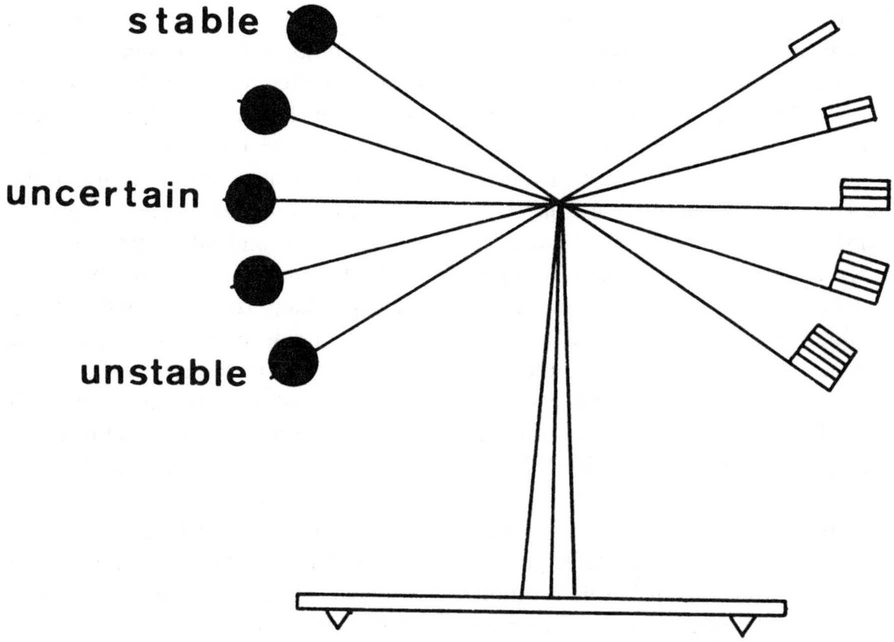

Figure 16–1 This imaginary balance should illustrate that even relatively small enhancements of different hurdles can result in a substantial improvement in the microbiological stability of a food. *Source:* Data from *Food Preservation by Combined Methods,* No. 25, pp. 151–158, © 1992, Food Research International.

believed to require refrigeration for storage, have become ambient stable. However, Combi-SSP do need strict rules for design and process control, which will be discussed in section 16.5.

16.5 PRODUCT DESIGN AND SHELF-LIFE EVALUATION OF MINIMALLY PROCESSED MEAT PRODUCTS

Hurdle technology as a concept has proved useful in the optimization in respect of the microbiological safety and stability of traditional foods as well as in the development of novel products. However, it is highly beneficial to combine hurdle technology in food design with predictive microbiology and in process control with Hazard Analysis Critical Control Point (HACCP) or good manufacturing practice (GMP). The integration of these concepts has been applied in a study for the German army, and in the establishment of a user guide for food product design.

16.5.1 Army study—a practical example

The Federal Centre for Meat Research (BAFF) in Kulmbach, Germany, has demonstrated the effectiveness of hurdle technology in an extensive study (supported by the Medical Corps of the German Army) of numerous minimally processed meat products with fresh product characteristics, which are ready-to-eat and stable at ambient temperatures (Hechelmann et al., 1991; Leistner and Hechelmann, 1993; Leistner, 1994b). The main requirements of the German army were as follows: meat products should be recommended for army provisions which have fresh product characteristics and remain tasty, stable, and safe for at least 6 days at 30°C; refrigeration storage should not be required because these rations are to be used during military exercises. Since large as well as small meat processing plants (i.e., the potential contractors) must be able to manufacture these rations to a tight specification, the processes had to be spelled out in detail. The Institute for Microbiology, Toxicology and Histology, in cooperation with the Institute of Technology of the BAFF, executed this project for the army using the following approach. German meat processors were asked to nominate meat products which taste like their counterparts available in delicatessen shops, and which do not require refrigeration. In response, 24 manufacturers put forward a total of 100 products. After storage trials, microbiological examination as well as sensory evaluation, 75 of these proved stable and safe. Ten samples of each of these products were purchased by the project team and scrutinized for their physical, chemical, microbiological, and technological characteristics. As a result of this work, eight categories of products were identified, with their stability and safety being based on different principles of hurdle technology (Table 16–6). Pilot plant work covering all these categories continued, which included the optimization of their microbiological safety and stability. Finally the products were reproduced in a commercial meat plant under industrial conditions. The manufacturers of most of these products did not know why their products were ambient stable; what they had was empirical knowledge. Information about the properties of each of the products and the principles behind their stability and safety were disseminated in the form of written reports to the manufacturers. This cooperation has been much appreciated and paved the way for the intentional application of hurdle technology in the German meat industry.

Table 16–6 Categories of minimally processed, ready-to-eat, ambient-stable meat products recommended for the German army

1. Quick-ripened fermented sausage
2. Mini-salami (two different types)
3. Bologna-type and liver sausage as F-SSP
4. Dried bologna-type sausage as a_w-SSP
5. Repasteurized bologna-type sausage as a_w-SSP
6. Bologna-type sausage and brawns as pH-SSP
7. Bologna-type sausages as Combi-SSP
8. Bologna-type sausage in autoclaved flat pouches

Source: Reprinted from L. Leistner and H. Hechelmann, Food Preservation by Hurdle-Technology, *Proceedings of Food Preservation 2000 Conference,* Vol. II, U.S. Army, Natick Research, Development and Engineering Center, Natick, Massachusetts, (1993), pp. 511–520.

Of the eight categories of minimally processed, stable and safe meat products evaluated in the German army study only two were raw meat products (quick-ripened fermented sausage, and raw mini-salami); the others were heat-processed products (F-SSP, a_w-SSP, pH-SSP, Combi-SSP, and flat pouches). It was necessary to define the manufacturing processes of these products in detail, and for this purpose a link between hurdle technology and HACCP was introduced (Leistner, 1992b, 1994b). Generic models for the processing of meats of each category were suggested (Hechelmann et al., 1991), which contained about 20 critical control points for the application of the HACCP concept. However, since quality aspects were included in these generic models, HACCP in its strictest sense was not used; the control points were really guidelines for GMP. The most prominent control points of the minimally processed meats encountered in the German army study are listed in Table 16–7. In the manufacturing plants processing the selected meats for the army, no microbiological tests are carried out, but processing time, temperature, pH, and a_w have to be strictly controlled. Monitoring of these parameters should be done on-line or at least close to the line. In the past this was difficult with respect to water activity measurements. However, an accurate a_w instrument has become available (Rödel et al., 1990), which allows a_w determinations for meat products to be completed within 10–20 minutes. The essence of the German army study has been published in a booklet illustrated with many color pictures (Leistner, 1994b), which has been distributed by request to more than 4000 food scientists and technologists in numerous countries worldwide.

16.5.2 User guide to food design

For the design, processing, and storage of stable, safe, and tasty foods, four concepts are

Table 16–7 Prominent control points for the processing of the eight categories of fresh-like, ambient-stable meats for the German army

1. Quick-fermented sausages	pH < 5.4; a_w < 0.95; starter cultures; smoke; sometimes modified atmosphere packaging
2. Mini-salami	As fermented sausages: a_w < 0.82; as dried bologna-type sausages: a_w < 0.85; aluminium pouches to exclude oxygen and light
3. F-SSP	F_0 > 0.4; a_w < 0.97 (bologna-type sausages) or < 0.96 (liver and blood sausages); pH < 6.2; casings or cans without headspace
4. a_w-SSP	Product core temperature > 75°C; a_w < 0.95; uncut surfaces; smoke
5. Repasteurized a_w-SSP	Temperature > 75°C; a_w < 0.95; pH < 6.2; after packaging heated for 45 min at 82–85°C
6. pH-SSP	pH < 5.2; hot filled and heated to core temperature > 72°C
7. Combi-SSP	Low initial spore count; nitrite > 100 ppm; core temperature > 72°C; a_w < 0.965; pH < 5.8; after packaging heated for 45–60 min at 82–85°C
8. Flat pouches	Laminated aluminium pouches with vacuum; diameter < 3 cm; F_0 > 3.0 ("botulinum cook")

Source: Reprinted from L. Leistner and H. Hechelmann, Food Preservation by Hurdle-Technology, *Proceedings of Food Preservation 2000 Conference,* Vol. II, U.S. Army, Natick Research, Development and Engineering Center, Natick, Massachusetts, (1993), pp. 511–520.

topical: hurdle technology (HT), predictive microbiology (PM), Hazard Analysis Critical Control Point (HACCP), and good manufacturing practice (GMP). These four concepts have related but different goals: hurdle technology is primarily used in food development; predictive microbiology for process refinement; HACCP to prevent microbiological, chemical, and physical hazards of foods; and quantitative GMP to establish standards for the processing and storage of a particular food. However, these four concepts could be usefully combined in a user guide to food design. A 10-step procedure which includes all four concepts is shown in Table 16–8. This approach has proved useful when solving real product development and processing problems in the food industry (Leistner, 1994c, 1994b).

Hurdle technology is the key to advanced food design. Efforts have been made to develop hurdle technology quantitatively by combining it with predictive modeling (McMeekin et al., 1999). Predictive microbiology (McMeekin et al., 1993; McClure et al., 1994) is a promising and powerful tool that involves computer-based and quantitative predictions of microbial growth, survival, and death in foods, and is clearly an integral part of advanced food design (Table 16.8). However, the predictive models constructed so far can only handle up to four factors (hurdles) simultaneously. There are numerous hurdles (> 60) in existence and several might be important for the stability, safety, and quality of a particular food product (see section 16.2). It is unlikely that all or even a majority of these hurdles can be covered by predictive modeling. Thus, predictive microbiology cannot yet be a quantitative approach to the totality of hurdle technology. However, it does allow quite reli-

Table 16–8 User guide to food design using an integrated approach comprising hurdle technology, predictive microbiology, and HACCP or GMP

1. For the modified or novel food product the desired sensory properties and the desired shelf life are tentatively defined.
2. A feasible technology for the production of this product is then outlined.
3. The food is manufactured according to this tentative technology, and the resulting product is analyzed for pH, a_w, preservatives, or other inhibitory factors. Temperatures for thermal processing (if intended) and storage as well as the expected shelf life are defined.
4. For preliminary microbiological stability testing of the food product, predictive microbiology might be employed.
5. Now the food product is challenged with food poisoning and spoilage microorganisms, using somewhat higher inocula and storage temperatures than would be "normal" for this product.
6. If appropriate, the hurdles in the product are modified, taking multitarget preservation and the organoleptic quality of the food (i.e., total quality) into consideration.
7. The food is now challenged again with relevant microorganisms, and if necessary the hurdles are modified once more. Predictive microbiology for assessing the safety of the food might be helpful at this stage too.
8. After the established hurdles of the modified or novel food product are exactly defined, including tolerances, the methods for monitoring the process are agreed on. Preferably physical methods for monitoring should be used.
9. The designed food product should now be produced under industrial conditions, because the feasibility for the scale-up of the proposed manufacturing process must be validated.
10. Finally, for the industrial process the critical control points (CCPs) and their monitoring must be established, and therefore the manufacturing process will be controlled by HACCP. If the HACCP concept is not appropriate, because food quality is controlled at the same time, guidelines for securing total quality of the product by application of quantitative GMP must be defined.

Source: Data from User Guide to Food Design, in *Food Preservation by Combined Processes,* L. Leistner and L.G.M. Gorris, eds., FLAIR Final report, EUR 15776 EN, European Commission DG XII, Brussels, pp. 25–28. (1994c).

able predictions of the fate of a number of microorganisms of significance in food systems, while considering few but the most important hurdles. Due to the fact that many hurdles (i.e., factors) have not been included in contemporary models, the predicted results from which have tended to err on the safe side, e.g., growth models giving failsafe predictions, i.e., the limits indicated for growth of pathogens in foods by the models are more prudent than the limits found in real foods. Nevertheless, predictive microbiology should be an important tool in advanced food design, because it can narrow down considerably the range over which challenge tests with relevant microorganisms need to be performed. Although predictive microbiology will never make challenge testing redundant, it may greatly reduce the time for and the cost of product development of which microbiological examination is a major part (Leistner, 1994a; 1995a).

After a food product has been properly designed, its manufacturing process must be effectively controlled, for which purpose HACCP may be applied. However, by definition, HACCP should only be used to control the safety hazards of foods, not their microbiological stability (spoilage) or organoleptic quality (Pierson and Corlett, 1992). In practice safety and quality issues of foods often overlap (Leistner, 1994b). And, for hurdle technology foods, microbiological safety and stability as well as organoleptic quality are all important, i.e., the total quality of the product (section 16.2.4) is essential. In this case, the HACCP concept might be too narrow as it relates only to microbiological, chemical, and physical hazards. Therefore, the HACCP concept should be broadened in order to cover the microbiological safety (food poisoning) and stability (spoilage) of foods as well as their organoleptic quality. If this is not practicable, the production process should be controlled by employing the principles of good manufacturing practice (GMP), and as far as possible, rules or guidelines must be quantitatively defined. For the main types of German meat products quantitative data based on GMP have been published (Wirth et al., 1990), and are widely used as guidelines in the processing of these products. Also, for food products of developing countries where small producers predominate, GMP guidelines are more appropriate because the application of HACCP often poses practical difficulties.

In food design, different kinds of researchers, including microbiologists and technologists, must work together. The microbiologist should determine the types and intensity of hurdles necessary for the safety and stability of a particular food product. The technologist should determine what ingredients or processes are suitable for establishing these hurdles in the product, taking into account the legal, technological, organoleptic, and nutritional requirements. Because the engineering, economic, and marketing aspects must be considered too, food design is indeed a multidisciplinary endeavor (Leistner, 1994b, 1995a).

16.6 CURRENT TRENDS IN HURDLE TECHNOLOGY FOODS

Food products based on combined preservation methods (hurdle technology) are very common in industrialized as well as developing countries. In the past, and even today, hurdle technology is often applied empirically without the knowledge of the governing principles in the preservation of a particular food product. But, with a better understanding of these principles and improved availability of suitable monitoring devices, the deliberate application of hurdle technology is advancing worldwide. Some examples and trends will be briefly mentioned.

16.6.1 Topical applications in industrialized countries

16.6.1.1 Minimally processed foods

Deliberate and intelligent hurdle technology was first used for the gentle preservation of mildly heated, fresh-like meats that could be stored without refrigeration (Leistner, 1985a); this has been discussed in section 16.4. Furthermore, a better understanding of the sequence of hurdles which lead to microbiological stability of fermented sausages (see section 16.3.1) has improved the safety and quality of these products (Leistner, 1995c), and paved the way for mildly fermented meats. However, advanced hurdle technology is applicable not only to minimally processed meats but also minimally processed fruits and vegetables, as has been discussed elsewhere (Leistner, in press).

16.6.1.2 Chilled foods

The advantages of hurdle technology are most obvious in high-moisture foods that are shelf stable, designed by the intelligent application of combined methods of preservation. However, the use of hurdle technology is appropriate for chilled foods too. In case of temperature abuse, which can easily happen during distribution, chilled foods may lose their microbiological safety and stability, especially if low-temperature storage is the only hurdle. Therefore, it is advisable to incorporate into chilled foods (meats, sous vide dishes, salads, fresh-cut vegetables, etc.) some additional hurdles that can act as a back-up in case of temperature abuse. This type of safety precaution for chilled foods has been called "invisible technology," meaning that additional hurdles are there to act as safeguards against temperature abuse in chilled foods ensuring that they remain microbiologically stable and safe during storage in retail outlets as well as in the home (Leistner, 1994b, 1999).

16.6.1.3 Healthy foods

A recent application of hurdle technology relates to the stabilization of novel health foods derived from meat, poultry, or fish, which contain less fat and/or salt and are therefore more prone to microbial deterioration. The reduction of salt and fat as well as use of substitutes and replacements for the traditional ingredients in these products cause microbiological instability as several hurdles (a_w, pH, preservatives, and possibly E_h and microstructure) will be affected. This instability can be overcome by the intelligent application of hurdle technology (Leistner, 1997, 1999).

16.6.1.4 Packaging

For most food products packaging is an important hurdle, since it contributes toward microbiological stability and safety, and helps to preserve organoleptic quality. Industrialized countries have a tendency to over-package foods, and this is especially true in Japan where, for example, "active" packaging (as scavengers, absorbers, emitters, antimicrobial or antioxidizing packaging materials, etc.) has been developed to perfection. These "smart" packaging systems are very sophisticated, but can be wasteful too. Thus, Japanese experts are aiming to use less food packaging (Ono, 1994). Future packaging shall aim to provide the necessary information only, together with some convenience to the consumer. The required shelf life of the product should not be so dependent on packaging, but that it should be based on superclean packaging procedures combined with either just-in-time delivery or the use of advanced hurdle technology that gives food products which are stable and safe despite minimal packaging (Ono, 1996; Leistner, 1999).

16.6.1.5 Futuristic technologies

Finally, it should be mentioned that some emerging nonthermal preservation methods (e.g., based on physical means such as high hydrostatic pressure, pulsed electric fields, oscillating magnetic fields, light pulses, etc.) are often most effective when used in conjunction with traditional methods of food preservation. It is likely that hurdle technology will play an increasingly important part in the application of these novel procedures of food preservation (Barbosa-Cánovas et al., 1998).

A summary of the possibilities outlined in this section is given in Table 16–9.

Table 16–9 Topical applications of advanced hurdle technology in industrialized countries

Minimal processing (of fresh-like or convenient foods)
Chilled foods (back-up hurdles as "invisible technology")
Healthy foods (with less fat, salt, and/or preservatives)
Less packaging (due to advanced hurdle technology foods)
Futuristic technologies (combined with additional hurdles)

16.6.2 Topical applications in developing countries

Many of the foods in developing countries, which should preferably be stored without refrigeration as electricity is expensive and sometimes not continuously available, are based on empirical use of hurdle technology. However, several of the foods have now been optimized by the intentional application of hurdles. Relevant examples include meat products of China and Taiwan as well as dairy and meat products of India. Even in Africa (e.g., South Africa, Nigeria) work based on the intentional application of hurdle technology in food and drink preservation has commenced. The most impressive advances have been made in South America. In several countries of Latin America (i.e., Argentina, Mexico, Venezuela), novel high-moisture fruit products (HMFP) have been developed using hurdle technology. In spite of their high water activity (a_w 0.98 to 0.93), these products can be stored at ambient and will remain in fresh-like condition for several months at room temperatures, and even become commercially sterile during storage due to metabolic exhaustion (see section 16.2.3) of the bacteria, yeasts, and molds originally present in them.

There is a general trend in developing countries to move gradually away from the traditional intermediate-moisture foods because they are often too salty or too sweet, contain too many preservatives, and have a less appealing texture and appearance than high-moisture foods. It may therefore be expected that high-moisture, fresh-like foods that nevertheless can be stored at ambient temperatures will be on the increase in developing countries once the application of intelligent hurdle technology has been mastered. However, deliberate hurdle technology should be applied to high-moisture foods without sacrificing microbiological stability and safety, especially of those destined for ambient storage. The state of the art in the use of intentional hurdle technology in countries of Latin America, Asia, and Africa has recently been presented in a comprehensive review (Leistner, 2000).

A summary of this section is given in Table 16–10.

16.7 CONCLUSIONS

The stability, safety, and quality of most preserved foods are based on empirical application of combined methods of preservation and more recently, on intentional hurdle technology. The deliberate and intelligent application of hurdle technology allows gentle but effective preservation of safe, stable, nutritious, and tasty foods and is advancing worldwide. Moreover, knowledge of the physiological mechanisms of growth, survival, and death of pathogenic and spoilage microorganisms in foods is increasing. The disturbance of the homeostasis of microorganisms is a key strategy in food preservation. Microbial stress reactions may hamper food preservation, whereas the metabolic exhaustion of microorganisms, existent in stable hurdle technology foods, clearly favors food preservation. An ambitious goal for the future in respect of optimal food preservation is the multitarget preservation of foods in which intelligently applied gentle hurdles will have a synergistic effect. After the targets of different preservation factors within the microbial cells have been elucidated, which should definitely be a major research topic of the future, the preservation of foods could progress far beyond the current state of knowledge of the hurdle technology as it is known today.

At present, interest in hurdle technology is mostly associated with minimally processed, fresh-like food products which are mildly heated or fermented, ready-to-eat, and ambient-stable. Meat products were the first foods to which the technology was applied. Today there is a large variety of advanced hurdle technology meat products on the German market, which have a good public health record and for several decades they have caused neither food poisoning nor spoilage problems, so long as the proposed guidelines are being followed. Experience of recent years has indicated that a similar approach, with due adaptation, will be suitable for other minimally processed foods besides meats. It is anticipated that intentional and intelligent hurdle technology will be increasingly used for the benefit of the general food supply.

If advanced hurdle technology foods were to become more popular and sophisticated, a thorough understanding of the principles involved as well as strict guidelines to underpin their production and distribution would be required. Advanced hurdle technology food products are in general less robust than traditional ones which are often over-processed and thus possess a larger margin of safety. Processes developed to produce advanced hurdle technology foods must be accurately defined and controlled. For the design of hurdle technology foods a 10-step procedure has been proposed, which comprises hurdle technology, predictive microbiology, and HACCP. However, the latter might be too narrow a concept since it is supposed to control only microbiological, chemical, and physical safety hazards, and exclude the organoleptic quality of foods. Ultimately, it is the latter which is the most important characteristic as far as the consumer is concerned. Thus, for the design, processing, and distribution of hurdle technology foods, quantitative guidelines based on GMP might be more appropriate than the HACCP system.

Table 16–10 Topical applications of advanced hurdle technology in developing countries

South America (new fruit products, meats, fish)
China, Taiwan (optimized Chinese meat products)
India (optimized dairy and meat products)
Africa (new fruit and juice items, fresh meat products)

REFERENCES

Alzamora, S.M., Cerrutti, P., Guerrero, S. and López-Malo, A. (1995) Minimally processed fruits by combined methods, in *Food Preservation by Moisture Control: Fundamentals and Applications* (eds G.V. Barbosa-Cánovas and J. Welti-Chanes), Technomics Publishing Co., Lancaster, PA, pp. 463–492.

Alzamora, S.M., Tapia, M.S., Argaiz, A. and Welti, J. (1993) Application of combined methods technology in minimally processed fruits. *Food Res. Internat.*, **26**, 125–130.

Barbosa-Cánovas, G.V., Pothakamury, U.R., Palou, E. and Swanson, B.G. (1998) *Nonthermal Preservation of Foods*, Marcel Dekker, Inc., New York, NY.

Denyer, S.P. and Hugo, W.B. (eds) (1991) *Mechanisms of Action of Chemical Biocides: Their Study and Exploitation*, Blackwell Scientific Publications, Glasgow, Scotland.

Gould, G.W. (1988) Interference with homeostasis—food, in *Homeostatic Mechanisms in Micro-organisms* (eds R. Whittenbury, G.W. Gould, J.G. Banks and R.G. Board), Bath University Press, Bath, England, pp. 220–228.

Gould, G.W. (1995) Homeostatic mechanisms during food preservation by combined methods, in *Food Preservation by Moisture Control: Fundamentals and Applications* (eds G.V. Barbosa-Cánovas and J. Welti-Chanes), Technomic Publishing Co., Lancaster, PA, pp. 397–410.

Gould, G.W. and Jones, M.V. (1989) Combination and synergistic effects, in *Mechanisms of Action of Food Preservation Procedures*, (ed G.W. Gould), Elsevier, London, England, pp. 401–422.

Häussinger, D. (ed.) (1988) *pH Homeostasis: Mechanisms and Control*, Academic Press, London, England.

Hechelmann, H. and Leistner, L. (1984) Mikrobiologische Stabilität autoklavierter Darmware. *Mitteilungsbl. Bundesanst. Fleischforsch. Kulmbach*, Nr. 84, 5894–5899.

Hechelmann, H., Kasprowiak, R., Reil, S., Bergmann, A. and Leistner, L. (1991) *Stabile Fleischerzeugnisse mit Frischprodukt-Charakter für die Truppe*, BMVg FBWM 91-11, Dokumentations- und Fachinformationszentrum der Bundeswehr, Bonn, Germany.

Hechelmann, H., Leistner, L. and Albertz, R. (1985) Ungleichmäßiger a_w-Wert als Ursache für mangelhafte Stabilität von F-SSP. *Jahresber. Bundesanst. Fleischforsch. Kulmbach,* 1985: C 27.

Katsaras, K. and Leistner, L. (1991) Distribution and development of bacterial colonies in fermented sausages. *Biofouling,* **5**, 115–124.

Leistner, L. (1978) Hurdle effect and energy saving, in *Food Quality and Nutrition* (ed W.K. Downey), Applied Science Publishers, London, England, pp. 553–557.

Leistner, L. (1985a) Hurdle technology applied to meat products of the shelf stable product and intermediate moisture food types, in *Properties of Water in Foods in Relation to Quality and Stability* (eds D. Simatos and J.L. Multon), Martinus Nijhoff Publishers, Dordrecht, pp. 309–329.

Leistner, L. (1985b) Empfehlungen für sichere Produkte, in *Mikrobiologie und Qualität von Rohwurst und Rohschinken* (ed L. Leistner), Institut für Mikrobiologie, Toxikologie und Histologie der Bundesanstalt für Fleischforschung, Kulmbach, Germany, pp. 219–244.

Leistner, L. (1986a) Allgemeines über Rohwurst. *Fleischwirtschaft*, **66**, 290–300.

Leistner, L. (1986b) Allgemeines über Rohschinken. *Fleischwirtschaft*, **66**, 496–510.

Leistner, L. (1987) Shelf-stable products and intermediate moisture foods based on meat, in *Water Activity: Theory and Applications to Food* (eds L.B. Rockland and L.R. Beuchat), Marcel Dekker, Inc., New York, NY, pp. 295–327.

Leistner, L. (1992a) Food preservation by combined methods. *Food Res. Internat.*, **25**, 151–158.

Leistner, L. (1992b) Linkage of hurdle-technology with HACCP. *Proceedings of 45th Reciprocal Meat Conference*, published by National Live and Meat Board, Chicago, IL, pp. 1–3.

Leistner, L. (1994a) Further developments in the utilization of hurdle technology for food preservation. *J. Food Engineering*, **22**, 421–432.

Leistner, L. (1994b) *Food Design by Hurdle Technology and HACCP*, printed and distributed by the Adalbert-Raps-Foundation, Kulmbach, Germany.

Leistner, L. (1994c) User guide to food design, in *Food Preservation by Combined Processes* (eds L. Leistner and L.G.M. Gorris), FLAIR Final Report, EUR 15776 EN, European Commission DG XII, Brussels, Belgium, pp. 25–28.

Leistner, L. (1995a) Principles and applications of hurdle technology, in *New Methods for Food Preservation* (ed G.W. Gould), Blackie Academic and Professional, Glasgow, Scotland, pp. 1–21.

Leistner, L. (1995b) Emerging concepts for food safety. *Proceedings of the 41st International Congress of Meat Science and Technology*, San Antonio, Texas, pp. 321–322.

Leistner, L. (1995c) Stable and safe fermented sausages world-wide, in *Fermented Meats* (eds G. Campbell-Platt and P.E. Cook), Blackie Academic and Professional, Glasgow, Scotland, pp. 160–175.

Leistner, L. (1997) Microbial stability and safety of healthy meat, poultry and fish products, in *Production and Processing of Healthy Meat, Poultry and Fish Products* (eds A.M. Pearson and T.R. Dutson), Blackie Academic and Professional, Glasgow, Scotland, pp. 347–360.

Leistner, L. (1999) Combined methods for food preservation, in *Handbook of Food Preservation*, (ed. M. Shafiur Rahman), Marcel Dekker, Inc., New York, NY, pp. 457–485.

Leistner, L. (2000) Use of combined preservative factors in foods of developing countries, in *The Microbiological Safety of Foods* (eds B.M. Lund, A.C. Baird-Parker and G.W. Gould), Aspen Publishers, Inc., Gaithersburg, MD, pp. 294–314.

Leistner, L. (in press) Hurdle technology in the design of minimally processed foods, in *Design of Minimal Processing Technologies for Fruits and Vegetables*, (eds S.M. Alzamora, M.S. Tapia and A. López-Malo), Aspen Publishers, Inc., Gaithersburg, MD.

Leistner, L. and Hechelmann, H. (1993) Food preservation by hurdle-technology. *Proceedings of Food Preservation 2000 Conference*, Vol II, U.S. Army Natick, Research, Development and Engineering Center, Natick, MA, pp. 511–520.

Leistner, L. and Karan-Djurdjic, S. (1970) Beeinflussung der Stabilität von Fleischkonserven durch Steuerung der Wasseraktivität. *Fleischwirtschaft*, **50**, 1547–1549.

Leistner, L., Rödel, W. and Krispien, K. (1981) Microbiology of meat and meat products in high- and intermediate-moisture ranges, in *Water Activity: Influences on Food Quality* (eds L.B. Rockland and G.F. Stewart), Academic Press, New York, NY, pp. 855–916.

Leistner, L., Wirth, F. and Vukovíc, I. (1979) SSP (Shelf Stable Products)—Fleischerzeugnisse mit Zukunft. *Fleischwirtschaft*, **59**, 1313–1318.

McClure, P.J., Blackburn, C. de W., Cole, M.B., Curtis, P.S., Jones, J.E., Legan, J.D., Ogden, I.D., Peck, M.W., Roberts, T.A., Sutherland, J.P. and Walker, S.J. (1994) Modelling the growth, survival and death of microorganisms in foods: the UK Food MicroModel approach. *Int. J. Food Microbiol.,* **23**, 265–275.

McKenna, B.M. (1994) Combined processes and total quality management, in *Food Preservation by Combined Processes* (eds L. Leistner and L.G.M. Gorris), FLAIR Final Report, EUR 15776 EN, European Commission DG XII, Brussels, Belgium, pp. 99–100.

McMeekin, T.A., Olley, J.N., Ross, T. and Ratkowsky, D.A. (1993) *Predictive Microbiology: Theory and Application*, Research Studies Press Ltd., Taunton, Sommerset, United Kingdom.

McMeekin, T., Presser, K., Ratkowsky, D., Ross, T., Salter, M. and Tienungoon, S. (1999) Quantifying the hurdle concept by modelling the bacterial growth/no growth interface, in *Book of Abstracts of International Symposium on Microbial Stress*, Quimper, France, p. 83.

Ono, K. (1994) *Packaging Design and Innovation*. Material for a Third Country Training Programme in the field of food packaging, conducted at Singapore.

Ono, K. (1996) Personal communication, Snow Brand, Tokyo, Japan.

Pierson, M.D. and Corlett Jr., D. (1992) *HACCP: Principles and Applications*, Van Nostrand Reinhold, New York, NY.

Rödel, W., Scheuer, R. and Wagner, H. (1990) A new method of determining water activity in meat products. *Fleischwirtschaft Int.*, **11**, 1–2.

Stanley, D.W. (1991) Biological membrane deterioration and associated quality losses in food tissues. *Crit. Rev. Food Sci. & Nutr.*, **30**, 487–553.

Tändler, K. and Rödel, W. (1982) Herstellung und Haltbarkeit von dünnkalibrigen Dauerwürsten. I. Herstellung. *Fleischwirtschaft*, **62**, 457–464.

Tändler, K. and Rödel, W. (1983) Herstellung und Haltbarkeit von dünnkalibrigen Dauerwürsten. II. Haltbarkeit. *Fleischwirtschaft*, **63**, 150–162.

Tapia de Daza, M.S., Alzamora, S.M. and Welti Chanes, J. (1996) Combination of preservation factors applied to minimal processing of foods. *Crit. Rev. Food Sci. & Nutr.*, **36**, 629–659.

Wirth, F., Leistner, L. and Rödel, W. (1990) *Richtwerte der Fleischtechnologie, 2. Auflage*, Deutscher Fachverlag GmbH, Frankfurt am Main, Germany.

Index

D

S